INTERNATIONAL SERIES OF MONOGRAPHS IN
NATURAL PHILOSOPHY

GENERAL EDITOR: D. TER HAAR

VOLUME 52

PROBLEMS IN OPTICS

PROBLEMS IN OPTICS

by

M. ROUSSEAU

Professor of the Faculty of Science, Rouen

and

J. P. MATHIEU

Professor of the Faculty of Science, Paris

Translated by

J. WARREN BLAKER

*Professor of the Department of Physics and Astronomy,
Vassar College, New York*

PERGAMON PRESS

OXFORD · NEW YORK · TORONTO
SYDNEY · BRAUNSCHWEIG

Pergamon Press Ltd., Headington Hill Hall, Oxford

Pergamon Press Inc., Maxwell House, Fairview Park, Elmsford, New York 10523

Pergamon of Canada Ltd., 207 Queen's Quay West, Toronto 1

Pergamon Press (Aust.) Pty. Ltd., 19a Boundary Street, Rushcutters Bay, N.S.W. 2011, Australia

Vieweg & Sohn GmbH, Burgplatz 1, Braunschweig

First edition 1973

Translated from the French edition *'Problemes d'optique'* by M. Rousseau and J.P. Mathieu. Copyright © 1966 Dunod Editeur Ltd., Paris. Incorporating corrections and additions supplied by the authors during translation.

Library of Congress Cataloging in Publication Data

Rousseau, Madeleine Gandeix.
   Problems in optics.

   (International series of monographs in natural philosophy, v. 52)
     1. Optics--Problems, exercises, etc.
   I. Mathieu, Jean Paul, joint author. II. Title.
   QC363.R613   535'.076   72-13018
   ISBN 0-08-016980-5

Printed in Hungary

CONTENTS

FOREWORD

COLLECTIONS of problems are useful both for faculty use in the evaluation of the state of a student's knowledge and for the student himself to use in self-evaluation. This collection of problems is at the level of the present state of knowledge expected of a student candidate for certification in optics and many of these problems are, in fact, drawn from certification examinations.

Physical optics is a traditional subject and a very large choice of problems is available in this area. An attempt has been made here to provide a broad selection of modern material using some of the newer experimental and theoretical results and, in addition, those areas of electromagnetic theory relevant to optics.

Quantum optics, which involves the elements of wave mechanics and its applications to atomic and molecular spectroscopy and, thus, to the propagation of electromagnetic radiation in material media, has only recently been introduced into optics courses. As a result of the relatively short experience in the presentation of these techniques, the problems in this area are generally presented at a somewhat lower level than the classical problems in spite of their significance in modern optical work.

An attempt has been made here to find a balance between extreme detail in solution and sufficient detail as to be of use. In general, whenever detail is not presented in the solution, reference is made to the general principle used. References are often given in the form § 8.3 (chapter 8, section 3) or § B.3 (Appendix B, section 3) and are keyed to the complementary volume *Optics*: Part 1, Electromagnetic Optics; Part 2, Quantum Optics, which forms part of this series. References to Appendices A and B of this volume are given in the form Appendix A (or B) and references to Problems (or parts thereof) as Problem 1 (or Problem 1, II. 1, etc.).

Many thanks are due our colleagues who provided us with a selection of problems, thus enhancing our coverage. To these individuals, MM. Boiteaux, Fert, Françon, Jacquinot, Kahane, Nikitine, Rouard, Rousset, Servant, Vienot, goes our gratitude. The solutions, however, are ours, and thus any error in detail or omission must remain with us.

We are also grateful to Professor J. W. Blaker for the accurate translation from the French.

M. R., J. P. M.

PRINCIPAL PHYSICAL CONSTANTS

(MKSA rationalized units)

Avogadro's number	$\mathscr{N}$	$= 6.025 \times 10^{26}$ molecules/kilomole
Volume of one kilomole of an ideal gas at standard conditions	V_m	$= 22,420$ m^3
Ideal gas constant	R	$= 8.3169 \times 10^3$ joules/kilomole-°K
Boltzmann constant	k	$= R/\mathscr{N} = 1.380 \times 10^{-23}$ joule/°K
Permittivity of free-space	ε_0	$= 8.834 \times 10^{-12}$ farads/m
Permeability of free-space	μ_0	$= 4\pi \times 10^{-7} = 1.257 \times 10^{-6}$ henrys/m
Faraday's constant	$\mathscr{F}$	$= 96.522 \times 10^6$ coul/kilomole
Electron charge	e	$= 1.602 \times 10^{-19}$ coul
Rest mass of the electron	m_e	$= 9.1083 \times 10^{-31}$ kg
Mass of the proton	M_{H}	$= 1.6724 \times 10^{-27}$ kg
Specific charge of the electron	e/m_e	$= 1.759 \times 10^{11}$ coul/kg
Planck's constant	h	$= 6.6252 \times 10^{-34}$ joule-sec
Speed of light in vacuum	c	$= 2.997\ 93 \times 10^8$ m/s
Rydberg constant for H	R_{H}	$= 10,967,758$ m^{-1}
Ground state radius of H	r_0	$= 0.5292 \times 10^{-10}$ m
Bohr magneton	μ_B	$= eh/4\pi m_e = 9.27 \times 10^{-24}$ A-m^2
Compton wavelength for the electron	λ_c	$= 2h/m_e c = 4.8524 \times 10^{-12}$ m

Energy conversion factors:

$$1 \text{ calorie} = 4.185 \text{ joules}$$

$$1 \text{ electron-volt} = 1.602 \times 10^{-19} \text{ joules}$$

$$= 8068 \text{ cm}^{-1} \ (\times hc)$$

Unless otherwise indicated, these constants will be used for the calculations which follow.

INTERFERENCE

PROBLEM 1

Visibility of Young's Fringes

In all of these problems assume that the source is monochromatic and radiates at a wavelength $\lambda = 0.55\ \mu$.

I

1. A point source S_0 illuminates two narrow, parallel slits, F_1 and F_2, ruled vertically in an opaque screen. The slits are separated by 2 mm. One observes the interference pattern in a plane π parallel to and at a distance of 1 m from the screen. A point M in the plane π is assigned the coordinates X and Y (Y parallel to the slits). Determine the expression governing the distribution of the illumination over the plane π.

2. How is the image modified when S_0 is replaced by a narrow slit F_0 parallel to F_1 and F_2? Calculate the interference pattern.

3. The observation of the fringes is made using a Fresnel eyepiece similar to a thin lens of focal length $f = 2$ cm. What are the advantages of observation with an eyepiece in comparison to observation with the naked eye? Indicate the positions of the eyepiece and the eye with respect to plane π for which the observation of the fringes is made under the best conditions.

II

Cover the slit F_1 with an absorbing screen (which introduces no phase-shift) of optical density $\Delta = 2$.

$\left(\text{The optical density is defined by: } \Delta = \log_{10} \dfrac{\text{incident intensity}}{\text{transmitted intensity}}.\right)$

Find the visibility, V, of the fringes defined by:

$$V = \frac{I_{\max} - I_{\min}}{I_{\max} + I_{\min}},$$

where $I_{\max}$ and $I_{\min}$ represent the maximum and minimum intensities respectively.

III

Here a large incoherent source is used.

1. The source slit has a height h (fixed) and a width a (variable). This is situated at a distance $d = 1$ m behind the plane of the slits F_1 and F_2. Under these conditions, what is the expression for the illumination at a point M in the plane π? How does the visibility of the fringes, V, vary as a function of a? Use this expression to describe the phenomenon observed when one progressively opens the source slit F_0. Determine the maximum width of the slit so that the loss in contrast does not exceed 10%.

2. To increase the luminosity of the image an incoherently illuminated grating is used as a source (slits parallel to F_1 and F_2). Determine the width a of the transparent intervals and the grating step p so that the visibility retains its preceding value.

IV

1. Assume that the source slit F_0 is sufficiently narrow that it can be considered as a line and replace the Fresnel eyepiece observing apparatus by a photocell. Place the slit of the cell in the plane π parallel to the fringes. The height of the slit is fixed; its width is variable. Assume that the intensity of the photocurrent is proportional to the luminous flux falling on the cell. Give the law for the variation of the current as a function of the abscissa X of the slit. Describe what happens when the slit is opened.

2. What is the expression for the intensity of the current assuming that the source slit is not vanishingly fine but has width a? Determine the visibility factor.

V

1. Take the width of the source slit as $a = 0.01$ mm and the width of the slit of the detector as $b = 0.02$ mm. Find the visibility.

This theoretical visibility V_t is greater than the experimental visibility V_r which has a value $V_r = 0.5$. Show that this can be explained by taking into account the parasitic current $\mathfrak{J}_0$ (dark current) found in the absence of all luminous flux. Calculate the ratio, $\mathfrak{J}_0/\mathfrak{J}_{max}$, of the dark current to the maximum signal intensity.

2. The width of the slit of the detector is fixed by its construction at a value $b = 0.02$ mm, while, on the other hand, the width a of the source slit can be altered.

Calculate V_r and present graphically its variation as a function of a. For what value of a will V_r be maximum? What can be concluded from this investigation?

SOLUTION

I. *Coherent illumination*

1. *Point source*

Designate by x and y the coordinates in the plane of the pupil and by X and Y the coordinates of a point M in the image plane (Fig. 1.1). The infinitely thin slits diffract uniformly in the plane perpendicular to Oy.

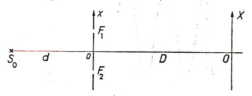

FIG. 1.1

Only the line Ox is illuminated with a light distribution

$$I = 4\cos^2(\pi u s) \tag{1}$$

where

$$u = \frac{\sin i}{\lambda} \approx \frac{i}{\lambda} = \frac{X}{D}\frac{1}{\lambda}. \tag{2}$$

One gets this result from the fact that, for coherent illumination, the distribution of the amplitude in the image is equal to the Fourier transform of the amplitude distribution in the pupil (see Appendix A).

The amplitude in the exit pupil is

$$f(x) = \delta\left(x+\frac{s}{2}\right) + \delta\left(x-\frac{s}{2}\right). \tag{3}$$

The amplitude in the image plane is

$$F(u) = \text{F.T.}[f(x)] \tag{4}$$
$$F(u) = \Delta(u)\left[e^{j\pi us} + e^{-j\pi us}\right] \tag{5}$$

with

$$\Delta(u) = \text{F.T.}[\delta(x)] = 1 \tag{6}$$

from which

$$F(u) = 2\cos\pi us \rightarrow \text{period } 2/s.$$

and

$$I(u) = |F(u)|^2 = 4\cos^2\pi us \rightarrow \text{period } 1/s. \tag{8}$$

2. *Linear source*

Here one observes no interference along the lines parallel to Oy. Each point on the source slit gives a light distribution centred on the geometric image and parallel to Ox. One then has fringes parallel to F_1 and F_2.

The period of these fringes is such that:

$$\Delta u = 1/s \tag{9}$$

giving a linear fringe spacing:

$$\Delta X = \lambda \frac{D}{s}. \tag{10}$$

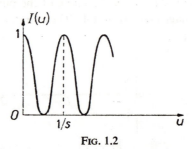

FIG. 1.2

Numerically:

$$\Delta X = 0.55 \times \frac{10^{-3}}{2} = 0.275 \text{ mm}.$$

3. Observation of the fringes

Naked eye. A normal eye working at its near point (25 cm) has difficulty in resolving the image. In effect the fringe spacing is seen at an angle:

$$\varepsilon = \frac{0.275}{250} \approx 10^{-3} \text{ rad}.$$

This value is only slightly larger than the angular limit of resolution of the eye which is of the order of 1 minute or 3×10^{-4} rad.

Eyepiece + eye. To avoid fatigue it is preferable that the eye does not accommodate. For this reason one uses an eyepiece whose focal plane coincides with the plane π; the image is then formed at infinity. This image is easily resolvable since the angular fringe spacing becomes

$$\varepsilon = \frac{0.275}{20} = 0.0135 \text{ rad}.$$

The magnification of the eyepiece is

$$G = \frac{\varepsilon'}{\varepsilon} = \frac{\text{angle at which the image is seen}}{\text{angle of the object when at the near point}}.$$

Note. In principle the slits diffract through an angle of 180° so that, even with large aperture, the eyepiece cannot collect all of the rays. The observer, in order to collect the maximum light, must place his pupil in the plane F_1', F_2' conjugate to the plane F_1, F_2 (Fig. 1.3).

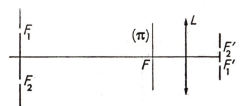

FIG. 1.3

The slits are at a distance ξ from the lens, their images are at a distance ξ', such that:

$$\frac{1}{\xi'} - \frac{1}{\xi} = \frac{1}{f}$$

$$\frac{1}{\xi'} = \frac{1}{100+f} + \frac{1}{f} = \frac{52}{102}$$

$$\xi' = 1.965 \text{ cm} \approx 2 \text{ cm}.$$

The magnification is equal to

$$\frac{\eta'}{\eta} = \frac{\xi'}{\xi} = \frac{1}{52}.$$

The image has dimension

$$\eta' = \frac{\eta}{52} = \frac{2}{52} \approx 0.04 \text{ mm}.$$

All of the rays which enter the eyepiece get to the eye since the value of η' is less than the minimum diameter of the pupil of the eye.

II. The vibrations passing through F_1 and F_2 are in phase but have different amplitudes

When the vibrations are out of phase by ϕ, the intensity at point M is given by

$$I(M) = A_1^2 + A_2^2 + 2A_1A_2 \cos \phi = I_1 + I_2 + 2\sqrt{I_1I_2} \cos \phi. \tag{11}$$

The maximum and minimum intensities are respectively equal to

$$I_{\max} = (\sqrt{I_1} + \sqrt{I_2})^2$$

$$I_{\min} = (\sqrt{I_1} - \sqrt{I_2})^2,$$

the visibility is

$$V = \frac{2\sqrt{I_1I_2}}{I_1 + I_2}. \tag{12}$$

Assuming that the optical density filter is placed in front of F_1 one has:

$$\log_{10} \frac{I_2}{I_1} = 2 \quad \text{where} \quad \frac{I_2}{I_1} = 100,$$

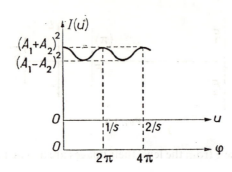

FIG. 1.4

from which $V = 0.2$ (Fig. 1.4).

The positions of the maxima and minima are the same with or without the filter. On the other hand the visibility, V, is not unity unless the amplitudes passing through the slits are equal.

III. *Large source. Incoherent illumination* (§ 6.7)

1. *The source is a large slit*

All the points lying on a line parallel to Oy give fringes parallel to Oy with period $\Delta u = 1/s$. Break the slit (width a) into an infinite number of vanishingly thin slits.

Let v be the reduced coordinate of a point in the source plane. The width of the slit can be characterized by

$$v_0 = a/\lambda d. \tag{13}$$

The intensity produced on M by an element of width dv is

$$dI = A \times h\{1 + \cos 2\pi[(u+v)s]\}\, dv. \tag{14}$$

A = constant, λvs = path difference between the disturbances arriving from F_1 and F_2.

Each elementary slit of infinitesimal width gives a system of fringes with period $\Delta u = 1/s$ and centred on the geometric image of the elementary slit.

Thus, the intensity transmitted to M by the slit source is

$$I = Ah \int_{-v_0/2}^{+v_0/2} [1 + \cos 2\pi(u+v)s]\, dv \tag{15}$$

$$I = I_0 \left[1 + \frac{\sin \pi v_0 s}{\pi v_0 s} \cos 2\pi us \right]. \tag{16}$$

One finds:

$$V = \frac{\sin \pi v_0 s}{\pi v_0 s}.$$

The graph of V is given in Fig. 1.5.

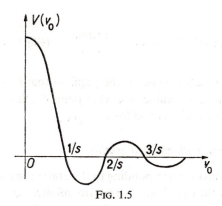

FIG. 1.5

Numerical application. One wants $V \geqslant 0.9$ so that

$$\frac{\sin \pi v_0 s}{\pi v_0 s} = 0.9 \rightarrow \pi v_0 s = \frac{\pi}{4} \rightarrow v_0 = \frac{1}{4s}.$$

From the definition of v_0 one gets

$$\frac{1}{4s} = \frac{a}{\lambda d} \quad \text{so that} \quad a = d\frac{\lambda}{4s} = 10^6 \times \frac{0.55 \times 10^{-3}}{4 \times 2}$$

$$V = 0.9 \quad \text{for} \quad a \approx 70 \ \mu.$$

The fringes vanish for $a = 275 \ \mu$.

The Van Cittert–Zernike theorem (Appendix B) gives this result immediately. The degree of coherence between the slits F_1 and F_2 is given by the Fourier transform of the intensity distribution in the source plane. Since the problem is one-dimensional, it is sufficient to assume that the source is a slit parallel to OY with a width a and that the pupil is formed by two points, P_1 and P_2, set in an opaque screen (P_1 and P_2 corresponding to the intersection of the slits F_1 and F_2 with the line Ox are separated by a distance s). The intensity distribution in the source can be represented by a rectangular function (Fig. 1.6).

$$I(v) = 0 \quad \text{for} \quad v < -v_0/2 \quad \text{and} \quad v > +v_0/2, \tag{17}$$

$$I(v) = 1 \quad \text{for} \quad -v_0/2 < v < +v_0/2.$$

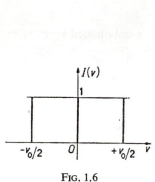

FIG. 1.6

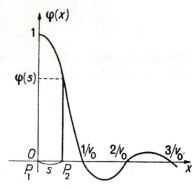

FIG. 1.7

One finds

$$F.T.[I(v)] = \phi(x) = \frac{\sin \pi v_0 x}{\pi v_0 x} \quad \text{(Fig.1.7)}. \tag{18}$$

Place each imaginary diffraction spot on the pupil so that its centre coincides with P_1. The fringe visibility is equal to the value of $\phi(x)$ at point P_2, that is, at $\phi(s)$ (Fig. 1.7). One can see that the fringe contrast is still good for $s = \frac{1}{4}v_0$.

2. The source is an incoherently illuminated grating

Call v_p the reduced coordinate corresponding to the grating spacing p.

(a) Assume initially that the illuminated strips are infinitely thin.

The intensity distribution in the source is a Dirac series (Fig. 1.8). Its Fourier transform is a Dirac series of period $1/v_p$ (Fig. 1.9).

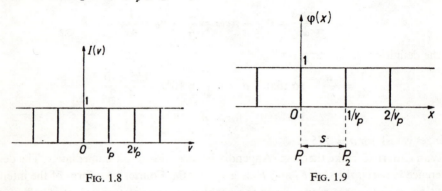

FIG. 1.8   FIG. 1.9

As before, place the imaginary diffraction spot $\phi(x)$ on the pupil so that $\phi(0)$ coincides with P_1. The fringe visibility will be unity if

$$1/v_p = s \quad \text{(Fig. 1.9)}$$

that is, if

$$s = \lambda d/p,$$

so that

$$p = \frac{\lambda d}{s} = 0.55 \times \frac{10^3}{2} = 275 \ \mu.$$

(b) The grating openings have a finite width a. $I(v)$ is a unbounded series of rectangular functions (Fig. 1.10) with period v_p and width v_0.

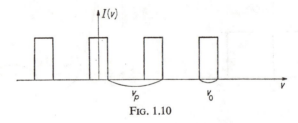

FIG. 1.10

The Fourier transform is shown in Fig. 1.11.

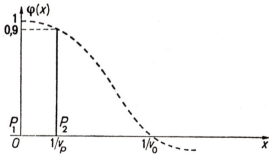

FIG. 1.11

To have an image well contrasted one needs

$$s = \frac{1}{v_p} = \frac{1}{4v_0}.$$

Numerical results

Grating spacing $p = 275$ μ.

Width of the grating openings $a = 70$ μ.

Note. One can also get these results by another simple process (Fig. 1.12).

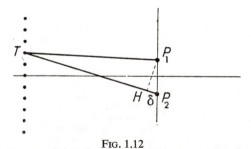

FIG. 1.12

(a) Fine grating openings: the fringes remain fixed if the vibrations transmitted by an opening T are phase-shifted by an integral multiple of 2π when arriving at P_1 and P_2.

(b) Grating with large openings: the vibrations transmitted from the edges of any window should produce at P_1 and P_2 a path difference lying between $k\lambda$ and $(k+\frac{1}{8})\lambda$ in which case the fringes do not overlap (the fringes produced by the extreme edges of an opening are shifted by a maximum of $\frac{1}{4}$ fringe).

IV. *The opening of the detector has finite width b*

1. *The source slit is infinitely thin*

The fringes on plane π have unit contrast (see question I). On the other hand, because of the finite width of the detector slit, the flux recorded by the receiver is never zero (Fig. 1.13). The illumination is the same at all points along a single vertical in the observing plane. Break the window of the receiving cell down into elements of width du and height l.

2*

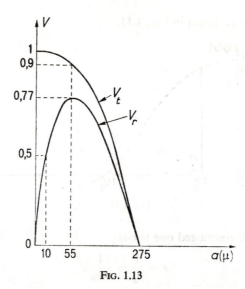

FIG. 1.13

Call u_c the reduced coordinate which corresponds to the linear width b of the slit. The flux which penetrates through the surface element at abscissa u' is

$$d\Phi = Bl(1+\cos 2\pi us)\,du, \qquad (19)$$

from which

$$\Phi(u') = \int_{u'-u_c/2}^{u'+u_c/2} d\Phi = Blu_c\left[1+\frac{\sin \pi u_c s}{\pi u_c s}\cos 2\pi u's\right]. \qquad (20)$$

As before, one can define a coefficient of visibility by

$$V = \frac{\sin \pi u_c s}{\pi u_c s}. \qquad (21)$$

As long as u_c is less than $\frac{1}{4}s$, the intensity of the photocurrent, proportional to the luminous flux, varies in a reasonably sinusoidal fashion. When one opens the slit, the difference between the maxima and minima lessens. Finally, for $u_c = 1/s$, the intensity of the photo-current does not vary regardless of the placement of the cell.

2. *The source slit has a finite width a*

One has

$$I(u) = I_0\left[1+\frac{\sin \pi v_0 s}{\pi v_0 s}\cos 2\pi us\right], \qquad (22)$$

from which

$$\Phi(u') = Blv_0\int_{u'-u_c/2}^{u'+u_c/2}\left[1+\frac{\sin \pi v_0 s}{\pi v_0 s}\cos 2\pi us\right]du, \qquad (23)$$

$$\Phi(u') = Blu_c v_0\left[1+\frac{\sin \pi u_c s}{\pi u_c s}\times\frac{\sin \pi v_0 s}{\pi v_0 s}\cos 2\pi u's\right]. \qquad (24)$$

One finds that

$$V = \frac{\sin \pi u_c s}{\pi u_c s} \times \frac{\sin \pi v_0 s}{\pi v_0 s}. \tag{25}$$

The visibility can be defined using an "instrument function".
   V = F.T. (source slit, width v_0)×F.T. (cell opening, width u_c)
   In the case of an infinitely fine source slit, the first term of the product reduces to a value of one since F.T. $[\delta(x)] = 1$.

V. *The effect of the dark current*

Recall equation (25) which gives the theoretical visibility.
   1. $a = 0.01$ mm, $b = 0.02$ mm (v_0 = const, u_c = const).
One has

$$V_t = 0.991 \times \frac{\sin \pi v_0 s}{\pi v_0 s} = 0.987.$$

Taking the dark current into account, the intensity of the real current becomes

$$\Im_r(u) = \Im(u) + \Im_0. \tag{26}$$

Hence, the experimental visibility is

$$V_r = \frac{\Im_{r\,\mathrm{max}} - \Im_{r\,\mathrm{min}}}{\Im_{r\,\mathrm{max}} + \Im_{r\,\mathrm{min}}} = \frac{\Im_{\mathrm{max}} - \Im_{\mathrm{min}}}{\Im_{\mathrm{max}} + \Im_{\mathrm{min}} + 2\Im_0}, \tag{27}$$

$$V_r = \frac{\dfrac{\sin \pi u_c s}{\pi u_c s} \times \dfrac{\sin \pi v_0 s}{\pi v_0 s}}{1 + \dfrac{\Im_0}{v_0}}. \tag{28}$$

(The constant coefficient Blu_c has been set equal to one.) Thus one has the relationship

$$V_r = \frac{V_t}{1 + \Im_0/v_0}. \tag{29}$$

Numerical application

$$1 + \frac{\Im_0}{v_0} = \frac{V_t}{V_r} = \frac{0.987}{0.5} = 1.974,$$

$$\frac{\Im_0}{v_0} = 0.974.$$

One has

$$\Im_{\mathrm{max}} = v_0[1 + V_t] = v_0[1.987],$$

hence

$$\frac{\Im_0}{\Im_{\mathrm{max}}} = \frac{0.974}{1.987} \approx \frac{1}{2} \quad \text{(in practice this value is much smaller).}$$

2. a variable, $b = 0.02$ mm (v_0 variable, u_c = const).
One has

$$V_t = 0.991 \frac{\sin \pi v_0 s}{\pi v_0 s}. \tag{30}$$

from which

$$V_r = \frac{0.991 \dfrac{\sin \pi v_0 s}{\pi v_0 s}}{1 + \Im_0/v_0} = 0.991 \times \pi s \frac{\sin \pi v_0 s}{\pi s (v_0 + \Im_0)}.$$

V_r has a maximum for $dV_r/dV_0 = 0$, so that

$$\tan \pi v_0 s = \pi v_0 s + \pi s \Im_0 = \pi v_0 s + 0.111.$$

This equation is satisfied for $\pi v_0 s \approx 35°$.

$$v_0 \approx \frac{36}{180} \times \frac{1}{2 \times 10^3} = 10^{-4} \, \mu^{-1}, \quad \text{since} \quad v_0 = \frac{a}{\lambda f},$$

hence

$$a = v_0 f \lambda = 10^{-4} \times 10^6 \times 0.55$$
$$a = 0.55 \, \mu.$$

The variations of V_t and V_r as a function of v_0 are shown in Fig. 1.13. For $a = 55 \, \mu$ one has

$$V_t = 0.991 \times \frac{\sin 35°}{\dfrac{3.14 \times 36}{180}} = 0.89.$$

hence

$$V_r = \frac{V_t}{1 + \Im_0/v_0} = \frac{0.9}{1 + 0.974 \times 10/55} = \frac{0.9}{1.1771}.$$
$$V_{r\,\text{max}} = 0.77.$$

Conclusion. In theory, to have the best contrast, it is advantageous to close the source slit to the smallest possible opening. In practice, in the presence of a dark current, it is necessary to give the source slit its optimal width.

PROBLEM 2

Young's Experiment. Achromatic Fringes

I

Monochromatic source ($\lambda = 0.55 \, \mu$).
Young's apparatus as illustrated in Fig. 2.1 has the following characteristics:
 Slit separation $a = 3.3$ mm.
 Distance from the pupil to the screen $D = 3$ m.

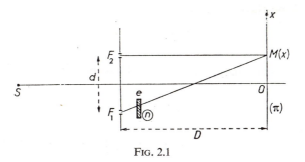

FIG. 2.1

1. Calculate the fringe spacing i.

2. Place a sheet of glass with plane parallel faces and thickness $e = 0.01$ mm in front of slit F_1.

(a) Determine the direction of the displacement of the fringes and the formula giving the relationship for their displacement.

(b) Knowing that the fringes are displaced by 4.73 mm, find the index of the glass. How precise is this value of n, if the displacement can be measured to 0.01 mm?

II

Nature of the fringes in white light.
The dispersion of glass is given by

$$n = n_0 + \frac{A}{\lambda^2} \quad \text{with} \quad \begin{cases} n_0 = 1.50, \\ A = 0.00605 \text{ for } \lambda \text{ in microns.} \end{cases}$$

Express x as a function of the interference order and the wave length.
Give $x = f(\lambda)$ for $p = \delta/\lambda = 1, 0, -1,$ and -2.

1. Describe the nature of the zero-order fringe.

2. Show that there exists a bright fringe for which x is stationary (λ between 0.4 and 0.75 μ). What is the interference order of this achromatic fringe?

SOLUTION

I. *Monochromatic source*

1. *Fringes spacing*

$$i = \lambda \frac{D}{a} = 0.55 \times 10^{-3} \times \frac{3 \times 10^3}{3.3} = 0.5 \text{ mm.}$$

2. (a) *Displacement Δx of the fringes*

The difference in path length for the rays which interfere on M is:
without the glass:

$$\delta_1 = F_1 M - F_2 M = x \frac{\lambda}{i} \tag{1}$$

after insertion of the glass:

$$\delta = x\frac{\lambda}{i} + (n-1)e. \tag{2}$$

The fringe of order p which had abscissa $x_1 = pi$ has the new abscissa:

$$x = \frac{i}{\lambda}[p\lambda - (n-1)e]. \tag{3}$$

The system of fringes is displaced toward negative x by an amount:

$$\Delta x = -\frac{i}{\lambda}(n-1)e. \tag{4}$$

(b) *Measurement of the index*

From equation (4) one gets

$$n = 1 - \frac{\lambda}{i}\frac{\Delta x}{e}. \tag{5}$$

Now $x = -4.73$ mm, thus

$$n = 1 + \frac{0.55 \times 10^{-3}}{0.5} \times \frac{4.73}{10^{-2}}$$

$$n = 1.5203.$$

Error determination:

$$\frac{d(\Delta x)}{\Delta x} = \frac{d(n)}{n-1} \quad \text{with} \quad d(\Delta x) = 2 \times 10^{-2} \text{ mm}$$

$$d(n) = 0.5 \times \frac{2 \times 10^{-2}}{4.73} \approx 2 \times 10^{-3},$$

$$n = 1.520 \pm 0.002.$$

II. *Fringes in white light*

In equation (3) replacing i by $\lambda D/d$ and $n-1$ by

$$0.50 + \frac{A}{\lambda^2} = 0.50 + \frac{0.00605}{\lambda^2},$$

one gets

$$x(p, \lambda) = -4.545 + 0.909p\lambda - \frac{0.055}{\lambda^2}. \tag{6}$$

with x in mm if λ is in microns.

1. $p = 0$. One finds a coloured fringe (Fig. 2.2). When one scans in the direction $x < 0$, the following tints are found:

red	for	$x = -4.64$ mm,
yellow-green	for	$x = -4.73$ mm,
blue	for	$x = -4.89$ mm.

2. $p = -1$. The spectrum bends back (Fig. 2.2).

One finds $dx/d\lambda = 0$ for λ near 0.5 μ.

The achromatic fringe spreads out about $x = -5.22$ mm.

Note. On Fig. 2.2 is traced a group of lines from the equation $x = -4.545 + 0.909 \ p\lambda$ corresponding to the fringes given by the insertion of a non-dispersive glass of index $n_0 = 1.50$.

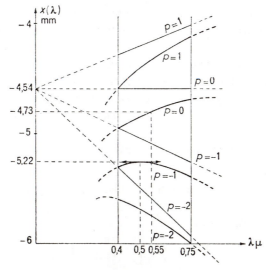

Fig. 2.2

PROBLEM 3

Fourier Spectroscopy

One wants to determine the spectral distribution of the radiance $B(\sigma)$ of a source. For this a Michelson interferometer is used as a modulator.

A point source P is placed at the focus of a collimator L_1. One of the mirrors is rigorously parallel to the image of the other formed in the beam splitter S_p. Mirror M_1 is moved with a constant speed starting from the position of zero path-length difference.

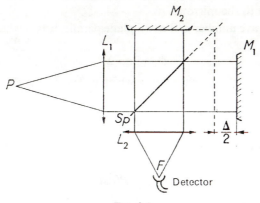

Fig. 3.1

I. *The path of the mirror is assumed to be unlimited*

1. Calling Δ the difference in path between the two mirrors, calculate the intensity received by the photocell R (placed at the focus of lens L_2):
(a) for a monochromatic source (σ_0), and
(b) for the case where the spectrum is comprised of σ_1 and σ_2.

2. The expression for the intensity may be thought of as the sum of a constant term (mean intensity) and a term dependent on Δ. These two terms when multiplied by 2 form the *interferogram: $I(\Delta)$.*

Show that $B(\sigma)$ and $I(\Delta)$ can be derived from one another by the Fourier transformation. To facilitate these calculations, it will be useful to introduce artificially a spectrum $B(-\sigma)$ composed of negative frequencies and symmetric with $B(\sigma)$. One can then use the following property: the F.T. of an even function is an even function. In all of these problems one normalizes the functions.

Applications. Describe and calculate the interferogram for the following cases:

(α) The source emits a monochromatic radiation $B(\sigma_0) = \delta(\sigma-\sigma_0)$.
(β) The ray is a doublet: $B(\sigma) = \alpha_1 \times \delta(\sigma-\sigma_1)+\alpha_2\times\delta(\sigma-\sigma_2)$ (α_1 and α_2 are constants less than one).
(γ) The ray has a gaussian profile:

$$B(\sigma) = \exp\left[-\pi\left(\frac{\sigma-\sigma_0}{\mathrm{d}\sigma}\right)^2\right].$$

II. *The movement of the mirror is limited. Resolving power*

Δ is allowed to vary only between 0 and a maximal value L. Spectroscopists call the "instrument profile" the spectral distribution one obtains if the instrument receives a rigorously monochromatic radiation of wave number σ_0. Starting with the interferogram limited by $\Delta = 0$ and $\Delta = L$, derive the instrument profile and represent it graphically.

SOLUTION

The rays are normal to the mirrors.

For radiation with wave number σ, the two plane parallel waves which interfere are out of phase by

$$\phi = 2\pi\sigma\,\Delta. \tag{1}$$

At L_2 one has a state of uniform interference which is detected at F.

I. *Consider the ideal case where the movement of the mirror M_1 is unlimited*

1. Let I_t be the total intensity received at R.
(a) Monochromatic source,

$$I_t = B(\sigma_0)\cos^2\pi\sigma_0\,\Delta = \frac{B(\sigma_0)}{2}(1+\cos 2\pi\sigma_0\,\Delta). \tag{2}$$

(b) Polychromatic source.

The element of intensity produced by every radiation in the interval $d\sigma$ is

$$dI_t = B(\sigma) \cos^2 \pi\sigma \, \Delta d\sigma$$

hence

$$I_t = \int_{\sigma_1}^{\sigma_2} \frac{B(\sigma)}{2} [1 + \cos 2\pi\sigma \, \Delta] \, d\sigma. \tag{3}$$

2. The interferogram thus has the form

$$I(\Delta) = \int_{\sigma_1}^{\sigma_2} B(\sigma) \cos 2\pi\sigma \, \Delta d\sigma.$$

Only the positive frequencies occur. One has, therefore

$$I(\Delta) = \int_0^\infty B(\sigma) \cos 2\pi\sigma \, \Delta d\sigma. \tag{4}$$

$B(\sigma)$, for example, is given in Fig. 3.2. Generate an artificial spectrum $B(-\sigma)$ composed of negative frequencies symmetric with the preceding spectrum. If $B_p(\sigma)$ is the even part of function $B(\sigma)$, one can write

$$B_p(\sigma) = \tfrac{1}{2}[B(\sigma) + B(-\sigma)] \qquad \text{(see Fig. 3.3)} \tag{5}$$

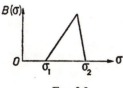

FIG. 3.2

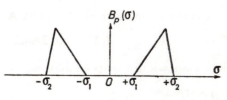

FIG. 3.3

Equation (4) can then be written

$$I(\Delta) = \int_{-\infty}^{+\infty} B_p(\sigma) \cos 2\pi\sigma \, \Delta d\sigma = \int_{-\infty}^{+\infty} B_p(\sigma) \, e^{j2\pi\sigma \Delta} \, d\sigma. \tag{6}$$

If one knows the interferogram precisely for Δ varying between 0 and ∞ (then, in fact between $-\infty$ and $+\infty$ since it is symmetric) the spectrum can be constructed exactly by the Fourier transform:

$$B_p(\sigma) = \int_{-\infty}^{+\infty} I(\Delta) \cos 2\pi\sigma \, \Delta \, d\Delta = \int_{-\infty}^{+\infty} I(\Delta) \, e^{-j2\pi\sigma \Delta} \, d\Delta,$$

$$B_p(\sigma) \xrightarrow{\text{F.T.}} I(\Delta). \tag{7}$$

Applications (see Appendix A dealing with the Fourier transformation)
Given $B(\sigma)$, one can get $B_p(\sigma)$ and from this the interferogram by Fourier transforming.

(α) $B(\sigma) = \delta(\sigma - \sigma_0)$,

    $B_p(\sigma) = \delta(\sigma + \sigma_0) + \delta(\sigma - \sigma_0) \rightarrow I(\Delta) = \cos 2\pi\sigma_0 \Delta$;

(β) $B(\sigma) = \alpha_1\delta(\sigma - \sigma_1) + \alpha_2\delta(\sigma - \sigma_2) \rightarrow I(\Delta) = \alpha_1 \cos 2\pi\sigma_1 \Delta + \alpha_2 \cos 2\pi\sigma_2 \Delta$.

(See Fig. 3.4.)

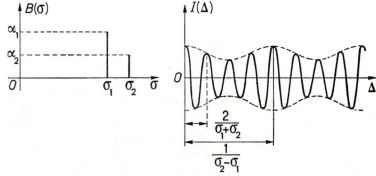

FIG. 3.4

In the special case where $\alpha_1 = \alpha_2$ the minima of the envelope are zero.

(γ) $B(\sigma) = \exp\left[-\pi\left(\dfrac{\sigma - \sigma_0}{d\sigma}\right)^2\right] = \delta(\sigma - \sigma_0) \otimes \exp\left[-\pi\left(\dfrac{\sigma}{d\sigma}\right)^2\right]$.

Applying the convolution theorem (Appendix A):

$$I(\Delta) = \cos 2\pi\sigma_0\Delta \times e^{-\pi(d\sigma \times \Delta)^2}.$$

The envelope has a width $1/d\sigma$ (Fig. 3.5).

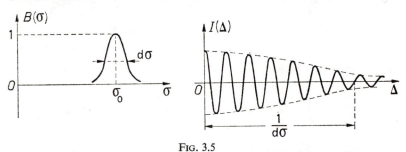

FIG. 3.5

II. *Resolving power*

1. *Limited interferogram* $(0 < \Delta_0 < L)$

This interferogram can be represented by the function $I'(\Delta)$ such that:

$$I'(\Delta) = I(\Delta) \times F(\Delta) \tag{8}$$

with

$$F(\Delta) = \begin{cases} 1 & \text{for} \quad -L < \Delta < +L \\ 0 & \text{for} \quad \Delta < -L \quad \text{and} \quad \Delta > L. \end{cases} \tag{9}$$

Using Parseval's theorem one can write

$$G_p(\sigma) = \text{F.T.}[I'(\Delta_0)] = \text{F.T.}[I(\Delta_0)] \otimes \text{F.T.}[F(\Delta_0)]. \tag{10}$$

One knows the F.T. of the slit function, namely,

$$\text{F.T.}[F(\Delta)] = \frac{\sin 2\pi\sigma L}{2\pi\sigma L}, \tag{11}$$

hence

$$G_p(\sigma) = \frac{\sin 2\pi(\sigma-\sigma_0)L}{2\pi(\sigma-\sigma_0)L} + \frac{\sin 2\pi(\sigma+\sigma_0)L}{2\pi(\sigma+\sigma_0)L}. \tag{12}$$

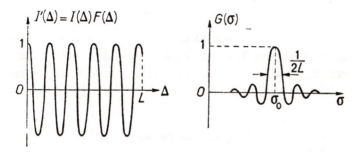

FIG. 3.6

The instrument profile is given in Fig. 3.6:

$$G(\sigma) = \frac{\sin 2\pi(\sigma-\sigma_0)L}{2\pi(\sigma-\sigma_0)L}. \tag{13}$$

$G(\sigma)$ is the spectrum obtained from a strictly monochromatic radiaton source. $G(\sigma)$ has width $\Delta\sigma = \frac{1}{2}L$. The resolving power for the radiation σ_0 is then

$$R = \frac{\sigma_0}{\Delta\sigma} = \sigma_0 \times 2L = \frac{2L}{\lambda_0} = 2N.$$

The resolving power is thus proportional to the number of fringes, N, recorded.
Numerically:

$$L = 10^3, \quad \lambda = 0.5 \ \mu,$$
$$R = \frac{2 \times 10^3}{0.5} = 4000.$$

PROBLEM 4

Mach Interferometer

Here one examines the interference of separate beams as shown in section in Fig. 4.1.

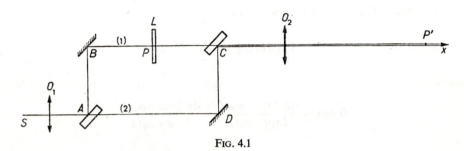

FIG. 4.1

B and D are two mirrors with unit reflectivity; A and C are two identical beam splitters placed as indicated in the figure. Take $AB = CD = d$ and $BC = AD = 2d$.

A point source, S, at the focus of the objective O_1, emits monochromatic radiation of wavelength $\lambda = 0.5\,\mu$.

I. *Equal path interferometer*

1. With the apparatus in adjustment, show that the two plane waves travelling along Cx are coherent and in phase. What intensity does each separate beam have? I_0 is the source intensity. The reflection and transmission coefficients for the beam splitters are exactly $\frac{1}{2}$.

Place an objective of focal length d at O_2 so that $CO_2 = d$. This objective images the plane P lying along BC and gives a real image at P'. What is the appearance of the plane P'?

2. Place at P a thin film (assume the thickness, absorption, and dispersion negligible and the phaseshift uniform). Describe the new appearance of the field P. Show that by photometric measurement, one can determine the phaseshift ϕ introduced by L. Take as the definition of the contrast:

$$\Gamma = \frac{I_{\max} - I_{\min}}{I_{\max}}.$$

Numerical application. Calculate the phaseshifts ϕ_1 and ϕ_2 caused respectively by two films L_1 and L_2 which give contrasts $\Gamma_1 = 1$ and $\Gamma_2 = 0.25$.

3. By what quantity Δy is it necessary to displace the beam splitter C parallel to itself in order to see a black field. Calculate the contrast and discuss the advantage of this method over the previous one.

II. *Interferometer with fringes*

1. Consider the apparatus in I.1. Rotate the mirror D through an angle $\alpha = 2'$ about an axis perpendicular to the figure. Describe the system of fringes and calculate their spacing.

2. What is the appearance of the field in white light?

3. Reinsert the film L. Show that the displacement of the central fringe gives sufficient information to permit the evaluation of the phase shift introduced by the film. Calculate the displacement for the values ϕ_1 and ϕ_2 treated in I.2. Take as the unit length the fringe spacing i corresponding to the wavelength $\lambda = 0.5\ \mu$.

4. Enlarge the source S and find the plane in which the fringes are localized.

SOLUTION

I. *Equal path interferometer* (§ 6.9)

1. The beams (1) and (2), coming from the same point source, propagate as coherent waves:

the geometric paths ABC and ADC are equal.
the reflections experienced at A and B on one hand, and D and C on the other are the same and each beam passes through a beam splitter once.
The optical paths are therefore equal and the waves moving along Cx are in phase.
Since the source is a point, the interference is not localized.
After reflection at A, ray (1) carries energy $I_0/2$, and after passing through the beam splitter C, its energy falls to $I_0/4$.
Both vibrations which interfere have amplitude $\sqrt{I_0/4}$. Thus, the field in the direction Cx is uniformly illuminated.

$$I_1 = \left(2\sqrt{\frac{I_0}{4}}\right)^2 = I_0. \tag{1}$$

Note. If one finds constructive interference along Cx, one finds destructive interference along Cy normal to Cx, since the reflections on the beam splitter C are of a different nature (air-glass for the ray Cx and glass-air for the ray Cy).

Planes P and P' are conjugate with unit magnification since these are the antiprincipal planes of the objective C_2. Later in this problem it will be found necessary to remove the plane of observation from the interferometer in order to make measurements.

2. The film L, when placed at P, produces a constant phaseshift for all the rays which traverse it. These rays have the amplitude:

$$\sqrt{\frac{I_0}{4}} \times e^{j\phi}$$

while the amplitude of the rays (2) remains as:

$$\sqrt{\frac{I_0}{4}}.$$

Thus, the illumination of the image of the film is:

$$I_2 = \frac{I_0}{2}[1 + \cos \phi] \leqslant I_0. \tag{2}$$

The object will appear more or less dark on a bright field (Fig. 4.2) with contrast:

$$\Gamma = \frac{I_{max} - I_{min}}{I_{max}} = \frac{I_1 - I_2}{I_1} = \frac{1 - \cos^2 \phi/2}{1}$$

$$\Gamma = \sin^2 \frac{\phi}{2}. \tag{3}$$

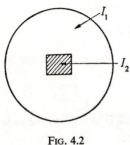

FIG. 4.2

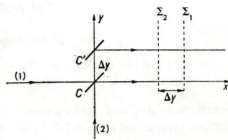

FIG. 4.3

Numerical application:

$\Gamma = 1 \rightarrow I_2 = 0 \rightarrow \phi = \pi, 3\pi, \ldots$ the image of the object is dark on a bright field,

$\Gamma = 0.25 \rightarrow I_2 = 0.75 I_0 \rightarrow \phi = \dfrac{\pi}{3}, \dfrac{5\pi}{3}, \ldots$

$\Gamma = 0 \rightarrow I_2 = I_0 \rightarrow \phi = 2\pi, 4\pi, \ldots$ the field will be uniformly bright and the object invisible

From the photometric measurement of I_1 and I_2 one can deduce ϕ.

Note. If the film introduces only a very slight phaseshift, the contrast can be written:

$$\Gamma = \frac{\phi^2}{4}$$

In this expression ϕ is squared and thus the value of Γ is very small. This apparatus does not lend itself to the detection of a small phaseshift.

3. It is necessary that the vibrations of rays (1) (before inserting the object) and (2) be completely out of phase. When C is displaced, the path (1) does not change but path (2) is increased by Δy (Fig. 4.3).

One wants to have

$$\Delta\phi = \frac{2\pi}{\lambda}\,\Delta y = \pi. \tag{4}$$

From which

$$\Delta y = \frac{\lambda}{2} = 0.25\ \mu.$$

Under these conditions

$$I_1 = 0,$$

$$I_2 = \frac{I_1}{2}[1 + \cos(\phi - \pi)] = \frac{I_0}{2}[1 - \cos\phi] = I_0 \sin^2\frac{\phi}{2}$$

Hence

$$\Gamma = \frac{I_{max} - I_{min}}{I_{max}} = \frac{I_2 - I_1}{I_2} = \frac{I_0 \sin^2\phi/2 - 0}{I_0 \sin^2\phi/2} = 1.$$

The object appears bright on a dark field with maximal contrast for any value of ϕ. Using this method extremely small variations in phase can be detected.

II. *Interferometer with fringes*

1. When mirror D is turned through α, the wave surface Σ_2 turns through 2α. One sees vertical, linear fringes normal to the plane of the figure. The fringe of order zero is on the axis Cx. The system of fringes with bright central lines has spacing:

$$i = \frac{\lambda}{2\alpha} = \frac{0.5 \times 10^{-3}}{2 \times 2 \times 3 \times 10^{-4}} = 0.42\ \text{mm}.$$

2. *Observation in white light.* The central fringe is white and achromatic. The fringes which surround it are rainbow-like with blue toward the centre and red to the outside.

3. The reference wave Σ_2 is not perturbed. Σ_1 has a shift (Fig. 4.4a). In the image of the film L, the white fringe is displaced from 1 to 2 (Fig. 4.4b). One can measure this displacement with an ocular micrometer.

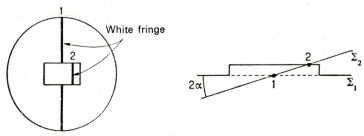

FIG. 4.4

Numerical application. For $\phi = 2k\pi$ (any k), the fringe displacement is equal to $d = ki$.

$$\phi = \pi \rightarrow k = \frac{1}{2} \rightarrow d = \frac{i}{2} = 0.21 \text{ mm}$$

$$\phi = \frac{\pi}{3} \rightarrow k = \frac{1}{6} \rightarrow d = \frac{i}{6} = 0.07 \text{ mm}.$$

To obtain ϕ two measurements are required:

the displacement of the central fringe,

the fringe spacing in monochromatic light of known wavelength.

4. *Localization*. When the source S is extended the fringes become localized.

THEOREM. *The surface of localization is at the point of intersection of the two rays generated from the single incident beam* (§ 6.6).

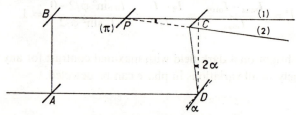

FIG. 4.5

In the case of the Mach interferometer, the localization surface coincides with the image of D formed in the beam splitter C. It is the plane π passing through P and inclined at 45° to BC (Fig. 4.5). (In practice the fringes are found to be localized in a somewhat more extended region surrounding P.)

PROBLEM 5

Michelson Interferometer

Consider the Michelson interferometer as shown in Fig. 5.1. The source S is placed at the focus of lens L_1. Initially, mirrors M_1 and M_2 are mutually perpendicular and are at 45° to the beam splitter C. One generally does not consider the effects related to reflection or transmission through G in this problem.

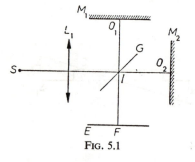

FIG. 5.1   FIG. 5.2

I

The point source S emits a monochromatic wave $\lambda = 5461$ Å.

1. The mirror M_1 remains in its initial position while M_2 is swung through an angle α on the axis O_2 normal to the figure (Fig. 5.2). What does one see?
   (a) Explain why the fringes are not localized and have the same separation everywhere.
   (b) Calculate the fringe spacing in a plane E normal to the direction IO_1, given $\alpha = 1'$.

2. M_1 is again a plane mirror, but M_2 is now replaced by a spherical mirror (convex or concave) with radius $R = 10$ m (Fig. 5.3). The center of M_2 is on IO_2. The vertex of M_2' (the image of M_2 in the beam splitter) is in contact with M_1. The observing plane E passes through O_1.

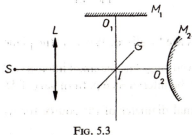

FIG. 5.3

   (a) What is the appearance of the centre of the rings? Calculate the radii of the first three bright rings observed under these conditions. Is the result the same for both concave and convex mirrors?
   (b) M_1 is moved toward the beam splitter. Comment on the displacement of the rings if M_2 is convex.
   (c) What happens in (b) if M_2 is concave?

II

The point source S emits white light, $0.4 < \lambda < 0.8$ μ. A spectrograph with dispersion proportional to wavelength has its slit in a plane conjugate to the plane E. The slit is parallel to the plane of the figure. Its centreline coincides with the extension of the axis IO_1. What does one observe in the exit focal plane of the spectrograph when the experiments described in I.1 and I.2 are performed.
   In both cases give the precise position of the bright lines. (The height of the slit is $l = 10$ mm.)

III

S is now taken to be a large monochromatic source ($\lambda = 5461$ Å). M_1 remains in its initial position. M_2 is again replaced by a plane mirror. Assume that M_2' is parallel to M_1 and at a distance very close to 1 cm from M_1 (Fig. 5.4).
   The lenses L_1 and L_2 both have focal length $f = 1$ m.

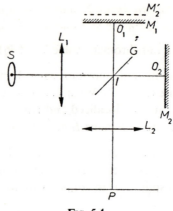

FIG. 5.4

1. Explain why the fringes are localized. Where is the plane of localization?

2. The centre of the observed rings has maximal intensity. Calculate the radius of the first three bright rings. How will the interference pattern vary if M_2' is brought toward M_1?

3. What will be the minimal diameter of the source for which three fringes will be visible?

4. What will occur if one moves the source S off centre by 12.8 mm? Precisely draw the appearance of the localization plane.

SOLUTION

I. *Monochromatic point source at infinity*

1. *The mirrors M_1 and M_2 are planar*

(a) With a point source the fringes are never localized. If the mirrors are at an angle α, one has two plane waves which make an angle of 2α everywhere. One gets equidistant rectilinear fringes normal to the plane of Fig. 5.2.

(b) Fringe spacing: $i = \dfrac{\lambda}{2\alpha} = \dfrac{0.5461}{2\times3\times10^{-4}}\times10^{-3} = 0.91$ mm.

2. *M_1 is planar, M_2 is spherical*

The interference is now produced by:
  a plane wave Σ_1,
  a spherical wave Σ_2 centred at the focus of the mirror at a distance of

$$f_{\text{mirror}} = \frac{R}{2} = 5 \text{ m} \qquad \text{(Fig. 5.5).}$$

(a) The relative position of the two waves, Σ_1 and Σ_2, is shown in Figs. 5.6a and 5.6b.

Regardless of the sign of the radius of M_2, one sees the same interference pattern (Fig. 5.6). These are *Newton's rings* with a bright centre (since Σ_1 and Σ_2 are in contact on the axis).

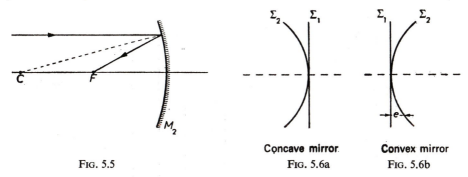

Concave mirror.   Convex mirror
FIG. 5.5   FIG. 5.6a   FIG. 5.6b

At a distance x from the axis, the path difference between the two rays is $\delta = e$, such that

$$x^2 = (2f-e)e \approx 2fe = Re. \tag{1}$$

The radii of the bright rings ($\delta = k\lambda$) are given by:

$$r = \sqrt{k}\,\sqrt{\lambda R} = \sqrt{k}\,\sqrt{0.5461 \times 10^{-3} \times 10 \times 10^3}$$
$$r_{mm} = 2.34\,\sqrt{k} \quad (k \text{ integer}). \tag{2}$$

The radii of the first three bright rings are:

$$k = 1 \quad r_1 = 2.34 \text{ mm,}$$
$$k = 2 \quad r_2 = 3.30 \text{ mm,}$$
$$k = 3 \quad r_3 = 4.04 \text{ mm.}$$

One now moves the mirror M_1 toward the beam splitter.

(b) M_2 *convex*. The relative position of the wave surfaces is shown on Fig. 5.7a. When M_1 is moved forward, the rings rise at the edges and contract into the centre. The number of rings is much greater than in the case of optical contact between the mirrors.

(c) M_2 *concave*. The rings rise at the centre and move outward. The central ring lies at the intersection of Σ_1 and Σ_2.

Note. The interferometer when modified in this way is called the Twyman interferometer. It is used to check the quality of objectives (Fig. 5.8).

One mounts the objective Ob in such a way that its image side focus coincides exactly with the centre, C, of M_2. If the objective is flawless, the rays returning along this path appear as plane waves and one observes the interference pattern of two plane waves. If the objective has flaws, the wave Σ_2 is no longer planar. This wave, after interference with the reference plane wave Σ_1 (reflected by M_1), gives deformed fringes.

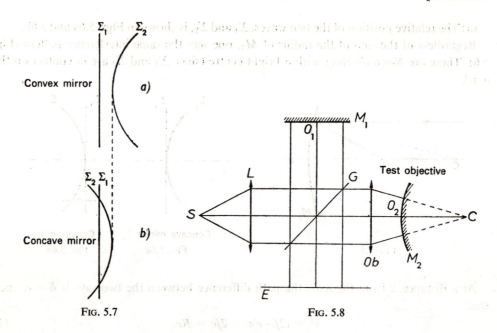

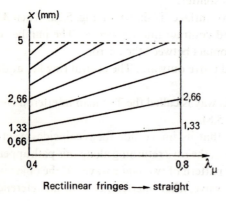

FIG. 5.7

FIG. 5.8

II. *Polychromatic source. Channelled spectra*

1. The slit of the spectrograph is perpendicular to the linear fringes and parallel to the face of M_1 (along the x-axis) in the plane of the figure.

One has constructive interference if $2\alpha x = k\lambda$ (k an integer). Since the dispersion of the spectrograph is proportional to λ (take the coefficient of proportionality equal to 1), the equation of the bright bands is

$$x = \frac{1}{2\alpha}k\lambda. \tag{3}$$

These are clusters of lines (Fig. 5.9a).

The interference of two waves gives wide bands.

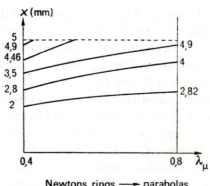

FIG. 5.9a

FIG. 5.9b

2. Expression (1) gives the equation of the bright bands

$$x^2 = R \times k\lambda. \tag{4}$$

These are parabolas (Fig. 5.9b) with vertices along the λ-axis coinciding with $\lambda = 0$.

III. *Large monochromatic source. Rings at infinity*

1. One observes fringes of equal inclination localized at infinity.

2. If i is the angle of incidence of the rays on mirrors M_1 and M_2, the path difference between the reflected rays is

$$\delta = 2e \cos i = 2e\left(1 - \frac{i^2}{2}\right). \tag{5}$$

The centre appears as a bright point. The interference order on the axis is an integer k_0. The bright rings are produced by rays making the angle i with the axis such that

$$i = \sqrt{k_0 - k}\,\sqrt{\frac{\lambda}{e}} \qquad (k \text{ integer}). \tag{6}$$

In the focal plane of L_2, the radii of the bright rings are given by

$$r_{mm} = fi = 10^3\sqrt{k_0 - k}\,\sqrt{\frac{0.5461}{10^4}} = \sqrt{k_0 - k}\,\sqrt{54.61}$$

$$k_0 - k = 1 \quad r_1 = 7.39 \text{ mm},$$
$$k_0 - k = 2 \quad r_2 = 10.45 \text{ mm},$$
$$k_0 - k = 3 \quad r_3 = 12.80 \text{ mm}.$$

3. These rings form an image of the source (L_1 and L_2 have the same focal length, the source and its image has the same dimension).

In order to be able to observe three rings, it is necessary that the source have a minimal dimension $D = 2r_3 = 2 \times 12.8$ mm.

$$D \text{ minimal} = 25.6 \text{ mm}.$$

4. When one moves the source off the centre line, one only sees the portions of the rings lying on the geometric image of the source. The centre of the rings coincides none the less with the axis of the instrument.

The rings are centred on the point S (Fig. 5.10) and the image of the source is centred on S' ($SS' = 12.8$ mm).

Note. Newton's rings and the rings at infinity have the same appearance. The first are due to variations in thickness (constant incidence). The second are due to variations in the incident angle (constant thickness). In the first case k increases when one lengthens the axis and in the second it decreases.

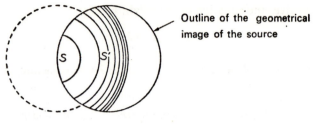

Outline of the geometrical image of the source

S   S'

FIG. 5.10

PROBLEM 6

Interference Filters

Two semi-metallized sheets of glass are separated by a fixed distance e with a constant index of refraction n (Fig. 6.1).

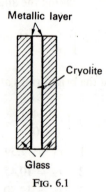

Metallic layer

Cryolite

Glass

FIG. 6.1

1. One wants this filter to have a transmission maximum for normally incident waves of wave length $\lambda = 5500$ Å. Given the fact that the spacing material is cryolite with index, $n = 1.35$, determine the possible values for the spacing e. Only one pass band between 4000 and 7500 Å is wanted (neglect the phase shift due to reflections on the metallized surfaces).

2. How is the wavelength of the transmission maximum changed when parallel rays fall on the filter at an angle of incidence i rather than at normal incidence?

SOLUTION

1. *Normal incidence* (§ 7.5)

The transmission maxima correspond to constructive interference, that is, to path differences

$$\delta = 2ne = k_0\lambda_0 = \ldots = k\lambda \tag{1}$$

$k_0, \ldots, k$ are integers giving the interference order for wavelengths $\lambda_0, \ldots, \lambda$.

One finds a transmission maximum for the filter for thicknesses of cryolite such that

$$e = k_0 \frac{\lambda_0}{2n} = k_0 e_0 \tag{2}$$

$$e = k_0 \times \frac{5500}{2 \times 1.35} = k_0 \times 2040 \text{ Å}.$$

Check for other pass bands in the visible spectrum.

e	λ_1	λ_0	λ_2	Number of pass bands
e_0	$k = 0.73$	$k_0 = 1$	$k = 1.37$	1
$2e_0$	$k = 1.46$	$k_0 = 2$	$k = 2.74$	1
$3e_0$	$k = 2.12$	$k_0 = 3$	$k = 4.11$	2

Only the spacings $e = e_0 = 2040$ Å and $e = 2e_0 = 4080$ Å give but one pass band.

2. *Oblique incidence*

The path difference becomes:
$$\delta = 2ne \cos r = k_0 \lambda_0'. \tag{3}$$

Compare this expression with equation (1). One finds the same interference order for shorter wavelengths.
$$\lambda_0' < \lambda_0.$$

When the filter is inclined, the pass bands shift toward shorter wavelengths.

PROBLEM 7

Fabry–Pérot Etalon. Use of Screens

The plates of Fabry–Pérot etalon are held strictly parallel at a distance of 1 cm by means of three invar wedges. This etalon is placed between two identical converging lenses L_1 and L_2 having focal length 15 cm. In the focal plane of L_1 one places a luminous source 1 cm in diameter (centred on the principal focus of L_1). This source emits monochromatic radiation of wavelength $\lambda = 0.49 \, \mu$ (Fig. 7.1).

Take the index of air equal to 1.

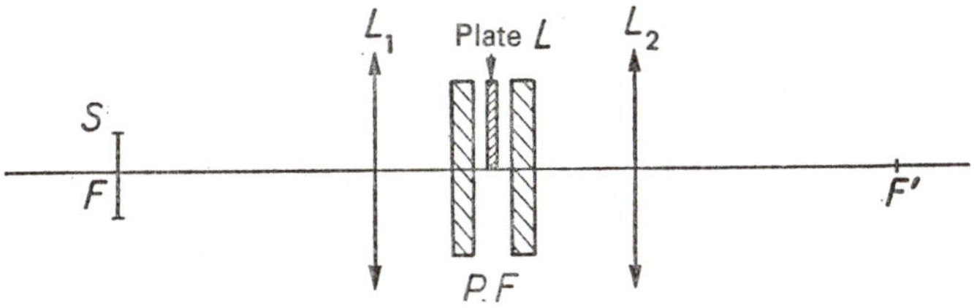

FIG. 7.1

1. Calculate the interference order at point F'.

How many bright rings can one observe in the focal plane of L_2?

What is the order and the radius of the largest of these rings?

2. Between the half-silvered plates place an opaque screen which covers half the surface of these plates.

What is observed in the focal plane of L_2?

3. Replace the opaque screen by a transparent 0.5 mm plate of index 1.5. Explain the appearance of the field. Give the radii of the bright rings.

4. What will one observe if this same glass plate is inserted in an apparatus which gives Newton's rings for normally incident light? (Fig. 7.2.)

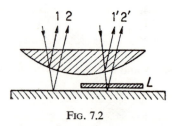

FIG. 7.2

SOLUTION

1. Let i be the angle of incidence. Two adjacent rays give a path difference:

$$\delta = 2ne \cos i = 2ne\left[1 - \frac{i^2}{2}\right]. \tag{1}$$

The bright rings, corresponding to constructive interference, are given by

$$2ne\left[1 - \frac{i^2}{2}\right] = k\lambda \tag{2}$$

where k is an integer.

The interference order at the centre,

$$k_0 = \frac{2e}{\lambda} = \frac{2\times10^4}{0.49} = 40{,}816.32. \tag{3}$$

The interference order at the edge of the field (i_M representing the maximum i) = $0.5/15 = 0.0333 = \frac{1}{30}$ rad.

$$k(i_M) = k_{(0)}\left[1 - \frac{i_M^2}{2}\right] = 40{,}793.65. \tag{4}$$

The order of interference for the largest bright ring is equal to: 40,794.

The angular radius corresponds to a value

$$i = \sqrt{2}\sqrt{\frac{k_{(0)} - k_{(i)}}{k_{(0)}}} = 0.03307 \text{ rad.}$$

From which the linear radius

$$r = fi = 0.4960 \text{ cm.}$$

The 22nd bright ring has a radius of 4.96 mm (it is essentially at the edge of the field).

2. Half the incident rays are intercepted. The useful surface of the etalon is halved. The position and radius of the bright rings remain unchanged but their illumination is halved (Fig. 7.3).

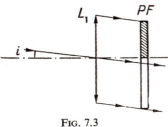

Fig. 7.3

3. Let a parallel bundle of rays make an angle i with the axis of the system. Two rays which pass through the lower part of the etalon have a path difference

$$\delta_1 = 2e \cos i.$$

Two rays which pass through the upper part of the etalon have a path difference δ_2 such that, if e' is the thickness of L,

$$\delta_2(i) = 2(e-e')\cos i + 2ne' \cos r = \delta_1(i) + 2e'[n \cos r - \cos i],$$

$$\delta_2(i) = \delta_1(i) + 2e'\left[n\left(1 - \frac{r^2}{2}\right) - \left(1 - \frac{i^2}{2}\right)\right],$$

$$\delta_2(i) = \delta_1(i) + 2e'\left[(n-1) + \frac{i^2}{2}\left(1 - \frac{1}{n}\right)\right] = \delta_1(i) + e'\left(1 + \frac{i^2}{3}\right).$$

Hence

$$k_2(i) = k_1(i) + \frac{500}{0.49}\left(1 + \frac{i^2}{3}\right).$$

One sees in the field two systems of bright rings centred on F' (Fig. 7.4).

	First ring system	Second ring system
Interference order at the centre	40,816.32	41,836.73
Interference order at the edge	40,793.65	41,814.43
Number of bright rings	23	22

4. At the same distance x from the axis of the system, the rays have path differences

$$\delta = \frac{x^2}{R} \qquad \text{for the rays 1 and 2,}$$

$$\delta' = \frac{x^2}{R} + 2e'(n-1) \quad \text{for the rays 1' and 2',}$$

Whatever value of k is taken, the variation in the interference ordered $[(2e'/\lambda(n-1) = 1020.41]$ differs by an integral number. One sees two ring systems as in Fig. 7.5.

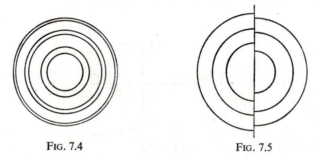

FIG. 7.4   FIG. 7.5

PROBLEM 8

Observation of Phase Objects by the Tolansky Method

Consider a Fabry–Pérot interferometer. The plates L_1 and L_2 are parallel. Their surfaces are separated by a distance e and may be thought of as half-reflecting. The index of the central medium is 1.5. The interferometer is illuminated by a source S situated at the focus of a collimator C. The eye is placed at the focus of an objective O which allows one to focus on any plane between L_1 and L_2 (Fig. 8.1).

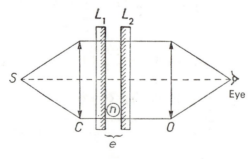

FIG. 8.1

I. *Monochromatic point source* ($\lambda = 0.5\ \mu$)

1. The image of the source is formed on the pupil of the eye and the field observed appears uniformly illuminated. Recall that the illumination is given by

$$I = I_M \frac{1}{1 + m \sin^2 \dfrac{\phi}{2}}$$

m being a characteristic of the apparatus taken as 2500. Find ϕ as a function of e and n.

While maintaining L_1 and L_2 parallel, vary the spacing e. Show graphically the variation of illumination as a function of e.

2. Place between the plates a small "phase-shifting object", that is, a transparent object of thickness e' which differs from the medium by only its index $n' \neq n$ (Fig. 8.2).

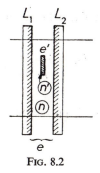

FIG. 8.2

For a given spacing e:

(a) Calculate the phase shift $\Delta\phi$ in the region occupied by the object.

(b) Derive the variation of illumination ΔI in this region. Calculate the contrast of the object with respect to the "background" illumination. Take as the definition of contrast

$$\Gamma = \Delta I / I.$$

(c) For what values of e is the contrast maximal?

(d) What is the smallest path difference detectable by this method if one can see a contrast $\Gamma = 0.1$?

II. *The influence of the size of the source*

The source is now a small luminous disc centred on S with diameter d. Each point on the source gives an illumination $I(i, e, n)$ as a function of spacing, index, and the angle of incidence of the rays on the etalon. Assume that the illumination remains unchanged for the eye when $\phi(i, e, n)$ and $\phi(o, e, n)$ have a maximum variation of $\pi/50$.

Derive the tolerances on the size of the source if the focal length of the collimator is $f = 50$ mm and $e = 1.5$ mm.

III. *The effect of the colour of the source*

Now consider a point source emitting radiation with a width $d\lambda$ and a coefficient of fineness $\lambda/d\lambda$. To this width corresponds a phase variation which should be, as before, less than $\pi/50$. Derive the coefficient of fineness for the source. Again take $e = 1.5$ mm.

SOLUTION

I. *Monochromatic point source*

1. Two adjacent parallel rays differ in phase by

$$\phi = \frac{2\pi}{\lambda} \times 2ne. \tag{1}$$

One has an equal path interferometer. The field has the colour of the source and a uniform illumination given by (§ 7.4)

$$I = I_M \frac{1}{1 + m \sin^2 \frac{\phi}{2}}. \tag{2}$$

When e varies, the illumination varies: it passes through a maximum for $e = k\lambda/2n$ (Fig. 8.3).

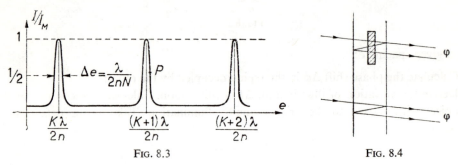

FIG. 8.3   FIG. 8.4

These maxima have widths

$$\Delta e = \frac{\lambda}{2nN}, \quad \text{with} \quad N = \frac{\pi \sqrt{m}}{2}. \tag{3}$$

2. (a) For two rays which have passed through the object, one has a phase difference

$$\phi' = \phi + \Delta\phi \tag{4}$$

such that

$$\Delta\phi = \frac{2\pi}{\lambda} \Delta\delta \tag{5}$$

and

$$\Delta\delta = 2e'(n' - n). \tag{6}$$

(b) By definition

$$\Gamma = \frac{\Delta I}{I} = \frac{-m \sin \phi/2 \cos \phi/2}{1+m \sin^2 \phi/2} \Delta\phi = \frac{-m \sin \phi}{2(1+m \sin^2 \phi/2)} \Delta\phi. \tag{7}$$

(c) The contrast is maximal for $d\Gamma/d\phi = 0$, so that

$$\cos \phi \left(1+m \sin^2 \frac{\phi}{2}\right) = 2m \sin^2 \phi/2 \cos^2 \phi/2$$

$$\sin^2 \phi/2 = \frac{1}{m+2}.$$

$$\sin \phi = \frac{2\sqrt{m+1}}{m+2} \approx \frac{2}{\sqrt{m}} = \frac{4}{100}. \tag{8}$$

The Fabry–Pérot etalon should have a spacing

$$e = \frac{\lambda}{2n}\left[k\pm\frac{1}{\pi\sqrt{m}}\right]. \tag{9}$$

By substituting the value of ϕ in (7) one has

$$\Gamma_{max} = \frac{m \Delta\phi}{2\sqrt{m+1}} \approx \frac{\sqrt{m}}{2}\Delta\phi = 25 \Delta\phi. \tag{10}$$

(d) If one takes $\Gamma_{max} = 0.1$, one has

$$\Delta\phi = \frac{1}{250} = 2\pi \frac{\Delta\delta}{\lambda},$$

hence

$$\Delta\delta = \frac{\lambda}{2\pi\times250} = \frac{5\times10^{-6}}{2\pi\times250} = 3.18 \text{ Å}.$$

Note. For the contrast to be good, it is necessary to place oneself at a point P on the curve 8.3 where the variations of I are large. The ideal point corresponds to

$$\frac{d}{d\phi}\left[\frac{I'(\phi)}{I}\right] = 0, \quad \text{or} \quad II'' - I'^2 = 0.$$

This does not coincide with the inflection point of the curve for which one has $I'' = 0$ but it is very close to it in the case where m is large.

II. *Influence of the size of the source*

One has

$$\delta_0 = 2ne,$$

$$\delta_i = 2ne \cos r = 2ne\left(1-\frac{r^2}{2}\right) = 2ne\left(1-\frac{i^2}{2n^2}\right), \tag{11}$$

$$\Delta\delta = \delta_0 - \delta_i = \delta_0 \frac{i^2}{2n^2} = e\frac{i^2}{n}. \tag{12}$$

One wants $\Delta\phi < \pi/50$ where $\Delta\delta < \lambda/100$. Thus,

$$i < \frac{1}{10}\sqrt{\frac{\lambda n}{e}} \quad \text{or} \quad d = 2fi = \frac{2\times50}{10}\times\sqrt{\frac{0.5\times1.5}{1.5\times10^3}}$$

$$d \leqslant 0.224 \text{ mm.}$$

III. *The influence of the spectral width of the source*

When the source is not monochromatic the phaseshift is not constant. From (1) one has

$$d\phi = -\frac{4\pi ne}{\lambda^2}\, d\lambda.$$

So that this variation is less than $\pi/50$, it is necessary that

$$\frac{\lambda}{d\lambda} \geqslant \frac{200ne}{\lambda} = \frac{200\times1.5\times1.5\times10^3}{0.5},$$

so that

$$\frac{\lambda}{d\lambda} \geqslant 9\times10^5.$$

This is a large coefficient of fineness.

PROBLEM 9

Fabry–Pérot (Interference) Spectroscopy

I

Consider the interference apparatus formed by a layer of air of thickness e with parallel faces limited by two plates of glass whose opposite faces P and Q have an improved reflection coefficient (Fabry–Pérot interferometer). The reflection of a light ray on each of these surfaces is accompanied by a phase shift which is taken to be zero for all radiation. Assume at the outset that the apparatus is such that reflection of light on faces other than P and Q does not occur. The index of air is taken equal to one. A lens L of focal length f whose optic axis is normal to the faces P and Q is situated behind the etalon (see Fig. 9.1). One wants to study the rings in the far-field resulting from transmission by the Fabry–Pérot interferometer.

1. Recall in which plane one should observe the rings so that contrast will not be lost when one uses an extended source.

2. Calculate the wavelengths of radiation λ_K for which the centre of the system of rings has maximum light intensity when using the interference order corresponding to K.

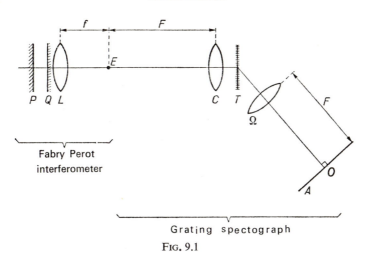

Fabry Perot
interferometer

Grating spectograph

FIG. 9.1

3. With the interferometer illuminated by radiation with wavelength λ slightly less than the preceding value λ_K, calculate the angular radius $\alpha_K(\lambda)$ of bright rings corresponding to interference order K as a function of only λ_K and λ. (One assumes that the difference $\lambda_K - \lambda$ is such that the the angle $\alpha_K(\lambda)$ must be thought of as small.)

II

Consider a grating spectrograph formed essentially of an entry slit E which is infinitely fine and a transmission grating T with a width (normal to the rulings) of $L = 5$ cm and which has 1000 rulings per millimetre. The collimator C and the objective Ω both have the same focal length $F = 3$ m. The optical axis of the collimator is normal to the grating and the optical axis of the objective is parallel to the diffracted rays in the first order for the wavelength $\lambda_0 = 5000$ Å (Fig. 9.1).

Each image of the slit formed by the spectrograph for a monochromatic radiation is recorded on a photographic plate A normal to the optical axis of the objective. The points on this plate are referenced by a system of rectangular axes Ox and Oy. The various monochromatic images from the slit are formed on Ox, O being the point where the optic axis of the objective intersects the plate.

1. What is the distance on the plate which separates the images of the slit corresponding to wavelengths $\lambda_1 = 4500$ Å and $\lambda_2 = 5500$ Å in the first diffraction order?

2. Calculate the values of the linear dispersion $D = dx/d\lambda$ in millimetres per Å for the wavelengths λ_0, λ_1, and λ_2 in the first diffraction order.

3. Calculate the resolving power of the grating when used in the first order.

III

The entry slit of the preceding spectrometer is placed along a diameter of the system of rings of the interferometer, the middle of the slit coinciding with the centre of the system of rings (Fig. 9.2). Assume throughout that the grating always acts in the first order and here the investigation will be limited to radiation with wavelengths between λ_1 and λ_2 so that one can consider the dispersion as being linear and equal to the value D calculated in II-2 for the wavelength λ_0.

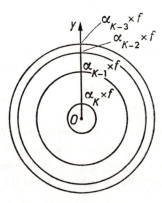

Fig. 9.2

1. When the interferometer is illuminated by light with wavelength λ, show that one sees points of maximum illumination on the plate. Find the coordinates of these points in the xOy system defined above.

2. When the interferometer is illuminated by two monochromatic waves with lengths λ and $\lambda + d\lambda$ which are very close together, find the difference dy of the ordinates of the two maxima corresponding to the same order of interference K. Derive the value $D' = dy/d\lambda$ of the dispersion due to the interferometer.

What happens to D' when the wavelength λ tends toward λ_K?

3. When the interferometer is illuminated with white light, show that a set of maxima appears on the photographic plate as lines and find the equation of these lines in the xOy coordinate system. Describe the nature of these curves.

Find the distance which separates the points of intersection of two successive curves with the Ox axis near $\lambda = \lambda_0$ for $e = 2$ mm.

4. Show that the difference $y_2 - y_1$ of the ordinates of two points M_2 and M_1 with maxima corresponding to the K and $K-1$ orders of interference for the wavelength λ, can be put in the form:

$$\Delta y = y_2 - y_1 = \frac{1}{y}\,\phi(\lambda, f, e),$$

where y represents an ordinate lying between y_2 and y_1. Assume that the difference Δy is small and derive the difference in ordinate δy which separates two points close to M_1 on the plate for which the illumination is equal to half the maximum illumination under the assumption that the reflection coefficients of P and Q are such that the coefficient of fineness has the value $N = 30$.

What is the resolving power R' for this device near the wavelength λ_0? Compare this with the resolution of the grating spectrograph. Indicate why it is necessary to use the spectrograph in conjunction with the Fabry–Pérot interferometer.

SOLUTION

I. *Rings in the far-field* (§ 7.4)

1. The very fine rings appearing bright on a dark background are localized at infinity. They can be observed in the focal plane E of the lens L.

2. One finds a bright centre for all wavelengths such that

$$\frac{2e}{\lambda_K} = K \rightarrow \lambda_K = \frac{2e}{K} \qquad (K \text{ integer}). \tag{1}$$

3. Rings of order K for wavelengths $\lambda < \lambda_K$. One has

$$2e \cos \alpha_K(\lambda) = K\lambda. \tag{2}$$

Since α_K is small

$$2e\left(1 - \frac{\alpha_K^2}{2}\right) = K\lambda,$$

hence, by using (1),

$$\alpha_K(\lambda) = \sqrt{2}\,\sqrt{1 - \frac{\lambda}{\lambda_K}}. \tag{3}$$

II. *Dispersion of the grating* (§ 7.8)

1. The incident rays are normal to the grating ($i = 0$).
Let P be the step of the grating. The principal maxima are given by

$$P \sin i' = p\lambda \qquad (p \text{ integer}).$$

The first-order spectra are formed in a direction i' such that

$$P \sin i' = \lambda \rightarrow \sin i' = \frac{\lambda}{P} = \frac{0.5}{1} = \frac{1}{2}, \tag{4}$$

$$i' = 30°.$$

By differentiating (4), one gets

$$P \cos i'\, \mathrm{d}i' = \mathrm{d}\lambda. \tag{5}$$

The images of the entry slit corresponding to λ_1 and λ_2 are separated by a distance of

$$dx = F \, di' = \frac{F \, d\lambda}{P \cos i'} = 3 \times 10^3 \times \frac{0.1}{1 \times \frac{\sqrt{3}}{2}} = 2 \sqrt{3} \times 10^2. \tag{6}$$

$$dx = 346.5 \text{ mm.}$$

2. Combining (4) and (5) one gets

$$\frac{di'}{d\lambda} = \frac{1}{P \cos i'} = \frac{1}{P} \frac{1}{\sqrt{1 - \lambda^2/P^2}} = \frac{1}{\sqrt{P^2 - \lambda^2}}.$$

for which the grating dispersion $D = dx/d\lambda$ in the first-order is

$$D = \frac{dx}{d\lambda} \frac{F}{\sqrt{P^2 - \lambda^2}}. \tag{7}$$

One uses F in mm and P and λ in Å.

Numerical application:

$$\lambda_1 = 4500 \text{ Å,} \quad D_1 = \frac{3 \times 10^3}{\sqrt{10^8 - (4.5)^2 10^6}} = \frac{3}{\sqrt{10^2 - (4.5)^2}} = 0.335 \text{ mm. Å}^{-1}$$

$$\lambda_2 = 5000 \text{ Å} \quad D_0 = \frac{3 \times 10^3}{\sqrt{10^8 - (5)^2 10^6}} = \frac{3}{\sqrt{10^2 - (5)^2}} = 0.346 \text{ mm. Å}^{-1}$$

$$\lambda_3 = 5500 \text{ Å} \quad D_2 = \frac{3 \times 10^3}{\sqrt{10^8 - (5.5)^2 10^6}} = \frac{3}{\sqrt{10^2 - (5.5)^2}} = 0.36 \text{ mm. Å}^{-1}$$

3. Resolving power of the grating in the first order.

For the two wavelengths λ and $\lambda + d\lambda$, two first-order spectra are separated by $di' = d\lambda/P \cos i'$. The width of each spectrum is

$$\delta i' = \frac{\lambda}{L \cos i'}. \tag{8}$$

The two spectra are resolved if

$$di' \geqslant \delta i'. \tag{8'}$$

The resolving power of the grating in the first-order is

$$R = \frac{\lambda}{d\lambda}, \tag{9}$$

so that, according to (8'),

$$R = \frac{L}{P} = \text{number of rulings} = nL,$$

$$R = 1000 \times 50 = 5 \times 10^4.$$

III. *Dispersion of the Fabry–Pérot etalon*

The objectives C and Ω have the same focal length. The ordinate of conjugate points is the same in plane E as in plane A (unit magnification). It is sufficient to examine the dispersion of the Fabry–Pérot etalon in the focal plane of L.

1. Using a monochromatic wave of length λ.
Bright rings occur for

$$\frac{2e}{\lambda}\cos\alpha_K = \frac{2e}{\lambda}\left(1-\frac{\alpha_K^2}{2}\right) = K \qquad (K \text{ integer}). \tag{10}$$

The entry slit of the spectrometer cuts these rings along a diameter. In the plane E the bright points have ordinates

$$y = f\alpha_K = f\sqrt{2-\frac{K\lambda}{e}} \qquad \text{(Fig. 9.2)}. \tag{11}$$

The entry slit coincides with Oy. The photographic plate is normal to the first-order diffracted rays for $\lambda_0 = 5000$ Å. For this spectrum the grating is taken to have a constant dispersion

$$D = \frac{dx}{d\lambda} = 0.346 \text{ mm Å}^{-1}.$$

In the plane A, one observes bright points with coordinates (Fig. 9.3)

$$\left. \begin{array}{l} x = D\lambda = \text{const.} \\ y = f\sqrt{2-\dfrac{K\lambda}{e}}. \end{array} \right\} \tag{12}$$

2. One has here two monochromatic waves λ and $\lambda+d\lambda$. Thus one finds two concentric systems of rings in plane E, hence two series of bright points on the entry slit. On the plate one sees two series of bright points shifted by $D\times d\lambda$ (Fig. 9.4). The derivation of (11) allows

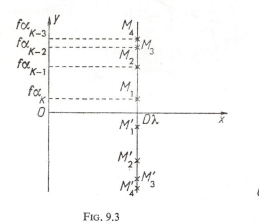

FIG. 9.3

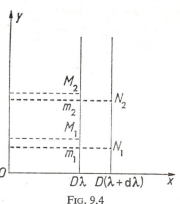

FIG. 9.4

one to write:

$$dy = -f\frac{K}{2e}\frac{1}{\sqrt{2-\dfrac{K\lambda}{e}}}\,d\lambda \qquad (13)$$

dy corresponding to the separation of two points such as Mm.

From (13) one derives the dispersion of the etalon:

$$D' = \frac{dy}{d\lambda} = -f\frac{K}{2e}\frac{1}{\sqrt{2-\dfrac{K\lambda}{e}}}. \qquad (14)$$

One can see that maximum dispersion is obtained for the maximum values of K, that is for rings of small diameter. At the limit, for $\lambda = \lambda_K = 2e/K$, D' becomes infinite.

3. *Band spectra.* If the source emits all the wavelengths lying between λ_1 and λ_2, the equation of the lines of maximal illumination in the plane xOy is given by (12). By eliminating λ one finds the expression

$$y^2 = f^2\left(2-\frac{K}{e}\frac{x}{D}\right). \qquad (15)$$

These are bright, very fine parabolas about the Ox axis.

One passes from one parabola to the next by making a unit change in K (Fig. 9.5). The apexes are on Ox and have as abscissa

$$x = \frac{2e}{K}D. \qquad (16)$$

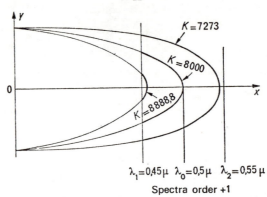

Spectra order +1

Fig. 9.5

Two successive apexes are separated by

$$|\Delta x| = \frac{2e}{K^2}D\times\Delta K = \frac{2e}{K^2}D = \frac{\lambda_0^2}{2e}D = \frac{(5\times10^3)^2\times0.346}{4\times10^7} \approx 0.2 \text{ mm.}$$

for
$$\lambda_1 = 4500 \text{ Å,} \quad \text{one has} \quad K_{\lambda_1} = 2e/\lambda_1 = 8888.8,$$
$$\lambda_0 = 5000 \text{ Å} \qquad\qquad K_{\lambda_0} = 2e/\lambda_0 = 8000.0,$$
$$\lambda_2 = 5500 \text{ Å} \qquad\qquad K_{\lambda_2} = 2e/\lambda_2 = 7272.7.$$

One sees then $(8888-7273)+1 = 1616$ "sections" of parabolas in the field. They are concave towards the blue.

4. Resolving power of the Fabry–Pérot. Wavelength λ. Orders K and $K-1$.

Let Δy be the distance between points M_1 and M_2 (Fig. 9.4). Expression (11) gives

$$\Delta y = -f \frac{\lambda}{2e} \frac{\Delta K}{\sqrt{2 - \frac{K\lambda}{e}}}.$$

If $\Delta K = 1$:

$$\Delta y = -f \frac{\lambda}{2e} \frac{1}{\sqrt{2 - \frac{K\lambda}{e}}} \tag{17}$$

$$\Delta y = -\frac{1}{y} \frac{f^2 \lambda}{2e} = M_2 M_1. \tag{18}$$

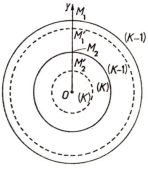

FIG. 9.6

The rings contract as one passes from the centre to the periphery (Fig. 9.6). For wavelength λ, the rings of order K and $K-1$ are represented as solid lines. For the wavelength $\lambda+d\lambda$, the rings are represented as dotted lines.

Consider a monochromatic wave λ. If Δy is the distance between two consecutive rings and δy the width of the rings, one has

$$N = \frac{\Delta y}{\delta y}. \tag{19}$$

N is called the coefficient of fineness. Its value $\pi \sqrt{R}/(1-R)$ depends only on the reflection factor R of the Fabry–Pérot plates.

If the source emits two close wavelengths λ and $\lambda+d\lambda$, one says that two points such as M_1 and M_1' are resolved if the distance $M_1 M_1' = dy$ is greater than δy (Fig. 9.7):

$$dy \geqslant \delta y. \tag{20}$$

Combining (19) and (20) one arrives at the condition

$$dy \geqslant \frac{1}{N} \Delta y. \tag{21}$$

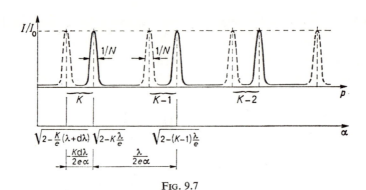

FIG. 9.7

Referring to the values of dy and Δy given by (13) and (17) and using them in the inequality (21) gives

$$\frac{\lambda}{d\lambda} \geqslant KN = 30 \, K.$$

The Fabry–Pérot resolving power is

$$R' = KN. \tag{22}$$

For λ_0, the highest order of interference is at the centre: $K = 8000$. One finds

$$R' = 30 \times 8000 = 24{,}000 \approx R/2.$$

The dispersion axes of the Fabry–Pérot (Oy) and of the grating (Ox) are crossed. One thus eliminates all ambiguity in spectral analysis for the case where one has overlapping of different interference orders at the same point of Oy.

ELECTROMAGNETIC OPTICS

PROBLEM 10

Interference of Hertzian Waves

The electrical properties of the water of the Atlantic Ocean with respect to hertzian (r.f.) waves are characterized by the following constants:

$$\varepsilon_r = 81, \quad \mu_r = 1, \quad \gamma = 4.3\,\Omega^{-1}\,\mathrm{m}^{-1}$$

1. Show that this water can be regarded as a good conductor for frequencies less than 10^8 Hz. In the following parts of the problem assume that the water is a perfect conductor.

2. Under these conditions, consider a horizontal dipole antenna situated at point S at a height H above the surface of the sea emitting monochromatic hertzian waves with wavelength λ. A receiver is placed at a point O at a height h above the water in the equatorial plane of the emission from the antenna and at a horizontal distance D from it. Assume that D is much larger than H and h and that the surface of the water, assumed planar, extends from S to O (Fig. 10.1).

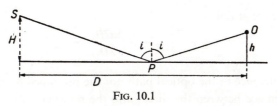

FIG. 10.1

Find the variations of the electric field as a function of H, h, and λ for a given value of D. Determine the minimum value of h for which there is optimum reception. Numerical application: $H = 300$ m, $D = 10$ km, $\lambda = 30$ m.

3. For small values of h, find the expression for the intensity of the wave at O as a function of D and compare this with the corresponding expression one would find in the absence of the ocean. Assuming that the waves propagate parallel to the surface of the sea (valid since H and h are small), calculate the mean power $\langle \mathcal{P} \rangle$ which passes normally through a unit surface area at O as a function of the total mean power $\langle \Phi \rangle$ radiated by the dipole.

Numerical application: $H = h = \lambda = 10$ m, $D = 10$ km, $\langle \Phi \rangle = 10$ W.

SOLUTION

1. For a good conductor, the conduction current exceeds the displacement current and the imaginary part of the complex index (§ 2.5) is larger than the real part. Now the ratio of these two variables for ocean water is

$$\frac{\gamma}{\varepsilon_0 \varepsilon_r \omega} = \frac{6.1 \times 10^9}{\omega} \approx \frac{10^9}{\nu}.$$

Thus this water is a conductor which insulates for $\nu < 10^9$ Hz ($\lambda = 0.3$ m) and is a good conductor for $\nu = 10^8$ Hz.

2. The antenna receives both the wave propagating along SO and the wave reflected at P by the surface of the water, which acts as a perfect mirror, at an angle close to $\pi/2$ since $D \gg H$ (here $i \approx 88°15'$). The reflection factor is essentially equal to one. The electric field of the waves is horizontal and the reflection introduces a phase shift of π, with the result that the tangential component of the electric field is zero at the surface of the conductor (§ 2.6.3).

The path difference of the waves reaching O is $SPO - SO$:

$$SO = \sqrt{D^2 + (H-h)^2} \approx D\left[1 + \frac{1}{2}\left(\frac{H-h}{D}\right)^2\right],$$

$$SPO = \sqrt{D^2 + (H+h)^2} \approx D\left[1 + \frac{1}{2}\left(\frac{H+h}{D}\right)^2\right],$$

so that

$$\delta = SPO - SO = \frac{2Hh}{D}$$

and the phase difference at O is

$$\phi = \frac{2\pi\delta}{\lambda} + \pi = \frac{4\pi Hh}{\lambda D} + \pi.$$

Neither the small difference in the optical path nor the presumed total reflection produces any reasonable difference between the direct and the reflected wave in amplitude (which varies as $1/r$). Since these fields are parallel, the resultant field is given by:

$$\frac{E_0}{D}\cos \omega t + \frac{E_0}{D}\cos (\omega t + \phi) = -\frac{2E_0}{D}\sin\left(\omega t + \frac{2\pi Hh}{\lambda D}\right)\sin \frac{2\pi Hh}{\lambda D}.$$

One then gets at every instant on the vertical at O a set of maxima and minima in the amplitude. At the surface of the water ($h = 0$) there is a zero minimum. The first maximum occurs at the height $h_1 = \lambda D/4H$.

Numerical application:

$$h_1 = \frac{30 \times 10^4}{4 \times 300} = 250 \text{ m.}$$

3. For $h \ll h_1$, one can replace $\sin 2\pi Hh/\lambda D$ by the angle. The amplitude of the electric field at O is then proportional to $1/D^2$ and the intensity of the resulting wave to $1/D^4$, while the intensity of the direct wave varies as $1/D^2$.

The power $\langle \mathcal{P} \rangle$ is given by the expression (§ 2.3)

$$\langle \mathcal{P} \rangle = \tfrac{1}{2}\varepsilon_0 c E_m^2. \tag{2}$$

E_m represents the field amplitude (1), so that

$$E \approx E_0 \frac{4\pi Hh}{\lambda D^2}. \tag{3}$$

The field E_0/D, produced by the dipole antenna at a distance D in its equatorial plane, has an amplitude (with $\theta = \pi/2$) given by equation (10.10) of § 10.3:

$$\frac{E_0}{D} = \frac{1}{4\pi\varepsilon_0 c^2} \frac{\omega_0^2 d_m}{D}. \tag{4}$$

On the other hand, the total mean power $\langle \Phi \rangle$ radiated by the sinusoidal dipole is given by equation (10.13) of § 10.3:

$$\langle \Phi \rangle = \frac{\omega_0^4 d_m^2}{12\pi\varepsilon_0 c^3} = \frac{4}{3}\pi\varepsilon_0 c E_0^2. \tag{5}$$

Using (2), (3), (4), and (5),

$$\langle \mathcal{P} \rangle = 6\pi \frac{H^2 h^2}{\lambda^2 D^4} \langle \Phi \rangle.$$

Numerical application:

$$\langle \mathcal{P} \rangle = 6\pi \times 10^{-8} \approx 0.2 \times 10^{-6} \text{ W}.$$

PROBLEM 11

Fresnel Formulas

I

The Fresnel equations give the reflection coefficients obtained by assuming that the magnetic permeabilities μ_1 and μ_2 of the dielectric are equal to that of free space μ_0. What happens to these equations if this assumption is set aside?

II

Consider the possibility of linear polarization of reflected light resulting from reflection if μ_1 and μ_2 are different from μ_0.

III

Show that under normal incidence the reflection coefficient is zero for a dielectric in vacuum where the relative permittivity ε_r is equal to the relative permeability μ_r.

SOLUTION

I

Maxwell's equations for a plane wave give the following general expression relating the magnetic field H to the electric field E (§ 2.4):

$$H = \sqrt{\frac{\varepsilon}{\mu}} E.$$

The equations of continuity relative to reflection and refraction at the surface between two transparent media for the component of the electric field normal to the plane of incidence is given by Fig. 11.1 (§ 3.2).

$$E_i + E_r = E_t, \quad \text{hence} \quad 1 + r = t;$$

$$(H_i - H_r) \cos i_1 = H_t \cos i_2, \quad \text{from which} \quad \sqrt{\frac{\varepsilon_1}{\mu_1}}(1-r)\cos i_1 = \sqrt{\frac{\varepsilon_2}{\mu_2}} t \cos i_2.$$

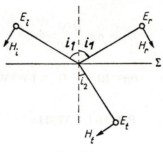

FIG. 11.1

The derived coefficient of reflection $r_\perp$ is

$$r_\perp = \frac{\sqrt{\dfrac{\varepsilon_1}{\mu_1}}\cos i_1 - \sqrt{\dfrac{\varepsilon_2}{\mu_2}}\cos i_2}{\sqrt{\dfrac{\varepsilon_1}{\mu_1}}\cos i_1 + \sqrt{\dfrac{\varepsilon_2}{\mu_2}}\cos i_2} \tag{1}$$

and the transmission coefficient $t_\perp$ is

$$t_\perp = \frac{2\sqrt{\dfrac{\varepsilon_1}{\mu_1}}\cos i_1}{\sqrt{\dfrac{\varepsilon_1}{\mu_1}}\cos i_1 + \sqrt{\dfrac{\varepsilon_2}{\mu_2}}\cos i_2}.$$

One finds for the coefficients relative to the case where the E fields are in the plane of incidence

$$r_{||} = \frac{\sqrt{\dfrac{\varepsilon_2}{\mu_2}}\cos i_1 - \sqrt{\dfrac{\varepsilon_1}{\mu_1}}\cos i_2}{\sqrt{\dfrac{\varepsilon_2}{\mu_2}}\cos i_1 + \sqrt{\dfrac{\varepsilon_1}{\mu_1}}\cos i_2},$$

$$t_{||} = \frac{2\sqrt{\dfrac{\varepsilon_1}{\mu_1}}\cos i_1}{\sqrt{\dfrac{\varepsilon_2}{\mu_2}}\cos i_1 + \sqrt{\dfrac{\varepsilon_1}{\mu_1}}\cos i_2}.$$

(2)

II

Incidence at Brewster's angle causes $r_{||}$ to vanish. This condition occurs, using (2), for

$$\sqrt{\frac{\varepsilon_2}{\mu_2}}\cos i_1 = \sqrt{\frac{\varepsilon_1}{\mu_1}}\cos i_2 = \sqrt{\frac{\varepsilon_1}{\mu_1}}\sqrt{1 - \sin^2 i_2}.$$

By using the law of refraction

$$\frac{\sin i_2}{\sin i_1} = \frac{v_2}{v_1} = \sqrt{\frac{\varepsilon_1\mu_1}{\varepsilon_2\mu_2}}$$

and by expressing $\sin i_1$ and $\cos i_1$ as a function of $\tan i_1$, one finds

$$\tan i_B = \sqrt{\frac{\varepsilon_2(\varepsilon_2\mu_1 - \varepsilon_1\mu_2)}{\varepsilon_1(\varepsilon_2\mu_2 - \varepsilon_1\mu_1)}}.$$

For $\mu_1 = \mu_2 = \mu_0$, this expression reduces to

$$\tan i_B = \sqrt{\frac{\varepsilon_2}{\varepsilon_1}} = \frac{n_2}{n_1},$$

which is the usual expression for the Brewster angle.

Equation (1) shows that under the assumptions made here, $r_\perp$ can also vanish. Proceeding as above, one finds that this occurs for an angle i'_B such that

$$\tan i'_B = \sqrt{\frac{\mu_2(\varepsilon_2\mu_1 - \varepsilon_1\mu_2)}{\mu_1(\varepsilon_1\mu_1 - \varepsilon_2\mu_2)}}.$$

Under ordinary conditions where $\mu_1 = \mu_2 = \mu_0$, one has $\tan i'_B = \sqrt{-1}$. There is no Brewster angle for the vibration perpendicular to the plane of incidence.

III

In the case of normal incidence, equations (1) and (2) become

$$r_\perp = \frac{\sqrt{\dfrac{\varepsilon_1}{\mu_1}} - \sqrt{\dfrac{\varepsilon_2}{\mu_2}}}{\sqrt{\dfrac{\varepsilon_1}{\mu_1}} + \sqrt{\dfrac{\varepsilon_2}{\mu_2}}} = -r_\| . \tag{3}$$

Since $\varepsilon = \varepsilon_0\varepsilon_r$ and $\mu = \mu_0\mu_r$, if $\varepsilon_r = \mu_r, r = 0$. One notes that the ratio $\sqrt{\mu/\varepsilon}$ is the intrinsic impedance of the medium (§§ 2.3 and 2.4). Realization of equation (3) at the surface of separation of two media is equivalent to the situation where the impedance is matched at the junction of two transmission lines.

PROBLEM 12

Fresnel Formulas. Thin Films

1. Starting with the Fresnel formulas for glassy reflection at normal incidence, show that the amplitude reflection coefficients r and r' for crossing the surface in both directions satisfy $r = -r'$ and the corresponding transmission coefficients t and t' satisfy $tt' = 1 - r^2$.

2. A layer of a transparent substance with index n_1, parallel faces, and thickness e covers a glass surface of index n_2. Its upper face is in contact with the air whose index is taken as unity. A plane monochromatic wave of length λ in air and unit amplitude intersects the layer from the air at normal incidence. Show that the reflected intensity, taking into account the multiple reflections is given by

$$R = \frac{r_1^2 + r_2^2 + 2r_1r_2 \cos \phi}{1 + r_1^2r_2^2 + 2r_1r_2 \cos \phi},$$

r_1 and r_2 being the reflection coefficients for air-layer and the layer-glass respectively and ϕ the phase difference between two successive reflected rays.

3. Show that if $1 < n_1 < n_2$, the transmission factor of the layer plus the glass is always greater than that of the glass alone for any spacing e. For $n_1 = 1.35$ and $n_2 = 1.50$, by how much will the reflection factor R be lessened (relative to the intensity) from that of the glass alone by the deposition of a layer having the optimal spacing.

Among the possible optimal spacings for the wavelength λ, why does one use only the smallest?

4. Show that if $n_1 > n_2$, the reflection factor at the glass is increased. For what thickness R is it maximal? Do the calculation for $n_1 = 2.30$.

5. Indicate the advantages and disadvantages which occur for partially reflecting metal and dielectric films.

SOLUTION

1. *Fresnel formulas* (§ 7.2).

$$r = \frac{n_1-n_2}{n_1+n_2}, \quad r' = \frac{n_2-n_1}{n_2+n_1} \rightarrow r = -r', \tag{1}$$

The continuity of the electric field requires, for normal incidence: $1+r = t$,

$$t = \frac{2n_1}{n_1+n_2}, \quad t' = \frac{2n_2}{n_1+n_2} \rightarrow tt' = 1-r^2. \tag{2}$$

2. *Reflection factor.* This can be found by the method used in Problem 14, part II. Here, one is required to take into account multiple reflections (Fig. 12.1).

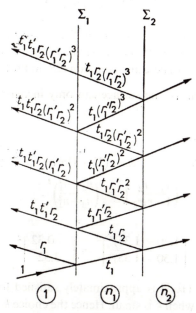

Fig. 12.1

To obtain the reflected amplitude, one can sum the amplitudes of the successive rays either in air or in the substance of index n_1. The electric field has the same value on either side of Σ_1 on the surface separating the media

Medium of index 1		Medium of index n_1	
$E = E_i+E_r$	$=$	E_t	(3)

Calling ϕ the phase shift introduced by a double pass through the substance

$$\phi = \frac{2\pi}{\lambda} \times 2n_1 e,$$

r being < 1 one always has $r^N \to 0$ (N is the number of reflections). Value of the resulting field in air:

$$E = 1 + r_1 + t_1 t_1' r_2 e^{-j\phi}[1 + r_1' r_2 e^{-j\phi} + \ldots] = 1 + r_1 + \frac{t_1 t_1' r_2 e^{-j\phi}}{1 - r_1' r_2 e^{-j\phi}}. \tag{4}$$

Value of the resulting field in the substance:

$$E = t_1[1 + r_1' r_2 e^{-j\phi} + \ldots] + t_1 r_2 e^{-j\phi}[1 + r_1' r_2 e^{-j\phi} + \ldots] = \frac{t_1[1 + r_2 e^{-j\phi}]}{1 - r_1' r_2 e^{-j\phi}}. \tag{5}$$

One can easily verify that equations (4) and (5) are identical, since $1 + r_1 = t_1$.

Given that $E = E_i + E_r = 1 + r$, one can immediately get from (4), using (1) and (2), the reflected amplitude:

$$r = r_1 + \frac{(1 - r_1^2) r_2 e^{-j\phi}}{1 + r_1 r_2 e^{-j\phi}}. \tag{6}$$

The reflected energy value $R = |r|^2$ is given in the text of the problem.

Note. As a further exercise, one can replace r_1 and r_2 by the values found from the Fresnel formulas and verify equation (6) here and equation (28) in Problem 14.

3. $1 < n_1 < n_2$. Treatment of the surface can only increase the transmission factor. In effect,

 (a) Treated surface film:

$$R_{\text{treated}} = R_{\text{minimum}} \text{ if } \phi = (2k+1)\pi \text{ or } n_1 e = (2k+1)\lambda_0/4,$$

$$R = \left(\frac{r_1 - r_2}{1 - r_1 r_2}\right)^2 = \left(\frac{n_2 - n_1^2}{n_2 + n_1^2}\right)^2.$$

Numerical application:

$$R = \left[\frac{1.50 - (1.35)^2}{1.50 + (1.35)^2}\right]^2 = \left[\frac{-0.32}{3.32}\right]^2 \approx 0.01.$$

The condition $n_1 e = (2k+1)\lambda_0/4$ is approximately satisfied for wavelengths about λ_0 and more so in a larger domain when k is small. Hence the choice $k = 0$.

 (b) Plain glass: $R_{\text{maximum}} = R_{\text{plain glass}}$ if $\phi = 2k\pi$ (or $n_1 e = k\lambda_0/2$):

$$R = \left(\frac{1 - n_2}{1 + n_2}\right)^2 = \left(\frac{1 - 1.5}{1 + 1.5}\right)^2 = 0.04.$$

4. $1 < n_1 > n_2$. Treatment of the surface can only increase the reflection factor.

 (a) $R_{\text{treated}} = R_{\text{maximum}}$ if $\phi = (2k+1)\pi$ or $n_1 e = (2k+1)\lambda_0/4$:

$$R = \left(\frac{r_1 - r_2}{1 - r_1 r_2}\right)^2 = \left(\frac{n_2 - n_1^2}{n_2 + n_1^2}\right)^2.$$

 (b) $R_{\text{plain glass}} = \left(\frac{1 - n_2}{1 + n_2}\right)^2.$

Numerical application:

$$R_{\text{treated glass}} = \left[\frac{1.50-(2.3)^2}{1.50+(2.3)^2}\right]^2 = \left(\frac{3.8}{6.8}\right)^2 = 0.31,$$

$$R_{\text{plain glass}} = 0.04.$$

5. Metallic films are absorbing. The dielectric films are selective.

PROBLEM 13

Newton's Rings in Polarized Light

One gets Newton's rings between a plano-convex glass lens with large radius of curvature and index n_1 and a glass flat whose index n_2 differs significantly from n_1. The incident light is parallel and linearly polarized. Describe qualitatively the effect of a variation of the angle of incidence on the visibility of the rings:

1. When the vibration is parallel to the plane of incidence.

2. When it is perpendicular.

SOLUTION

1. The transmitted intensity crossing the upper face of the lens increases uniformly with the incident angle. The amplitude reflected on the lower face vanishes for an angle of incidence i_1 such that $\tan i_1 = n_1$. There can no longer be interference with the rays reflected on the glass flat and the rings vanish. They also disappear for an angle of incidence i_2 such that $\tan i_2 = n_2$, the rays no longer being reflected on the glass plate.

2. The visibility undergoes some small uninteresting variations, but never vanishes.

PROBLEM 14

Propagation of Waves in a Stratified Dielectric Medium

I

A monochromatic plane wave whose amplitude can be taken as unity falls with normal incidence on a plane surface which separates two transparent media with indices n_1 and n_2.

1. Set up the Fresnel formulas giving the transmitted amplitude t and the reflected amplitude r.
Numerical application: $n_1 = 1$, $n_2 = 1.5$.

2. By using the conservation of energy, write the expression which connects r and t.

II

Consider a thin film of a transparent substance with index n, thickness d deposited on a support formed by a plate of plane glass L with index $n_s > n$. The thickness of the glass is sufficiently large that it can be thought of as infinite. The set up is a "treated plate" (Fig. 14.1). A monochromatic plane wave (wavelength λ_0 in vacuum) propagating in the direction Oz falls on the treated plate under normal incidence.

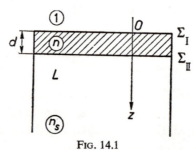

FIG. 14.1

1. Write the equations of continuity for the electric and magnetic fields on the surface Σ_I (air-film) and on Σ_{II} (film-glass). What expression relates the components of E and H before and after traversing the thin film? Show that this thin film can be characterized by a four-element square matrix. Assume $q = \sqrt{\varepsilon/\mu_0}$.

2. Determine, as a function of the indices n_s and n and the thickness d, the following characteristics of the treated plate: transmitted amplitude t; reflected amplitude r; transmission factor T; and reflection factor R.

3. What should the thickness and index of the film be so that the treated surface is non-reflecting?
Numerical application: $n_s = 1.60$, $\lambda_0 = 0.5\ \mu$.

III

Consider now a system of p thin films characterized by:

    their respective thicknesses $d_1, d_2, \ldots, d_p$.
    their indices $n_1, n_2, \ldots, n_p$.

The same illumination used above is used here.

1. Determine the characteristic matrix for this stratified medium. Derive the transmitted amplitude and the reflected amplitude for the system of p films.

2. One can make a mirror by using a system of thin films by using alternate high and low index films (the film in contact with the air being high index). Call the indices of these films n_h (high) and n_l (low). Assume that all the films have the same optical thickness $\lambda_0/4$. Justify thus choice of thickness.

(a) Determine the characteristic matrix relative to one period (two films), to $2p$ films and to $(2p+1)$ films.

(b) Calculate the reflection factor R of a mirror having $2p$ and $(2p+1)$ films.

Numerical application: $n_0 = 1$; $n_h = 2.3$ (zinc sulphide); $n_l = 1.38$ (magnesium fluoride); $n_s = 1.52$ (glass support). Number the films, 1, 2, ..., 11. Recall that the elements (c_{ij}) of a matrix $[C]$, equal to the product of matrices $[A]$ and $[B]$, are obtained by the following equation:

$$c_{ij} = \sum_{k=1}^{p} a_{ik} \times b_{kj}.$$

SOLUTION

I. Glassy reflection. Fresnel formulas

1. It is unnecessary to make a distinction between parallel and perpendicular vibration on the plane of incidence. In effect, for the case of normal incidence, all planes through which the rays pass are planes of incidence (Fig. 14.2).

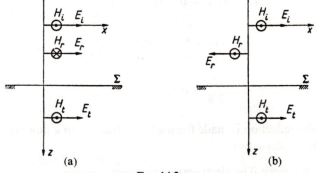

(a)   (b)

FIG. 14.2

Take the incident fields E_i and H_i oriented as in Fig. 14.2. The sense of the transmitted fields remains the same. For the reflected fields one has the choice between the orientations respresented on Fig. 14.2a and 14.2b. *Arbitrarily* take the case of Fig. 14.2a. The components of the electric and magnetic fields on the surface satisfy the continuity conditions

$$\left. \begin{array}{c} E_i + E_r = E_t \\ H_i - H_r = H_t \end{array} \right\}. \tag{1}$$

Between the E and H fields of each sinusoidal plane wave one has the relationships

$$\left. \begin{array}{c} \dfrac{H_i}{E_i} = \sqrt{\dfrac{\varepsilon_1}{\mu_0}} = \dfrac{H_r}{E_r} \\[3mm] \dfrac{H_t}{E_t} = \sqrt{\dfrac{\varepsilon_2}{\mu_0}} \end{array} \right\} \tag{2}$$

5*

With a dielectric medium one has $\mu_1 = \mu_0$ and

$$\left.\begin{array}{l}\sqrt{\dfrac{\varepsilon_1}{\varepsilon_0}} = n_1 = \dfrac{c}{v_1}, \\[4mm] \sqrt{\dfrac{\varepsilon_2}{\varepsilon_0}} = n_2 = \dfrac{c}{v_2}. \end{array}\right\} \tag{3}$$

Eliminating H between equations (1) and (2) one gets the Fresnel formulas

$$\left.\begin{array}{l} r = \dfrac{E_r}{E_i} = \dfrac{n_1-n_2}{n_1+n_2} = \dfrac{1-n}{1+n}, \quad \text{with} \quad n = \dfrac{n_2}{n_1}; \\[4mm] t = \dfrac{E_t}{E_i} = \dfrac{2n_1}{n_1+n_2} = \dfrac{2}{1+n}. \end{array}\right\} \tag{4}$$

Note

(a) t is always positive; E_t and$_i$ E have the same sense,

(b) r has the sign of n_1-n_2,

$$\text{if } n_1 > n_2, r > 0 \text{ (Fig. 14.2a)};$$

$$\text{if } n_1 < n_2, r < 0 \text{ (Fig. 14.2b)}.$$

Numerical application:

$$r = \frac{1-1.5}{1+1.5} = -\frac{0.5}{2.5} = -0.20,$$

$$t = \frac{2}{2.5} = 0.8.$$

In the case where the reflection is **made from a** less refracting to a more refracting medium, the electric field shifts phase by π.

2. *Conservation of energy.* The electromagnetic energy density is

$$w = \varepsilon E^2 = \mu_0 H^2. \tag{5}$$

From this the energy contained in the volume $d\tau$ is

$$dW = \varepsilon E^2 \, d\tau. \tag{6}$$

The amount of energy which passes through a surface element in time dt parallel to the plane of the wave is that amount of energy which is contained at time t in a cylinder of base dS and height $v \, dt$ (v being the velocity of propagation of the wave in the medium).

Equation (6) can be rewritten

$$dW = \varepsilon E^2 \, dS \times v \times dt. \tag{7}$$

The radiant energy flux passing the surface dS is then

$$d\Phi = \frac{dW}{dt} = \varepsilon v E^2 \, dS. \tag{8}$$

The conservation of energy flux is written

$$d\Phi_i = d\Phi_r + d\Phi_t. \tag{9}$$

After simplification one gets

$$v_1\varepsilon_1 E_i^2 = v_1\varepsilon_1 E_r^2 + v_2\varepsilon_2 E_t^2. \tag{10}$$

By dividing both sides of this equation by E_i^2 and using the definitions of t and r, one gets

$$1 = r^2 + nt^2. \tag{11}$$

This expression could also have been found directly using the Fresnel formulas (4).

Note. In the case of normal incidence, r and t are related by

$$\left. \begin{array}{l} 1+r = t, \\ 1 = r^2 + nt^2. \end{array} \right\} \tag{12}$$

For the energy, one has

$$1 = R + T. \tag{13}$$

Thus:

$$\left. \begin{array}{l} R = r^2 \\ T = nt^2 \neq t^2. \end{array} \right\} \tag{14}$$

In effect, if the reflected beam propagates in the same medium as the incident beam, it differs from the transmitted beam: *in general, the transmitted energy is not equal to the square of the transmitted amplitude.*

II. *A ntireflection coatings*

1. *Continuity conditions.* The fields E and H are the result of two waves which propagate in opposite directions.

Take

$$\left. \begin{array}{l} E = E^+ + E^-, \\ H = H^+ + H^-. \end{array} \right\} \tag{15}$$

The positive exponent designates a wave travelling in the positive Oz direction while the negative exponent designates the wave traveling in the opposite direction.

The second equation (15) can be written, using (2),

$$H = \sqrt{\frac{\varepsilon}{\mu}}(E^+ - E^-) = \sqrt{\frac{\varepsilon}{\mu_0}}(E^+ - E^-) = q(E^+ - E^-). \tag{16}$$

If A and B designate the amplitudes within the film, one has, taking $k_0 = 2\pi/\lambda_0$

$$\left. \begin{array}{l} E_{\text{film}} = A\,e^{-jk_0nz} + B\,e^{+jk_0nz}, \\ H_{\text{film}} = q[A\,e^{-jk_0nz} - B\,e^{+jk_0nz}]. \end{array} \right\} \tag{17}$$

When passing from one medium to another, Maxwell's equations require the continuity of the tangential components of the field vectors. Here, in the special case of normal incidence, E and H which are perpendicular to Oz should remain continuous.

On the surfaces Σ_I and Σ_{II} one has then

$$\text{on} \quad \Sigma_I \left\{ \begin{array}{l} E_I = A + B, \\ H_I = q(A - B), \end{array} \right. \tag{18}$$

$$\text{on} \quad \Sigma_{II} \left\{ \begin{array}{l} E_{II} = A\,e^{-jk_0nd} + B\,e^{+jk_0nd}, \\ H_{II} = q[A\,e^{-jk_0nd} - B\,e^{+jk_0nd}]. \end{array} \right. \tag{19}$$

Eliminate A and B from equations (18) and (19). One gets

$$E_{II} = \frac{1}{2q}[(qE_I + H_I)\,e^{-jk_0nd} + (qE_I - H_I)\,e^{+jk_0nd}]$$

$$H_{II} = \tfrac{1}{2}[(qE_I + H_I)\,e^{-jk_0nd} - (qE_I - H_I)\,e^{+jk_0nd}].$$

One is led to the following linear relationships:

$$\left. \begin{array}{l} E_{II} = E_I \cos k_0nd - \dfrac{j}{q} H_I \sin k_0nd, \\[2mm] H_{II} = -jqE_I \sin k_0nd + H_I \cos k_0nd. \end{array} \right\}$$

Also,

$$\left. \begin{array}{l} E_I = E_{II} \cos k_0nd + \dfrac{j}{q} H_{II} \sin k_0nd, \\[2mm] H_I = jqE_{II} \sin k_0nd + H_{II} \cos k_0nd, \end{array} \right\} \tag{20}$$

or, in matrix form,

$$\begin{bmatrix} E_I \\ H_I \end{bmatrix} = \begin{bmatrix} \cos k_0nd & \dfrac{j}{q} \sin k_0nd \\[2mm] jq \sin k_0nd & \cos k_0nd \end{bmatrix} \begin{bmatrix} E_{II} \\ H_{II} \end{bmatrix} = [M_1] \begin{bmatrix} E_{II} \\ H_{II} \end{bmatrix}. \tag{21}$$

The matrix $[M_1]$ characterizes the film. It is a unimodular matrix, that is, its determinant is unity.

2. In the vacuum, on surface Σ_I, equations (15) can be written

$$\left. \begin{array}{l} E_I = E_i + E_r \\ H_I = q_0(E_i - E_r). \end{array} \right\} \tag{22}$$

In the support, thought of as infinite, there exists no wave propagating in the negative direction. On the surface Σ_{II}, the field components are

$$\left. \begin{array}{l} E_{II} = E_t \\ H_{II} = H_t = q_s E_t. \end{array} \right\} \tag{23}$$

Using equations (22) and (23), equation (20) becomes

$$\left. \begin{array}{l} E_i + E_r = E_t \left[\cos k_0nd + j\,\dfrac{q_s}{q} \sin k_0nd \right. \\[2mm] q_0(E_i - E_r) = E_t[jq \sin k_0nd + q_s \cos k_0nd]. \end{array} \right\} \tag{24}$$

From this one gets:

(a) The transmitted amplitude t:

$$t = \frac{E_t}{E_i} = \frac{2q_0q}{q(q_0+q_s)\cos k_0 nd + j(q^2 + q_0q_s)\sin k_0 nd}, \tag{25}$$

$$t = \frac{2}{(1+n_s)\cos k_0 nd + j\left(n + \frac{n_s}{n}\right)\sin k_0 nd}. \tag{26}$$

(b) The reflected amplitude:

$$r = \frac{E_r}{E_i} = \frac{q(q_0-q_s)\cos k_0 nd + j(q_0q_s - q^2)\sin k_0 nd}{q(q_0+q_s)\cos k_0 nd + j(q_0q_s + q^2)\sin k_0 nd} \tag{27}$$

$$r = \frac{(1-n_s)\cos k_0 nd + j\left(\frac{n_s}{n} - n\right)\sin k_0 nd}{(1+n_s)\cos k_0 nd + j\left(\frac{n_s}{n} + n\right)\sin k_0 nd}. \tag{28}$$

(c) The transmission factor:

$$T = \frac{q_s}{q_0}|t|^2 = \frac{q_s}{q_0}tt^* = \frac{n_s}{n_0}tt^*. \tag{29}$$

Since the outer media are not identical, one has $T \neq t^2$. Equation (25) allows one to write

$$T = \frac{4q_0q_sq^2}{q^2(q_0+q_s)^2\cos^2 k_0 nd + (q^2+q_0q_s)^2\sin^2 k_0 nd}, \tag{30}$$

$$T = \frac{4q_0q_sq^2}{q^2(q_0+q_s)^2 + (q^2-q_0^2)(q^2-q_s^2)\sin^2 k_0 nd}, \tag{31}$$

or finally, as a function of the indices,

$$T = \frac{4n_s}{(1+n_s)^2 + (n^2-1)\left[1 - \left(\frac{n_s}{n}\right)^2\right]\sin^2 k_0 nd}. \tag{32}$$

(d) The reflection factor:
One gets immediately

$$R = 1 - T = \frac{(1-n_s)^2 + (n^2-1)\left[1 - \left(\frac{n_s}{n}\right)^2\right]\sin^2 k_0 nd}{(1+n_s)^2 + (n^2-1)\left[1 - \left(\frac{n_s}{n}\right)^2\right]\sin^2 k_0 nd}. \tag{33}$$

3. Go back to equation (32). Since $1 < n < n_s$, the second term in the denominator D is always negative. To have T maximum, D must be a minimum, that is, $\sin k_0 nd$ maximum, or finally,

$$nd = \lambda_0/4. \tag{34}$$

For this optical thickness in the treatment of the surface, the transmission factor will be

$$T = \frac{4n^2 n_s}{n^2(1+n_s)^2 + (n^2-1)(n^2-n_s^2)} = \frac{4n^2 n_s}{(n^2+n_s)^2}. \tag{35}$$

This value is a maximum for

$$n^2 = n_s. \tag{36}$$

Conclusion. The treated plate will be perfectly transparent if the thickness and index of the film satisfies the conditions

$$\begin{aligned} n &= \sqrt{n_s}, \\ nd &= \lambda_0/4. \end{aligned} \tag{37}$$

Numerical application:

$$n = \sqrt{1.6} = 1.265,$$

$$d = \frac{\lambda_0}{4n} = \frac{0.5}{5.06} \approx 0.1 \ \mu.$$

Note. There is no solid with an index less than 1.3. By depositing a single film one cannot get a perfect non-reflector, but rather only a close approximation.

III. *Multiple dielectric films*

1. Writing the continuity equations on surfaces Σ_{II} and Σ_{III}:

$$\begin{bmatrix} E_{\mathrm{II}} \\ H_{\mathrm{II}} \end{bmatrix} = \begin{bmatrix} \cos k_0 n_2 d_2 & \dfrac{\mathrm{j}}{q_2}\sin k_0 n_2 d_2 \\ \mathrm{j}q_2 \sin k_0 d_2 & \cos k_0 n_2 d_2 \end{bmatrix} \begin{bmatrix} E_{\mathrm{III}} \\ H_{\mathrm{III}} \end{bmatrix} = [M_2] \times \begin{bmatrix} E_{\mathrm{III}} \\ H_{\mathrm{III}} \end{bmatrix}. \tag{38}$$

Using equation (21), one can write

$$\begin{bmatrix} E_{\mathrm{I}} \\ H_{\mathrm{I}} \end{bmatrix} = [M_1] \times [M_2] \begin{bmatrix} E_{\mathrm{III}} \\ H_{\mathrm{III}} \end{bmatrix}. \tag{39}$$

It follows that the relationship between E_{I} and H_{I} (the values of E and H on the plane $z = 0$) and E_{P+1} and H_{P+1} (the values of E and H on the plane $z = d_1 + d_2 + \ldots + d_P$) is simply

$$\begin{bmatrix} E_{\mathrm{I}} \\ H_{\mathrm{I}} \end{bmatrix} = [M_1] \times [M_2] \times \ldots \times [M_p] \times \begin{bmatrix} E_{P+1} \\ H_{P+1} \end{bmatrix} \tag{40}$$

or finally

$$\begin{bmatrix} E_{\mathrm{I}} \\ H_{\mathrm{I}} \end{bmatrix} = [M] \times \begin{bmatrix} E_{P+1} \\ H_{P+1} \end{bmatrix}. \tag{41}$$

With a system of p thin films whose characteristic matrices are $[M_i]$, one has

$$[M] = \prod_{i=1}^{p} [M_i]. \tag{42}$$

Note. The matrix product is not commutative. The product should be taken in the order in which the incident wave falls on the films.

$[M]$ is always a unimodular matrix. This property is a result of the conservation of the energy transported by an electromagnetic wave.

Equation (40) can be written

$$\begin{bmatrix} E_I \\ H_I \end{bmatrix} = \begin{bmatrix} m_{11} & m_{12} \\ m_{21} & m_{22} \end{bmatrix} \begin{bmatrix} E_{P+I} \\ H_{P+I} \end{bmatrix}. \tag{43}$$

Transmitted amplitude t. Equation (22) is not modified. Equation (23) becomes:

$$\left. \begin{aligned} E_{P+I} &= E_t \\ H_{P+I} &= H_t = q_s E_t . \end{aligned} \right\} \tag{44}$$

Combining (22), (43), and (44), one can write

$$t = \frac{E_t}{E_i} = \frac{2q_0}{q_0(m_{11}+q_s m_{12})+(m_{21}+q_s m_{22})} . \tag{45}$$

Reflected amplitude r. In the same way as above one gets

$$r = \frac{E_r}{E_i} = \frac{q_0(m_{11}+q_s m_{12})-(m_{21}+q_s m_{22})}{q_0(m_{11}+q_s m_{12})+(m_{21}+q_s m_{22})} . \tag{46}$$

2. (a) *Characteristic matrices.* Since one has quarter-wave $(\lambda_0/4)$ films, the matrices $[M_1]$ and $[M_2]$ become

$$[M_h] = \begin{bmatrix} 0 & \dfrac{j}{q_h} \\ jq_h & 0 \end{bmatrix}, \tag{47}$$

$$[M_l] = \begin{bmatrix} 0 & \dfrac{j}{q_l} \\ jq_l & 0 \end{bmatrix}. \tag{48}$$

The characteristic matrix for one period is

$$[M_{1 \text{ period or 2 films}}] = \begin{bmatrix} -\dfrac{q_l}{q_h} & 0 \\ 0 & -\dfrac{q_h}{q_l} \end{bmatrix} = \begin{bmatrix} -\dfrac{n_l}{n_h} & 0 \\ 0 & -\dfrac{n_h}{n_l} \end{bmatrix}. \tag{49}$$

For p periods one gets

$$[M_{p \text{ periods or } 2p \text{ films}}] = \begin{bmatrix} \left(-\dfrac{n_l}{n_h}\right)^p & 0 \\ 0 & \left(-\dfrac{n_h}{n_l}\right)^p \end{bmatrix}. \tag{50}$$

For $(2p+1)$ the characteristic matrix is

$$[M_{(2p+1) \text{ films}}] = [M_{2p \text{ films}}] \times [M_h] \tag{51}$$

$$[M_{(2p+1) \text{ films}}] = \begin{bmatrix} \left(-\dfrac{n_l}{n_h}\right)^p & 0 \\[2ex] 0 & \left(-\dfrac{n_h}{n_l}\right)^p \end{bmatrix} \begin{bmatrix} 0 & \dfrac{j}{q_h} \\[2ex] jq_h & 0 \end{bmatrix} \tag{52}$$

$$[M_{2p+1 \text{ films}}] = \begin{bmatrix} 0 & \left(-\dfrac{n_l}{n_h}\right)^p \dfrac{j}{q_h} \\[2ex] \left(-\dfrac{n_h}{n_l}\right)^p jq_h & 0 \end{bmatrix}. \tag{53}$$

(b) *Calculation of the reflection factor* $R = |r|^2$. Return to equation (46) giving r and replace the m's by the elements of matrix (50) or those of (53).

Case with 2p films:

$$r_{2p} = \frac{q_0\left(-\dfrac{n_l}{n_h}\right)^p - q_s\left(-\dfrac{n_h}{n_l}\right)^p}{q_0\left(-\dfrac{n_l}{n_h}\right)^p + q_s\left(-\dfrac{n_h}{n_l}\right)^p} \tag{54}$$

hence

$$R_{2p} = \left[\frac{1 - \dfrac{n_s}{n_0}\left(\dfrac{n_h}{n_l}\right)^{2p}}{1 + \dfrac{n_s}{n_0}\left(\dfrac{n_h}{n_l}\right)^{2p}}\right]^2. \tag{55}$$

Case with $(2p+1)$ films:

$$r_{2p+1} = \frac{q_0 q_s\left(-\dfrac{n_l}{n_h}\right)^p \dfrac{1}{q_h} - \left(-\dfrac{n_h}{n_l}\right)^p q_h}{q_0 q_s\left(-\dfrac{n_l}{n_h}\right)^p \dfrac{1}{q_h} + \left(-\dfrac{n_h}{n_l}\right)^p q_h}, \tag{56}$$

$$R_{2p+1} = \left[\frac{1 - \left(\dfrac{n_h}{n_l}\right)^{2p} \dfrac{n_h^2}{n_0 n_s}}{1 + \left(\dfrac{n_h}{n_l}\right)^{2p} \dfrac{n_h^2}{n_0 n_s}}\right]^2. \tag{57}$$

Numerical application. The results found in the following table are obtained from the publications of Abeles. The films do not absorb and $R+T = 1$.

Number of films	1	2	3	4	5	6	7	8	9	10	11
T	0.693	0.619	0.340	0.289	0.138	0.116	0.0522	0.0432	0.0191	0.0158	0.0069
R	0.307	0.381	0.660	0.711	0.862	0.884	0.9478	0.9568	0.9809	0.9842	0.9931

Note. The choice of $\lambda_0/4$ as the optical thickness for the films is easily explained (Fig. 14.3). If two consecutive reflections are of different kinds, the two reflected rays have a path difference of $2nd + \lambda_0/2$.

For these rays to produce constructive interference, it is necessary that $nd = \lambda_0/4$. Hence

$$n_h d_h = \lambda_0/4 \quad \text{and} \quad n_l d_l = \lambda_0/4.$$

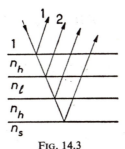

FIG. 14.3

It is necessary that the reflections experienced by the rays 1 and 2 are of a different kind as well as the reflections experienced by the last two rays. In other words, one is interested in having an odd number of films, zinc sulphide being in contact with the air on one hand and the glass on the other.

In the modern Fabry–Pérot interferometer the outer faces are treated with multiple dielectric films. Note that the selectivity increases with the number of films (§ 7.4).

PROBLEM 15

Electromagnetic Waves in a Resonant Cavity

A resonant cavity in the form of a cube has one corner at O and the three edges are oriented along the Ox, Oy, and Oz of a right tetrahedron. The cavity is evacuated and one assumes that the walls are perfectly conducting.

I

Produce an electromagnetic field in the cavity having frequency ν so that the excitation produces an electric field parallel to the Oz axis. Starting with the electromagnetic wave equation and the conditions imposed on the wave field at the walls, show that one can obtain stationary states in which the field E is parallel to Oz and has a modulus independent of z for which there is a relationship between the length of the cavity, L, and the vacuum wavelength λ_0 for a plane wave with frequency ν. Take

$$E_z(x, y) = X(x) \times Y(y).$$

Determine the minimum value of L and do the calculation for $\nu = 3 \times 10^9$ Hz.

II

Assign L the minimum value found above and let E_0 be the maximum amplitude of E_z. Express the fields E_z, H_x, H_y, and H_z as a function of x, y, z, t, and of the parameters L, E_0 and $\omega = 2\pi\nu$. Find the mean energy contained in the cavity as a function of L and E_0.

Application:

$$\nu = 3 \times 10^9 \text{ Hz}, \quad E_0 = 10^7 \text{ V/m}.$$

SOLUTION

I

For a monochromatic wave the wave equation is (§ B.2)

$$\Delta E + \sigma^2 E = 0 \qquad \left(\sigma = \frac{2\pi\nu}{c} = \frac{2\pi}{\lambda_0}\right). \tag{1}$$

For the desired field

$$E_x = 0, \quad E_y = 0, \quad E_z = E_z(x, y)$$

equation (1) becomes

$$\frac{\partial^2 E_z}{\partial x^2} + \frac{\partial^2 E_z}{\partial y^2} + \sigma^2 E_z = 0.$$

Substituting the solution suggested above $E_z(x, y) = X(x)\,Y(y)$

$$\frac{1}{X}\frac{d^2 X}{dx^2} + \frac{1}{Y}\frac{d^2 Y}{dy^2} + \sigma^2 = 0. \tag{2}$$

The general solution is

$$X = A_1 \sin(\sigma_1 x + \phi_1), \quad Y = A_2 \sin(\sigma_2 y + \phi_2) \tag{3}$$

and equation (2) requires

$$\sigma_1^2 + \sigma_2^2 = \sigma^2. \tag{4}$$

The conditions imposed by the walls of a perfect conductor are that the tangential component of E and the normal component of H must be zero, hence

$$X(0) = X(L) = 0, \quad Y(0) = Y(L) = 0.$$

Solutions (3) become

$$X = A_1 \sin\frac{K_1\pi x}{L}, \quad Y = A_2 \sin\frac{K_2\pi y}{L},$$

K_1 and K_2 are integers which, using (4), satisfy

$$(K_1^2 + K_2^2)\frac{\pi^2}{L^2} = \sigma^2$$

or

$$K_1^2 + K_2^2 = \frac{4L^2}{\lambda_0^2}.$$

The eigenfrequencies of the cavity in the mode under consideration correspond to wavelengths

$$\lambda_0 = \frac{2L}{\sqrt{K_1^2 + K_2^2}}.$$

The minimum value of L is obtained for $K_1 = K_2 = 1$

$$L_m = \lambda_0/\sqrt{2}.$$

For $\nu = 3 \times 10^9$ Hz and $\lambda_0 = 0.01$ m, $L_m = 7.07$ cm.

II

For $L = L_m$, the electric field is given by:

$$E_z = E_0 \sin \frac{\pi x}{L} \sin \frac{\pi y}{L} \cos \omega t.$$

One can find the magnetic field from

$$\text{curl } E = -\mu_0 \frac{\partial H}{\partial t}$$

which gives

$$-\frac{\partial H_x}{\partial t} = \frac{E_0}{\mu_0} \frac{\pi}{L} \sin \frac{\pi x}{L} \cos \frac{\pi y}{L} \cos \omega t,$$

$$-\frac{\partial H_y}{\partial t} = -\frac{E_0}{\mu_0} \frac{\pi}{L} \cos \frac{\pi x}{L} \sin \frac{\pi y}{L} \cos \omega t,$$

and, for the minimum value of L,

$$H_x = -\frac{E_0 \pi}{\mu_0 L \omega} \sin \frac{\pi x}{L} \cos \frac{\pi y}{L} \sin \omega t = -\sqrt{\frac{\varepsilon_0}{2\mu_0}} E_0 \sin \frac{\pi x}{L} \cos \frac{\pi y}{L} \sin \omega t,$$

$$H_y = \frac{E_0 \pi}{\mu_0 L \omega} \cos \frac{\pi x}{L} \sin \frac{\pi y}{L} \sin \omega t = \sqrt{\frac{\varepsilon_0}{2\mu_0}} E_0 \cos \frac{\pi x}{L} \sin \frac{\pi y}{L} \sin \omega t.$$

The mean energy contained in the cavity is obtained from the mean value of the energy density

$$w = \tfrac{1}{2}(\varepsilon_0 E^2 + \mu_0 H^2)$$

taken over time. Since one is dealing with sinusoidal functions, $\langle E_z^2 \rangle = E_0^2/2$. Finally, one must take the mean values over the volume

$$\langle E_z^2 \rangle = \frac{E_0^2}{2} \frac{1}{L} \int_0^L \sin^2 \frac{\pi x}{L} \, dx \times \frac{1}{L} \int_0^L \sin^2 \frac{\pi y}{L} \, dy = \frac{E_0^2}{2} \times \frac{1}{4},$$

$$\langle H_x^2 \rangle = \frac{\varepsilon_0}{2\mu_0} \times \frac{E_0^2}{2} \times \frac{1}{4}, \quad \langle H_y^2 \rangle = \frac{\varepsilon_0}{2\mu_0} \times \frac{E_0^2}{2} \times \frac{1}{4},$$

$$\langle W \rangle = L^3 \frac{E_0^2}{4} \left[\frac{\varepsilon_0}{4} + \frac{\varepsilon_0}{2} \left(\frac{1}{4} + \frac{1}{4} \right) \right] = \frac{\varepsilon_0 E_0^2 L^3}{8}.$$

Application.

$$\langle W \rangle = \frac{10^{14} \times 10^{-1}}{8 \times 4 \times 3.14 \times 9 \times 10^9 \times 2 \times \sqrt{2}} \simeq 4.0 \text{ joules.}$$

PROBLEM 16

Radiation Pressure

Give the expression for the radiation pressure exerted by a monochromatic plane wave of frequency v, containing N photons per unit volume, on a plane surface in vacuum when falling on it at an angle of incidence i. Consider the following cases: (a) the surface is a black body; (b) the surface reflects specularly with a reflection factor R; and (c) the surface is a perfect radiation scatterer.

Numerical application. Calculate the radiation pressure exerted by the sun's radiation on the earth assuming the earth is a perfect scatterer. The parameters are given in Problem 19.

SOLUTION

The momentum transported in one second by the incident photons contained in a cylinder of length c and cross-section $S \cos i$ (Fig. 16.1) i s

$$N \frac{hv}{c} cS \cos i$$

FIG. 16.1

since the momentum of a photon is equal to $h\nu/c$ (§ 11.2). The force exerted on the mirror in the incident direction is

$$F = Nh\nu S \cos i$$

and the radiation pressure normal to the surface is

$$\bar{\omega} = \frac{F \cos i}{S} = Nh\nu \cos^2 i = w \cos^2 i \tag{1}$$

since the radiant energy density is equal to $Nh\nu$, $h\nu$ being the energy of a photon.

(a) If the surface is totally absorbing, the radiation pressure is given by (1).

(b) If the surface is a mirror with reflection factor R, a fraction R of the incident photons leave the mirror in a direction symmetric with the normal to the mirror and these photons impart a momentum

$$R \frac{Nh\nu}{c} cS \cos i.$$

The corresponding pressure is

$$\bar{\omega}' = RNh\nu \cos^2 i = Rw \cos^2 i.$$

and the total radiation pressure becomes

$$\bar{\omega} + \bar{\omega}' = w(1+R) \cos^2 i. \tag{2}$$

(c) If the surface is a perfect scatterer, the incident photons are scattered from the surface in all directions with equal probability. The probability that a photon is scattered into the solid angle $d\Omega$ is then, $d\Omega/2\pi = \sin i \, di$, taking for the solid angle that angle which lies between two cones with half-angles i and $i+di$ respectively. The mean value of the projection of the impulse of a photon leaving the surface at angle i on the normal to the surface is

$$\int_0^{\pi/2} \frac{h\nu}{c} \cos i \sin i \, di = \frac{h\nu}{2c}.$$

This corresponds to a pressure

$$\bar{\omega}'' = Nc \cos i \frac{h\nu}{2c} = \frac{w}{2} \cos i,$$

to which is added the pressure of the incident photons given by (1). The total is

$$\bar{\omega} = w(\cos^2 i + \tfrac{1}{2} \cos i). \tag{3}$$

Numerical application. Using the parameters of Problem 19, 1 m² of the earth's surface receives at normal incidence a flux density $\Phi = 1.35 \times 10^3$ W. The energy density is $w = \Phi/c$ and for $i = 0$ equation (3) yields

$$\bar{\omega} = \frac{3}{2} w = \frac{3 \times 1.35 \times 10^3}{2 \times 3 \times 10^8} = 0.675 \times 10^{-5} \text{ N/m}^2$$

or about $\tfrac{3}{4}$ mg/m².

PROBLEM 17

Antennas

I

Plot, in a plane passing through a Hertzian dipole oriented along the $z'Oz$ axis, the polar diagram representing the modulus of the electric field E and the radiant flux Φ. Show that one can put the flux in a form analogous to the Joule power dissipated by a sinusoidal current, with maximum value I_m, which passes through the dipole. Calculate the resistance introduced thereby (radiation resistance of the antenna). Express the field as a function of the instantaneous intensity of the current, I.

II

Consider a linear antenna oriented along $z'Oz$ which has a slightly smaller length than the emitted wavelength and which is insulated at the ends. For $l = \lambda/2$, derive the stationary state of the current which is developed in the antenna. Give the expression $I(z)$ for the current intensity as a function of the ordinate z of a point on the antenna. Take the origin at the centre of the antenna. Starting with the expression obtained in the first part of the problem for the field E which can be now thought of as giving the field dE radiated by an element along the length of the antenna, find the radiated field at a large distance r_0 from O and plot the polar diagram in a plane passing through $z'z$.

III

Consider now a long linear antenna formed on N segments of length $\lambda/2$ between which are inserted $N-1$ identical self-inductances of negligible dimension and whose value is such that they phase shift the current by π. Establish the direction of radiation on a polar diagram in a plane through $z'z$.

SOLUTION

I

The radiant electric field at a distance r from a Hertzian dipole in a direction making an angle θ with the axis of the dipole is given by (§ 10.3)

$$E_\theta = \frac{1}{4\pi\varepsilon_0 c^2} \frac{\omega^2 d_m \sin\theta}{r} \sin\omega\left(t - \frac{r}{c}\right) \tag{1}$$

d_m being the amplitude of the sinusoidal dipole which has angular frequency ω. A magnetic field $H = \varepsilon_0 c E_0$ corresponds to the field E_θ in the electromagnetic wave.

The radiant energy flux which crosses a surface element normal to the direction OP at a distance r from O is given by (Fig. 17.1)

$$d\Phi = S\,d\Sigma = E_\theta H\,d\Sigma = \varepsilon_0 c E_\theta^2\,d\Sigma = \frac{\omega^4 d_m^2 \sin^2\theta\,d\Sigma}{16\pi^2\varepsilon_0 c^3 r^2}\sin^2\omega\left(t - \frac{r}{c}\right),\qquad(2)$$

$S = EH$ being the Poynting vector.

The polar diagram of (1)—varying as $\sin\theta$—is given in Fig. 17.2. That of (2)—varying as $\sin^2\theta$—is in Fig. 17.3.

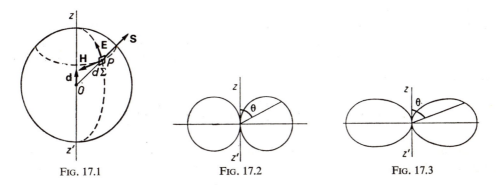

FIG. 17.1   FIG. 17.2   FIG. 17.3

The sinusoidal dipole moment d may be thought of as due to two oscillating charges $\pm q_m \sin\omega t$ separated by a small distance z. One has

$$d_m \sin\omega t = zq_m \sin\omega t$$

and the equivalent current is

$$I = \frac{dq}{dt} = \omega q_m \cos\omega t = I_m \cos\omega t,$$

hence

$$zI_m = \omega d_m.\qquad(3)$$

Furthermore, the total flux radiated over all space is obtained by integration of (2) with $d\Sigma = 2\pi r^2 \sin\theta\,d\theta$, so that

$$\Phi = \frac{\omega^4 d_m^2}{6\pi\varepsilon_0 c^3}\sin^2\omega\left(t - \frac{r}{c}\right) = \frac{\omega^2 z^2 I_m^2}{6\pi\varepsilon_0 c^3}\sin^2\omega\left(t - \frac{r}{c}\right).$$

The instantaneous power is in the form of a Joule power since it is proportional to I^2. One can then write $\mathcal{P} = RI^2$ where R is the radiation resistance given by

$$R = \frac{\Phi}{I^2} = \frac{\omega^2 z^2}{6\pi\varepsilon_0 c^3} = \frac{2\pi}{3\varepsilon_0 c}\left(\frac{z}{\lambda}\right)^2 = 789\left(\frac{z}{\lambda}\right)^2 \text{ ohms,}$$

using the wavelength expression $\lambda = 2\pi c/\omega$.

II

A real linear antenna differs from a Hertzian dipole in two ways: the high-frequency current which runs through it does not have the same value at every point at a given instant; and its length is not small with respect to the wavelength. In calculations one replaces the antenna by a line of dipoles the moment of each depending upon its position and proportional to the current intensity at the point where it is found. The field at large distances is obtained by adding up the elementary fields and taking into account the phase differences resulting from the two points discussed above.

In an antenna insulated at the ends the current is of necessity zero at these points at every instant. The sinusoidal current with angular frequency ω propagating along the length of such a conductor satisfies these condions at the ends and in the steady state establishes standing waves. The intensity is of the form (§ 3.7)

$$I = I_m \sin \omega t \sin \left(\frac{2\pi z}{\lambda} + \phi\right) = I_0 \sin \left(\frac{2\pi z}{\lambda} + \phi\right)$$

I_0 being the intensity at the centre O of the antenna. The conditions at the ends are $I = 0$ for $z = \pm \lambda/4$ so that

$$I = I_0 \cos 2\pi \frac{z}{\lambda}.$$

Using (3), the expression for the field radiated by an element of the antenna of length dz obtained from (1) and the total field is given by

$$E = \frac{I_m \sin \theta}{2\varepsilon_0 c \lambda r} \int_{-\lambda/4}^{+\lambda/4} \cos \frac{2\pi z}{\lambda} \, dz \sin \omega \left(t - \frac{r}{c}\right). \qquad (4)$$

The distance r, from an element dz situated at point A where $OA = z$ (Fig. 17.4) to point P, at a distance r_0 large with respect to λ and thus to OA, is given by a close approximation by

$$r = r_0 - z \cos \theta$$

which when introduced into (4) yields

$$E = \frac{I_0 \sin \theta}{2\varepsilon_0 c \lambda r_0} \int_0^{\lambda/4} \left[\sin \omega \left(t - \frac{r_0}{c} - \frac{z \cos \theta}{c}\right) + \sin \omega \left(t - \frac{r_0}{c} + \frac{z \cos \theta}{c}\right)\right] \cos \frac{2\pi z}{\lambda} \, dz$$

hence:

$$E = \frac{I_0 \sin \theta}{\varepsilon_0 c \lambda r_0} \sin \omega \left(t - \frac{r_0}{c}\right) \int_0^{\lambda/4} \cos \frac{2\pi z}{\lambda} \times \cos \frac{2\pi z \cos \theta}{\lambda} \, dz$$

$$E = \frac{I_0 \sin \theta}{\varepsilon_0 c \lambda r_0} \sin \omega \left(t - \frac{r_0}{c}\right)$$

$$\times \left[\frac{\lambda}{2\pi(1+\cos \theta)} \sin \frac{2\pi z}{\lambda}(1+\cos \theta) + \frac{\lambda}{2\pi(1-\cos \theta)} \sin \frac{2\pi z}{\lambda}(1-\cos \theta)\right]_0^{\lambda/4}.$$

$$E = \frac{I_0 \sin\theta}{2\pi\varepsilon_0 c r_0} \left\{ \frac{\cos\left(\frac{\pi}{2}\cos\theta\right)}{1+\cos\theta} + \frac{\cos\left(\frac{\pi}{2}\cos\theta\right)}{1-\cos\theta} \right\} \sin\omega\left(t - \frac{r_0}{c}\right)$$

$$E = \frac{I_0}{2\pi\varepsilon_0 c r_0} \frac{\cos\left(\frac{\pi}{2}\cos\theta\right)}{\sin\theta} \sin\omega\left(t - \frac{r_0}{c}\right) = 60\frac{I_m}{r_0} \frac{\cos\left(\frac{\pi}{2}\cos\theta\right)}{\sin\theta} \sin\omega\left(t - \frac{r_0}{c}\right)$$

Figure 17.5 shows the polar diagram of the radiation.

FIG. 17.4   FIG. 17.5

III

The proposed antenna is equivalent to a set of N antennas of length $l = \lambda/2$ placed in series and with their currents in phase thanks to the presence of the inductances. In direction θ the path difference between two successive elements is $(\lambda/2)\cos\theta$ (cf. Fig. 17.4). The calculation of the resulting field is made in exactly the same way as the diffraction calculation for a series of N identical, equidistant slits radiating in phase (§ 7.7). The resulting field is

$$E = \frac{60I_0}{r_0} \times \frac{\cos\left(\frac{\pi}{2}\cos\theta\right)}{\sin\theta} \times \frac{\sin\left(\frac{N\pi}{2}\cos\theta\right)}{\sin\left(\frac{\pi}{2}\cos\theta\right)} \times \sin\omega\left(t - \frac{r_0}{c}\right).$$

Figure 17.6 represents the variations of

$$\frac{\sin\left(\frac{N\pi}{2}\cos\theta\right)}{\sin\left(\frac{\pi}{2}\cos\theta\right)}$$

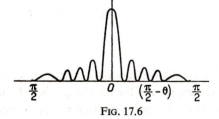

FIG. 17.6

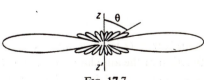

FIG. 17.7

as a function of θ for $N = 10$. Note that the difference in these relatively similar diagrams is that in the optical grating case there is only one principal maximum ($\theta = 0$) since the grating step here is $\lambda/2$.

Figure 17.7 is a polar diagram of the radiation.

PROBLEM 18

Hertzian Dipoles

I

1. Treat the Fresnel mirror experiment from the point of view of electromagnetic theory. One can regard the source S as a Hertzian oscillator vibrating parallel to $\varDelta$, the line of intersection of the mirrors and then the images of the sources S_1 and S_2 act as synchronous oscillators at separation l. Find the electric field, the magnetic field, and the Poynting vector of the resulting electromagnetic wave as a function of r, l, and α in a plane normal to $\varDelta$ and at a point P at a distance $CP = r_0$ from the centre C of S_1S_2, r_0 being large with respect to l and making with l the angle α.

2. A light source which will be compared to a Hertzian oscillator is placed at the centre O of the line II' joining two small plane dielectric mirrors M and M' (Fig. 18.1). The normals

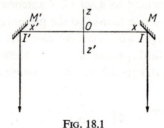

FIG. 18.1

to the mirrors IN and $I'N'$ make the same angle $\pi/4 - \varepsilon$ (ε being a very small angle) with II'. Discuss qualitatively the possibility of interference between the reflected rays in each of the following cases:

(a) dipole vibrating along $z'z$;
(b) dipole vibrating along $y'y$, normal to the figure;
(c) both dipoles above, assuming them to be identical, synchronous, and coherent. Study the state of polarization in the interference field.

Assume that the solid angle subtended from O by the mirrors M and M' is very small and neglect the difference in the reflection coefficients for the two principal vibrations.

II

A light source O is made up of a set of Hertzian oscillators randomly oriented. Write, as a function of the angular coordinates, θ and ϕ, the expression for the electric field E and its components E_x and E_z at a large distance from O for radiation emitted by the source:

(a) in direction Oy;

(b) in the direction Oy' on the xOy plane making an angle γ with Oy. Calculate, as a function of γ, the contrast of the fringes obtained through the interference of the radiation emitted along Oy and along Oy'.

SOLUTION

1. The oscillator images S_1 and S_2 are normal to the plane of the figure which constitutes the equatorial plane for both of them.

The electric field of the electromagnetic wave emitted by each of them is normal to the Fig. 18.2 at P and is given by (§ 10.3)

$$E = \frac{1}{4\pi\varepsilon_0 c^2}\frac{\omega d_m}{r_0}\sin\omega\left(t-\frac{r_0}{c}\right). \tag{1}$$

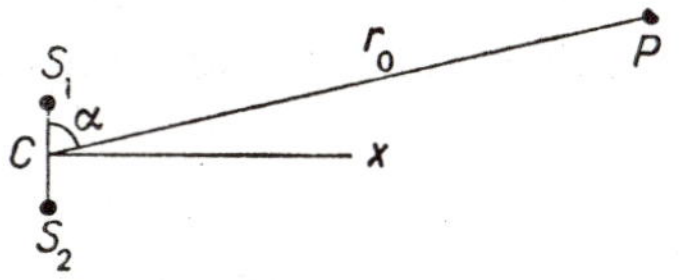

FIG. 18.2

The fields E_1 and E_2, parallel at P, have a phase difference at P due to the path difference $\delta = |r_2 - r_1|$. If this path difference is small enough so that the amplitude difference due to the $1/r$ factor is negligible, the fields have the same amplitude

$$a = \frac{\omega d_m}{4\pi\varepsilon_0 c^2 r_0}.$$

On the other hand, one can take

$$\delta = l\cos\alpha,$$

from which

$$\phi = \frac{2\pi}{\lambda}l\cos\alpha.$$

The resulting E field is given by the summation of two parallel vibrations with amplitude a and phase difference ϕ. Its intensity is

$$A^2 = 4a^2\cos^2\frac{\phi}{2}.$$

Thus

$$E = 2a\cos\frac{\pi l\cos\alpha}{\lambda}\sin\omega\left(t-\frac{r_0}{c}\right). \tag{2}$$

The H field is the resultant of the fields H_1 and H_2 which are in phase with E_1 and E_2 respectively, since the distance CP is large and for this same reason these fields are practically parallel. Since for all electromagnetic plane waves in free space

$$H = \sqrt{\frac{\varepsilon_0}{\mu_0}} E,$$

One gets from (2)

$$H = 2a \sqrt{\frac{\varepsilon_0}{\mu_0}} \cos \frac{\pi l \cos \alpha}{\lambda} \sin \omega \left(t - \frac{r_0}{c} \right). \qquad (3)$$

The Poynting vector:

$$S = E \times H = 4a^2 \sqrt{\frac{\varepsilon_0}{\mu_0}} \cos^2 \frac{\pi l \cos \alpha}{\lambda} \sin^2 \omega \left(t - \frac{r_0}{c} \right).$$

2. Since the mirrors give images of O separated by a very small angle, one can assume that the radiation from O forms a quasi-parallel bundle of rays. The electric fields of the emitted waves are parallel to the dipole and thus are either in the plane of incidence to the mirrors or normal to it.

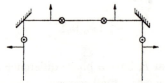

Fig. 18.3

For incidence close to $\pi/4$, less than the Brewster angle for all dielectrics, the vibrations parallel to $z'z$ or to $y'y$ undergo a phase advance of π by reflection. Their arrangement is as shown in Fig. 18.3. When the reflected bundles combine—which occurs at a large distance from O—there is, on the $z'z$ axis, constructive interference for the vibrations parallel to $y'y$ and destructive interference for the vibrations emitted by O parallel to $z'z$ (these are practically antiparallel at their point of recombination and are directed along $x'x$).

When one moves off the centre of the interference field parallel to $x'x$, the vibrations parallel to $y'y$ are:

$$E'_y = E_m \cos \omega t, \quad E''_y = E_m \cos (\omega t + \phi)$$

ϕ represents the phase difference due to the difference in the optical paths which vary linearly as a function of x. The resultant amplitude is

$$E_y = 2E_m \cos \frac{\phi}{2} \cos \left(\omega t + \frac{\phi}{2} \right). \qquad (4)$$

The vibrations emitted parallel to $z'z$ are

$$E'_x = E_m \cos \omega t, \quad E''_{x'} = E_m \cos (\omega t + \pi + \phi).$$

The amplitude E_m is the same as before since the reflection coefficients are assumed equal. The resultant amplitude is

$$E_x = 2E_m \sin \frac{\phi}{2} \sin \left(\omega t + \frac{\phi}{2}\right). \tag{5}$$

The vibrations (4) and (5) are perpendicular to each other and the ratio of their amplitudes varies with ϕ, that is, with the position of observational point along a line parallel to $x'x$. At each point the components give rise to a vibrational ellipse whose axes are parallel to $y'y$ and $x'x$ with variable dimensions respectively equal to $2E_m \cos \phi/2$ and $2E_m \sin \phi/2$. All of these ellipses can be inscribed in a square of edge $E_m \sqrt{2}$ (Fig. 18.4).

For $\phi = 0$, the ellipse reduces to a line parallel to Y, for $\phi = \pi$ to a line parallel to X, and for $\phi = \pi/2$ and $\phi = 3\pi/2$, one finds circles.

When the vibrations are squared, the resultant intensity is

$$I = E_y^2 + E_x^2 = 4E_m^2.$$

This is constant. In the absence of an analyser, the interference field is uniformly illuminated.

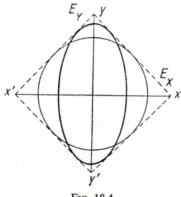

FIG. 18.4

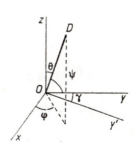

FIG. 18.5

II

Let θ and ϕ (Fig. 18.5) be the angles which define the orientation of a dipole OD in the rectangular system $Oxyz$. For an observer on the Oy axis, the electric field of the wave emitted by the dipole is proportional to $\sin \psi$, ψ being the angle $D\widehat{O}y$ (§ 10.3). The field E is in the plane xOz since the free electromagnetic waves are transverse. The components are

$$E_x = E \sin \theta \cos \phi, \quad E_z = E \cos \theta.$$

In the direction Oy' which makes the angle γ with Oy, the electric field is in the plane $x'Oz$ (Ox' normal to Oy') and its components are given by

$$E_x' = E \sin \theta \cos (\phi \pm \gamma), \quad E_z' = E \cos \theta.$$

The vibrations E_x and E'_x can interfere as can E_z and E'_z. But the first do not interfere with the second since they are perpendicular. The fields emitted by the various dipoles, which are incoherent, do not interfere. The intensity maxima are given by

$$I_M = \sum \{(E_x + E'_x)^2 + (E_z + E'_z)^2\}$$

(the sum being taken over all the dipoles) and the minima by

$$I_m = \sum \{(E_x - E'_x)^2 + (E_z - E'_z)^2\}.$$

Defining the contrast by

$$\Gamma = \frac{I_M - I_m}{I_M + I_m},$$

one has

$$\Gamma = \frac{\sum \{2(E_x E'_x + E_z E'_z)\}}{\sum (E_x^2 + E_x'^2 + E_z^2 + E_z'^2)} = \frac{N}{D}.$$

It is now necessary to take into account the random orientation of the dipoles whose axes are uniformly distributed over all the solid angle elements $d\Omega = \sin \theta \, d\theta \, d\phi$. The sums N and D become the integrals

$$N = 2 \int_{\phi=0}^{2\pi} d\phi \int_{\theta=0}^{\pi/2} d\theta (E_x E'_x + E_z E'_z)$$

$$= 2 \int_{\phi=0}^{2\pi} \cos(\phi \pm \gamma) \cos \phi \, d\phi \int_{\theta=0}^{\pi/2} \sin^3 \theta \, d\theta + 2 \int_{\phi=0}^{2\pi} d\phi \int_{\theta=0}^{\pi/2} \cos^2 \theta \sin \theta \, d\theta.$$

But

$$\int_0^{\pi/2} \sin^3 \theta \, d\theta = \frac{1}{2} \int_0^{\pi/2} (1 - \cos 2\theta) \sin \theta \, d\theta$$

$$= \frac{1}{2} - \frac{1}{2} \int_0^{\pi/2} (2 \cos^2 \theta - 1) \sin \theta \, d\theta = \frac{2}{3}.$$

$$\int_0^{\pi/2} \cos^2 \theta \sin \theta \, d\theta = \int_0^{\pi/2} d\left(-\frac{\cos^3 \theta}{3}\right) = \frac{1}{3}.$$

$$N = \frac{4}{3} \int_0^{2\pi} \cos(\phi \pm \gamma) \cos \phi \, d\phi + \frac{4\pi}{3}$$

$$= \frac{4}{3} \int_0^{2\pi} \cos \gamma \cos^2 \phi \, d\phi \mp \int_0^{2\pi} \sin \gamma \sin \phi \cos \phi \, d\phi + \frac{4\pi}{3}$$

$$N = \frac{4\pi}{3} (1 + \cos \gamma).$$

$$D = \int_{\phi=0}^{2\pi} d\phi \int_{\theta=0}^{\pi/2} \{\sin^2 \theta [\cos^2 \phi + \cos^2(\phi \pm \gamma)] + 2 \cos^2 \theta\} \sin \theta \, d\theta$$

$$= \int_{\phi=0}^{2\pi} [\cos^2 \phi + \cos^2(\phi \pm \gamma)] \, d\phi \int_{\theta=0}^{\pi/2} \sin^3 \theta \, d\theta + 2 \int_{\phi=0}^{2\pi} d\phi \int_{\theta=0}^{\pi/2} \cos^2 \theta \sin \theta \, d\theta.$$

The last integral has been previously calculated. Since

$$\int_0^{\pi/2} \sin^3 \theta \, d\theta = \tfrac{2}{3},$$

$$D = \frac{2}{3} \int_0^{2\pi} \cos^2 \phi \, d\phi + \frac{2}{3} \int_0^{2\pi} \cos^2 (\phi \pm \gamma) \, d(\phi \pm \gamma) + \frac{4\pi}{3} = \frac{4\pi}{3}.$$

Then

$$\Gamma = \frac{N}{D} = \frac{1 + \cos \gamma}{2}.$$

For $\gamma = 0$, $\Gamma = 1$. The Fresnel mirror experiment approximates this case since one has interference of waves emitted by a single source along approximately parallel directions.

For $\gamma = \pi$, $\Gamma = 0$. Finally, for $\gamma = \pi/2$, $\Gamma = \tfrac{1}{2}$. This latter result is similar to Selenyi's experiment.

EMISSION AND ABSORPTION

PROBLEM 19

Photometry. The Earth–Sun System

One square meter of the earth's surface illuminated by the sun at normal incidence receives a flux of 1.35 kW if one neglects the absorption by the atmosphere.

1. Calculate the flux emitted by 1 m² of the sun's surface assuming that it radiates according to Lambert's law. Recall that the apparent diameter of the sun when viewed from the earth is $2\alpha = 32'$.

2. Calculate the sun's mass loss per second due to radiation given the earth–sun distance as 15×10^7 km.

3. Assume that the surface of the earth uniformly scatters a fraction ϱ of the incident radiation flux. Calculate the luminance of the earth.

4. Calculate the amplitude of the electric and magnetic fields due to solar radiation at the surface of the earth.

SOLUTION

1. The sun radiates according to Lambert's law and its luminance $\mathcal{L}$ is a constant. The flux emitted by a surface elements dS into a solid angle $d\Omega$ whose axis makes an angle θ with the normal to dS is

$$d^2\Phi = \mathcal{L}\, dS \cos\theta\, d\Omega. \tag{1}$$

Taking as $d\Omega$ the solid angle lying between two cones with apex on dS and axis normal to dS having an aperture 2θ (Fig. 19.1), one has

$$d^2\Phi = \mathcal{L}\, dS \cos\theta \times 2\pi \sin\theta\, d\theta.$$

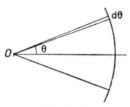

FIG. 19.1

The flux radiated by dS into all exterior space is:

$$d\Phi = \pi\mathcal{L}\,dS \int_0^{\pi/2} 2\sin\theta\cos\theta\,d\theta = \pi\mathcal{L}\,dS \int_0^{\pi/2} 2\sin\theta\,d(\sin\theta)$$

$$d\Phi = \pi\mathcal{L}\,dS \left[\sin^2\theta\right]_0^{\pi/2} = \pi\mathcal{L}\,dS.$$

The ratio $d\Phi/dS = B$ is the *emittance* or *emissive power* of the surface. For an emitter which follows Lambert's law, one has

$$B = \pi\mathcal{L}. \tag{2}$$

Note that a sphere which emits according to Lambert's law appears to be a plane disc, the factor $\cos\theta$ in (1) compensating exactly for the inclination of the surface when moving away from the normal (§ 1.8). This is the way the sun appears.

The flux (1) emitted by the sun at normal incidence and falling normally on a surface dS' of the earth at a distance r from the sun, can be written

$$d^2\Phi = \mathcal{L}\,dS\,\frac{dS'}{r^2}.$$

The illumination intensity produced is, by definition,

$$d\mathcal{E} = \frac{d^2\Phi}{dS'} = \mathcal{L}\frac{dS}{r^2}.$$

The illumination intensity due to the solar disc viewed through its angle

$$\frac{S}{r^2} = \pi\alpha^2$$

is

$$\mathcal{E} = \mathcal{L}\pi\alpha^2.$$

Thus, the emittance of the sun has the value

$$B = \pi\mathcal{L} = \frac{\mathcal{E}}{\alpha^2} = \frac{1.35\times10^3}{(16\times3\times10^{-4})^2} = 5.8\times10^7 \text{ W/m}^2.$$

2. The mass-energy equivalence (§ 9.11) allows one to write

$$\Delta m = \frac{\Delta W}{c^2},$$

c being the free-space velocity of light.

One can calculate the total power lost by the sun by noting that it is equal to the power received by a unit area of the earth's surface multiplied by the surface area of the sphere with radius equal to the earth–sun distance, that is

$$\Phi = 1.35\times10^3\times4\pi\times(15)^2\times10^{20} = 3.815\times10^{26} \text{ W}.$$

Hence, the mass loss per second is

$$\Delta m = \frac{\Phi}{c^2} = \frac{3.815 \times 10^{26}}{(3 \times 10^8)^2} = 4.24 \times 10^9 \text{ kg,}$$

which corresponds to an annual loss of 1.4×10^{13} tons. However, the mass of the sun is 2×10^{27} tons.

3. A surface area of the earth which receives a flux $d\Phi$, reradiates to all space a flux $d\Phi' = \varrho\, d\Phi$. Thus, the emittance of the earth is

$$B' = \frac{d\Phi'}{dS'} = \varrho \frac{d\Phi}{dS'} = \varrho\mathcal{E}$$

and its luminance is

$$\mathcal{L}' = \frac{\varrho}{\pi} \mathcal{E}.$$

4. The mean illumination produced by a plane electromagnetic wave is related to the amplitude of the electric field by (§ 2.3)

$$\langle \mathcal{E} \rangle = \varepsilon_0 c\, \frac{E_m^2}{2},$$

hence

$$E_m^2 = \frac{2\langle \mathcal{E} \rangle}{\varepsilon_0 c} = \frac{2 \times 1.35 \times 10^3 \times 36\pi \times 10^9}{3 \times 10^8} = 101.8 \times 10^4$$

$$E_m = 1010 \text{ V/m.}$$

The magnetic field of the wave has the amplitude (§ 2.3)

$$H_m = \frac{1}{c\mu_0} E_m = \frac{1010}{3 \times 10^8 \times 1.26 \times 10^{-6}} = 2.7 \text{ A/m.}$$

PROBLEM 20

The Spectra and Energy of a Laser

Suitably excited, a ruby laser can emit giant light pulses of wavelength $\lambda = 6935.9$ Å (wave number $\tilde{\nu} = 14{,}418$ cm^{-1}). Assume that each pulse can be ascribed to a linearly polarized plane wave train of constant amplitude, duration $\tau = 0.1$ milliseconds and carrying the energy $W = 0.3$ joule. The cross-section of the beam is circular with a diameter of 5 mm. The pulses propagate in air with the index taken as 1.

I

Calculate the number, N, of photons carried in a pulse. Knowing that the fluctuation of the number of photons in a wave is equal to $\sqrt{N}$, derive the corresponding fluctuation in the phase ϕ of the wave associated with the N photons. What conclusion can be drawn about its preponderant appearance—particle or wave?

II

Calculate the frequency spectrum $G(\nu)$ of each pulse. Derive the spectral width $\Delta\nu$, defined as half the separation of the two zeros of $G(\nu)$ which enclose the central maximum. Derive an expression between the corresponding width in wave numbers and the length L of the wave train.

Numerically find $\Delta\tilde{\nu}$ in millikaysers and $\Delta\lambda$ in milliangstroms.

If this pulse is injected into a Michelson interferometer, show, without new calculations what path difference will be required before one will no longer be able to observe the interference. Is this physically possible?

III

1. Calculate the volume energy density, w, carried by a pulse (to calculate the volume occupied by the wave train, neglect enlargement of the beam by diffraction).

2. Derive a numerical value for the electric field of the wave.

3. Calculate the pressure exerted on a plane screen perpendicular to the beam in the following cases:

(α) the screen is totally absorbing;
(β) the screen is totally reflecting; and
(γ) the screen has a reflection factor $R = 0.9$ and an absorption factor $A = 0.1$.

IV

Place on the trajectory of the beam an aberration free lens L with focal length $F = 5$ cm whose diameter is sufficiently large that it will not act as a pupil for the system. A film of steel 0.1 mm thick is placed in the plane of the focus of L (Fig. 20.1).

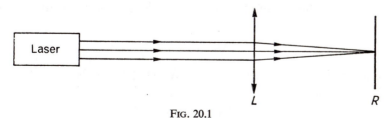

FIG. 20.1

1. Calculate the radius ϱ of the central diffraction spot. Assume that this spot receives 75% of the energy contained in the pulse (take into account the transmission factor of the lens).

2. The intensity absorption factor for the steel film R is equal to 0.1. The absorbed energy is transformed into heat and diffused about the spot isotropically. What amount of heat will be necessary to raise to the melting temperature a half-sphere of steel of radius 0.1 mm (Fig. 20.2)? Compare this value to the quantity of heat carried by one pulse. Conclusion?

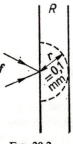

FIG. 20.2

Numerical values:

$$h = 6.62 \times 10^{-34} \text{ mksa,}$$
$$c = 3 \times 10^{8} \text{ m,}$$
$$\varepsilon_0 = \frac{1}{36\pi \times 10^{9}} \quad \text{(in rationalized mksa)}$$

1 kayser $= 1 \text{ cm}^{-1} = 100 \text{ m}^{-1}$,
density of steel $= 7.83 \text{ g/cm}^3$,
specific heat of steel $= 0.11 \times 4.18 \text{ J/g}$,
melting temperature of steel $t = 1525°C$.

SOLUTION

The energy of one pulse is equal to N times that of one of the photons carried so that

$$W = Nh\nu = Nhc\tilde{\nu}$$

$$N = \frac{0.3}{6.62 \times 10^{-34} \times 3 \times 10^{8} \times 14{,}418 \times 10^{2}} \simeq 1.05 \times 10^{18}.$$

The fluctuations in the number of photons and the phase of the wave are tied together by the uncertainty relation

$$\Delta W \, \Delta t \geqslant h.$$

In effect, the uncertainty in W is due to that in N, the quantum being well defined by the frequency. Thus, one has

$$\Delta Nh\nu \, \Delta t \geqslant h.$$

However, the uncertainty in the time Δt is related to the uncertainty $\Delta\phi$ in the phase. One finds

$$\frac{\phi}{2\pi} = \frac{t}{T} = vt,$$

hence

$$\Delta\phi = 2\pi v\,\Delta t$$

and

$$\Delta N\,\Delta\phi \geqslant 2\pi.$$

If $\Delta N = \sqrt{N}$,

$$\Delta\phi \geqslant 2\pi/\sqrt{N}.$$

N is large and $\Delta\phi$ is very small. For the frequencies corresponding to relatively long (red) wavelengths, the wave aspect is preponderant.

II

The pulse is represented by

$$s(t) = a\cos 2\pi v_0 t \qquad (-\tau/2 < t < \tau/2).$$

This has complex magnitude

$$s(t) = a\exp(-j\,2\pi v_0 t).$$

The corresponding frequency spectrum is

$$G(v) = a\int_{-\infty}^{+\infty} \exp[j\,2\pi(v-v_0)t]\,dt = \frac{a}{2\pi(v-v_0)}\left[\exp[j\,2\pi(v-v_0)t]\right]_{-\tau/2}^{+\tau/2}$$

$$G(v) = a\tau\,\frac{\sin 2\pi(v-v_0)\tau/2}{2\pi(v-v_0)\tau/2}.$$

The first two zeros of the well-known function $G(v)$ (§ A.7) arise for $2\pi(v-v_0)\tau/2 = \pm\pi$.
The half-interval between these values corresponds to a frequency domain

$$v - v_0 = 1/\tau$$

so that, since $\tilde{v} = v/c$ and the length $L = c\tau$;

$$\Delta\tilde{v} = \tilde{v} - \tilde{v}_0 = \frac{1}{L}.$$

Numerically

$$\Delta\tilde{v} = \frac{1}{c\tau} = \frac{1}{3\times10^8\times10^{-4}} = 0.33\times10^{-4}\ \text{m}^{-1} = 3.3\ \text{mkaysers},$$

$$|\Delta\lambda| = \lambda^2\,\Delta\tilde{v} = 48.10\times10^{-14}\times3.3\times10^{-5} = 1.59\times10^{-17}\ \text{m} = 1.59\times10^{-4}\ \text{mÅ},$$

$$L = \frac{1}{\Delta\tilde{v}} = 3\times10^4\ \text{m}.$$

One no longer observes interference in the Michelson interferometer when the distance between the mirrors is of the order of $L/2$ or 10 km. One cannot obtain homogeneous optical paths on this distance.

III

The volume occupied by the wave train is

$$v = \frac{\pi d^2}{4} L = 59 \times 10^{-2} \text{ m}^3.$$

1. The density of energy

$$w = \frac{W}{v} = \frac{0.3}{59 \times 10^{-2}} = 0.508 \text{ J/m}^3.$$

2. The expression for the energy density in electromagnetic theory is (§ 2.3.2)

$$w = \varepsilon_0 E^2,$$

hence:

$$E = \sqrt{575 \times 10^8} = 2.4 \times 10^5 \text{ V/m}.$$

3. The radiation pressure under normal incidence is (§§ 2.7 and 3.11.5)

(α) $\bar{\omega} = w = 0.508 \text{ N/m}^2$,
(β) $\bar{\omega} = 2w = 1.016 \text{ N/m}^2$,
(γ) $\bar{\omega} = (1+R)w = 1.9w = 0.965 \text{ N/m}^2$.

IV

1. $\varrho = 1.22\lambda F/d = 8.5 \times 10^{-4}$ cm (§ 5.11).

2. The mass of the hemisphere of steel is

$$M = \tfrac{2}{3}\pi r^3 \delta$$

δ being the density and r the radius

$$M = \tfrac{2}{3} \times 3.14 \times 10^{-6} \times 7.83 = 16.9 \times 10^{-6} \text{ g}.$$

The amount of heat necessary to raise the ordinary temperature to the melting point is

$$MC \, \Delta t = 16.9 \times 10^{-6} \times 0.11 \times 4.18 \times (1525 - 25) = 1.14 \times 10^{-2} \text{ J}.$$

During one pulse the film receives an amount of energy equal to

$$0.3 \times 0.75 \times 0.1 = 2.25 \times 10^{-2} \text{ J}.$$

The film will thus be melted at the point where it is struck by the radiation.

PROBLEM 21

Optical Constants of Germanium

The index for $\lambda_0 = 0.5\ \mu$ (in free-space) is given by

$$n = 3.47 - 1.40j \qquad (j = \sqrt{-1}).$$

1. Calculate the reflection factor at normal incidence for a polished germanium surface.

2. Calculate the phase shift ϕ_n introduced by reflection at normal incidence.

3. Calculate the depth which a plane wave penetrates into germanium when its intensity falls to 1/1000 of the incident intensity.

SOLUTION

The index of germanium is given in its complex form $n = n - jk$, n being the index of refraction and k the absorption index. The Fresnel formula is applicable to the complex index under normal incidence. The reflection coefficient for the light amplitude is complex (§ 3.5)

$$r_n = r_n \exp (j\phi) = \frac{n-1}{n+1} = \frac{n-jk-1}{n-jk+1}. \tag{1}$$

From which one gets the reflection factor

$$R_n = r_n r_n^* = \frac{(n-1)^2 + k^2}{(n+1)^2 + k^2} \tag{2}$$

and the phase advance of the reflected wave ϕ_n

$$\tan \phi_n = \frac{2k}{1 - n^2 - k^2}. \tag{3}$$

1. With the given values one finds

$$R_n = \frac{(3.47-1)^2 + (1.40)^2}{(3.47+1)^2 + (1.40)^2} = \frac{6.10 + 1.96}{19.98 + 1.96} = 0.37.$$

2.

$$\tan \phi_n = \frac{2.80}{1 - (3.47)^2 - (1.40)^2} = \frac{-2.80}{13.00} = -0.216.$$

$$\phi_n = 180° - 12.20° = 167.80°$$

3. The decrease in the light intensity as a function of the depth x is exponential

$$I = I_0 \exp (-2Kx)$$

with

$$K = \frac{2\pi k}{\lambda_0}.$$

One must have

$$\frac{I}{I_0} = 10^{-3},$$

$$\exp\left(-\frac{4\pi kx}{\lambda_0}\right) = 10^{-3},$$

$$\frac{4\pi kx}{\lambda_0} = 6.907,$$

$$x = \frac{6.907 \times 0.5}{4 \times 3.14 \times 1.40} \approx 0.2 \ \mu.$$

PROBLEM 22

Absorption. Black Bodies and Coloured Bodies

I

A small plane disc receives solar radiation at close to normal incidence. Of the two sides of the disc, only the side F turned towards the sun will be considered, the other side does not play a role. Assume that the disc is placed in a vacuum far removed from all other objects and that its temperature is always uniform. Assume that the sun radiates as a black body at 6000°K and call H its emittance. Its apparent diameter when viewed from a point D is small and will be taken as 2α. Calculate the equilibrum temperature of the disc in the following cases:

1. The disc emits and absorbs like a black body on the face F. Take 2α as 10^{-2} rad and then as 10^{-4} rad.

2. Repeat question 1 but assume here that the solar rays fall obliquely on the face F. The cosine of the angle of incidence can be taken to be 0.25.

3. Repeat question 1 replacing the disc D by a small sphere whose entire surface is a black body.

II

The disc has spectral energy emittance and an absorption factor which is zero for all radiation except for wavelengths very close to 0.40 μ. In this interval the disc acts like a black body. The angle 2α will be given the successive values 10^{-2} and 10^{-4} rad. Assume that near 0.40 μ the emittance of a black body is given in good approximation by the expression:

$$\log H_\lambda = a - \frac{b}{T}, \quad \text{with} \quad \frac{6000}{b} = 0.385,$$

a being a constant. Calculate the equilibrium temperature for 1 and 2 of part I.

III

A black body with sufficiently small dimension that its temperature will always be uniform and with heat capacity M is placed in the experimental arrangement indicated above. Initially it was protected from radiation and its temperature highly depressed. It was then exposed to solar radiation. According to what law will its absolute temperature rise as a function of time? How does this law behave near the equilibrium temperature?

SOLUTION

According to the definition of the emittance energy H, the energetic flux given off by a surface S into all exterior space is:

$$\Phi = HS.$$

In the case of the sun of radius R which emits like a black body:

$$S = 4\pi R^2 \quad \text{and} \quad H = \sigma T_0^4 \quad (\sigma = 5.672 \times 10^{-8} \text{ W m}^{-2} \text{deg}^{-4}).$$

This flux travels through spheres of increasing radius and that which reaches an area s on the sphere of radius r is:

$$\frac{\Phi s}{4\pi r^2} = s \frac{R^2}{r^2} \sigma T_0^4 = s\alpha^2 \sigma T_0^4 \tag{1}$$

for $\alpha^2 = R^2/r^2$.

On the other hand, the disc with area S at temperature T' radiates like a black body and its emittance is $H' = \sigma T'^4$. At equilibrium the incoming and outgoing fluxes are equal.

1. Normal disc, $s = S$.

$$S\alpha^2 \sigma T^4 = S\sigma T'^4, \tag{2}$$

hence,

$$T'^4 = \alpha^2 T_0^4, \quad T' = \sqrt{\alpha} T_0.$$

For $\alpha = 0.5 \times 10^{-2}$, $\quad T' = 0.071 \times 6000 = 426°\text{K}$.

$\quad \alpha = 0.5 \times 10^{-4}$, $\quad T' = 42.6°\text{K}$.

2. Oblique disc. This presents as above a cross-section of area S to the radiation but now the area $s = S \cos i$ (Fig. 22.1). However, it always radiates from the entire surface area S,

7*

thus at equilibrium:

$$S \cos i\alpha^2 \sigma T_0'^4 = S\sigma T''^4,$$

$$T'' = \sqrt{\alpha} \sqrt[4]{\cos i} T_0 = T'/\sqrt{2}.$$

For $\alpha = 0.5 \times 10^{-2}$,　$T'' = 300°K$.

For $\alpha = 0.5 \times 10^{-4}$,　$T'' = 30°K$.

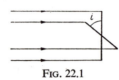

FIG. 22.1

3. *A sphere of radius ϱ.* This intercepts from the rays a section with area $s = \pi\varrho^2$ and radiates from its entire surface area $S = 4\pi\varrho^2$, hence at equilibrium:

$$\pi\varrho^2\alpha^2\sigma T_0^4 = 4\pi\varrho^2\sigma T'^4 \tag{4}$$

$$T' = \sqrt{\alpha} \sqrt[4]{\tfrac{1}{4}} T_0.$$

As a result of the numerical value $\cos i = \frac{1}{4}$ selected in (b) the temperature of the sphere in (c) is the same as that of the disc in (b).

II

Let $\Delta\lambda$ be the width of the spectral band in which the disc absorbs and emits, $H_{\lambda T_0}$ the spectral emittance of the sun, $H_{\lambda T'}$ the spectral emittance of a black body at the equilibrium temperature T' of the absorbing body, and the absorption factor A_λ. Calling S the surface area of the absorber and s the area which receives radiation the spectral flux energy received is

$$s\alpha^2 H_{\lambda T_0} \Delta\lambda.$$

The body absorbs a fraction A_λ of this. On the other hand, its spectral emittance is $A_\lambda H_{\lambda T'}$ according to Kirchhoff's law. The flux radiated is

$$SA_\lambda H_{\lambda T'} \Delta\lambda.$$

Hence, at equilibrium,

$$H_{\lambda T'} = \frac{s}{S} \alpha^2 H_{\lambda T_0}.$$

Taking the log of both sides and using the expression given for H_λ, this becomes

$$a - \frac{b}{T'} = \log \frac{s\alpha^2}{S} + a - \frac{b}{T_0}$$

from which

$$T' = \frac{6000}{1 + 0.385 \log \dfrac{S}{s\alpha^2}}.$$

1. *Normal disc: $S = s$.*

For $\alpha = 0.5 \times 10^{-2}$, $T' = 2165°K$.
$\alpha = 0.5 \times 10^{-4}$, $T' = 1392°K$.

2. *Oblique disc or sphere: $S = 4s$.*

For $\alpha = 0.5 \times 10^{-2}$, $T' = 1998°K$.
$\alpha = 0.5 \times 10^{-4}$, $T' = 1320°K$.

III

The black body receives flux [see eqn. (2)] from the sun. When its temperature is T, it radiates the flux $S\sigma T^4$. If its temperature is raised by dT in time dt, the energy balance is written

$$S\sigma(T'^4 - T^4)\, dt = M\, dt.$$

Taking $T/T' = x$ and $B = S\sigma T'^3/M$ this becomes

$$\frac{dx}{1-x^4} = B\, dt.$$

This fraction can be broken down into more simple elements. One has in effect

$$\frac{4}{1-x^4} = \frac{2}{1+x^2} + \frac{1}{1+x} + \frac{1}{1-x}.$$

Hence

$$4Bt + C = 2\arctan x + \log(1+x) - \log(1-x),$$

C being an integration constant.

When the temperature approaches the equilibrium value, $x \to 1$, $\arctan x \to \pi/4$, and $\log(1+x) \to \log 2$, thus

$$\log \frac{T'-T}{T'} = C - 4Bt.$$

$T'-T$ varies according to a decreasing exponential in t.

PROBLEM 23

Absorption. Kirchhoff's Law

I

1. A parallel beam of monochromatic light propagates in an absorbing liquid. Calling I_0 the intensity at a point taken as the origin and K the absorption coefficient, find the expression which will give the intensity I_x after the beam has travelled through a region of thickness x and also find the optical density D of this region.

2. Find the order of magnitude of the error in K and D for a relative error ε in the ratio I_0/I_x.

3. The absorbing medium under consideration is a plate of coloured glass with parallel faces and thickness x. This is placed normally across a beam of intensity I_0 and one measures a new intensity I_1.

The reflections which occur at the faces of the plate weaken the beam so that I_1 cannot be used as I_x without introducing an error ε' in the ratio I_0/I_x. Evaluate ε' by taking $\frac{3}{2}$ as the index of the glass and unity for the air index. How can one measure the optical density of this absorbing plate in such a way as to prevent this error? Neglect beams arising from multiple internal reflections.

II

1. Consider a homogeneous flame whose radiation obeys Kirchhoff's law and which is initially assumed to radiate monochromatically. Show that the spectral luminescence l_x of this flame with thickness x tends toward a limit as x increases without bound. Call $l_1 \, dx$ the luminous energy, $a_1 \, dx$ the absorption factor of an infinite thin film of thickness dx for radiation at the wavelength λ and call $\mathcal{L}$ the luminance of a black body at the temperature of the flame.

2. What occurs when the radiation is not rigorously monochromatic?

SOLUTION

I

1. The intensity of a parallel beam measured by the illumination which it produces on a surface and which in this case is constant, is proportional to the radiation flux so that the law for the variation of the intensity in an absorbing medium is (§ 1.11)

$$-dI = KI_0 \, dx,$$

hence

$$I_x = I_0 \exp(-Kx). \tag{1}$$

The ratio I_x/I_0 is the transmission factor of the substance with thickness x. The absorption factor is $(I_0-I_x)/I_0$. The optical density is given by

$$D = -\log_{10}\frac{I_x}{I_0} = \log_{10}\frac{I_0}{I_x} = \frac{1}{2.3}Kx. \tag{2}$$

The optical density is proportional to the distance travelled in the medium.

2. Taking

$$A = I_0/I_x = \exp(Kx): \tag{3}$$

$$dA = x\exp(Kx)\,dK, \qquad \varepsilon = \frac{dA}{A} = x\,dK,$$

$$dD = \frac{1}{2.3}\frac{dA}{A} \quad , \qquad \frac{dD}{D} = \frac{\varepsilon}{\log A} = \frac{dK}{K}.$$

3. The reflection factor for one face of the plate under normal incidence is (§ 3.4)

$$R = \left(\frac{n-1}{n+1}\right)^2.$$

The intensity which penetrates across the entry face is

$$I_0' = I_0(1-R).$$

At the exit face this has reduced to

$$I' = I_0' \exp(-Kx).$$

It then undergoes a second reflection and the emerging intensity is

$$I_1 = I'(1-R) = I_0(1-R)^2 \exp(-Kx). \tag{4}$$

Comparing (1) and (4) one sees that

$$I_1 = I_x(1-R)^2.$$

As a result of the reflections, the value A defined by (3) is reduced by

$$\Delta A = \frac{I_0}{I_1} - \frac{I_0}{I_x},$$

hence

$$\varepsilon' = \frac{\Delta A}{A} = \frac{1}{(1-R)^2} - 1 = \frac{R(2-R)}{(1-R)^2} \approx 2R.$$

These values approach one another when R is small. In effect,

$$R = \left(\frac{1.5-1}{1.5+1}\right)^2 = \frac{1}{25},$$

$$\varepsilon' = 0.08.$$

To eliminate the error caused by the reflections, it is necessary to make two intensity measurements on the transmitted ray for different thicknesses x_1 and x_2. Let D' be the optical density due to the reflections. One has

$$D_1 = D' + \frac{1}{2.3} Kx_1, \quad D_2 = D' + \frac{1}{2.3} Kx_2,$$

hence

$$D_1 - D_2 = \frac{1}{2.3} K(x_1 - x_2).$$

II

1. Compare this flame to an isothermal volume limited at O by a plane normal to Ox (Fig. 23.1). The variation of l_x in passing through a thin film of thickness dx is due to the emission $l_1\, dx$ and the absorption $-a_1 l_x\, dx$ following Beer's law. From this

$$dl_x = -a_1 l_x\, dx + l_1\, dx = a_1(\mathcal{L} - l_x)\, dx,$$

since $l_1 = a_1 \mathcal{L}$,

$$\frac{dl_x}{\mathcal{L} - l_x} = a_1\, dx.$$

For $x = 0$, $l_x = l_1$, hence $\quad \log \dfrac{\mathcal{L} - l_1}{\mathcal{L} - l_x} = a_1 x.$

$$\mathcal{L} - l_x = (\mathcal{L} - l_1) \exp(-a_1 x)$$
$$l_x = \mathcal{L}[1 - \exp(-a_1 x)] + l_1 \exp(-a_1 x).$$

For $x \to \infty$, $l_x \to \mathcal{L}$.

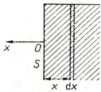

FIG. 23.1

2. If one is dealing with a set of radiations for which the absorption factor is not constant, the value of the total absorption factor evidently depends on the distribution of the energy in the radiation source. Now assume that this source follows Kirchhoff's law, $l_\lambda = a_\lambda \mathcal{L}_\lambda$, the the total luminance is

$$l = \int l_\lambda\, d\lambda = \int a_\lambda \mathcal{L}_\lambda\, d\lambda$$

and the total absorption factor is

$$a = \frac{l}{\mathcal{L}} = \frac{\displaystyle\int a_\lambda \mathcal{L}_\lambda\, d\lambda}{\displaystyle\int \mathcal{L}_\lambda\, d\lambda},$$

hence

$$\frac{l}{a} = \mathcal{L} = \int \mathcal{L}_\lambda\, d\lambda.$$

In the case where the incident radiation arises in a body which completely absorbs all radiation in the spectral interval under consideration. Kirchhoff's law applies to this set of radiations.

BIREFRINGENCE

PROBLEM 24

Stokes' Parameters. Poincaré Representation. Muller Matrices

I

Stokes' parameters

The state of polarization of a monochromatic light wave can be characterized by four quantities, all having the same dimensions, known as Stokes' parameters. These are

$$S_0 = a^2 + b^2, \quad S_1 = a^2 - b^2, \quad S_2 = 2ab \cos \phi, \quad S_3 = 2ab \sin \phi. \tag{1}$$

a and b are the amplitudes along two perpendicular directions Oy and Oz in the plane of the wave and ϕ is the difference in phase between them.

1. These four parameters are not independent. Find the relationship which exists between their squares.

2. What does the parameter S_0 represent? What are the four parameters relative to a linearly polarized wave along the Oy direction, along Oz, and at 45°; to right circular polarization, left circular polarization, and natural light (in this last case, in (1) use the mean values of the amplitudes) (§ 4.8.2)?

3. Prove the relationships

$$S_1 = S_0 \cos 2\beta \cos 2\theta, \quad S_2 = S_0 \cos 2\beta \sin 2\theta, \quad S_3 = S_0 \sin 2\beta, \tag{2}$$

β being the angle whose tangent is equal to the ratio of the axes of the elliptically polarized vibration and θ the angle which the major axis of the ellipse makes with Oy.

II

Poincaré sphere

The preceding equations show that the polarization state of a monochromatic vibration of given intensity can be represented on the surface of a sphere with radius S_0 by a point M whose latitude is 2β and whose longitude is 2θ. S_1, S_2, and S_3 are the cartesian coordinates of this point (Fig. 24.1).

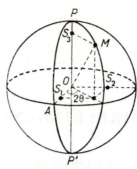

FIG. 24.1

1. The amplitudes a and b relative to the point M can be thought of as the components of a rectilinear vibration making angle α with Oy. How are the angles α and ϕ represented on the sphere (§ 8.2)?

2. What points on the sphere correspond to the directions Oy and Oz? What is the locus of points representing linear polarization, circular polarization? What do points situated on the same parallel represent?

3. Derive from II.1 a simple geometric construction which will allow one to get M, being given a birefringent medium with phase retardation ϕ and with known axes, for an incident linearly polarized wave coming in at angle α. Allow ϕ to vary and examine the results given by § 8.3.

4. Generalize the preceding construction representing the action of a birefringent medium with phase retardation ϕ on a vibration M. Look into a quarter wave plate as a special case (§ 8.6.4).

III

Muller matrices

The Stokes' parameters can be thought of as the four components of a column vector. One can represent the action of a polarizer or of a retarding system with known retardation and orientation on a light wave by a square matrix $[M]$ which when multiplied by the incident Stokes' vector $[V]$ gives the Stokes' vector $[V']$ of the outgoing vibration:

$$[V'] = [M][V]. \tag{3}$$

Here are several examples of the $[M]$ matrices:

$$\frac{1}{2}\begin{vmatrix} 1 & 1 & 0 & 0 \\ 1 & 1 & 0 & 0 \\ 0 & 0 & 0 & 0 \\ 0 & 0 & 0 & 0 \end{vmatrix} \qquad \frac{1}{2}\begin{vmatrix} 1 & 0 & 0 & 1 \\ 0 & 0 & 0 & 0 \\ 0 & 0 & 0 & 0 \\ 1 & 0 & 0 & 1 \end{vmatrix} \qquad \begin{vmatrix} 1 & 0 & 0 & 0 \\ 0 & 1 & 0 & 0 \\ 0 & 0 & 0 & 1 \\ 0 & 0 & -1 & 0 \end{vmatrix} \qquad \begin{vmatrix} 1 & 0 & 0 & 0 \\ 0 & 0 & 0 & -1 \\ 0 & 0 & 1 & 0 \\ 0 & 1 & 0 & 0 \end{vmatrix}$$

$$\qquad\quad (M_1) \qquad\qquad\qquad (M_2) \qquad\qquad\qquad (M_3) \qquad\qquad\qquad (M_4)$$

M_1: linear polarization with transmission direction Oy,
M_2: right circular polarization,

M_3: quarter-wave plate with its advancing axis along Oy,
M_4: quarter-wave plate with its axis at $45°$ to Oy.

With this information we want to use these methods to generate the following known results:

1. The action of a linear polarizer on natural light.

2. The action of a right circular polarizer on natural light.

3. The action of a quarter wave plate with axis along Oy and Oz on right circular polarized light.

4. Repeat question 3 with the axis at $45°$.
In each case, find the components of the vectors $[V]$ and $[V']$ and verify equation (3).
(The advantage of this method of calculation is that the action of a succession of polarizers and phase shifters on a light wave reduces to the multiplication of the Stokes' vector by a unique matrix which is the product of the matrices appropriate to the successive devices.)

SOLUTION

I

1.
$$S_0^2 = S_1^2 + S_2^2 + S_3^2. \tag{4}$$

2. S_0 represents the intensity of the vibration. For a linear polarization along Oy, $b = 0$, hence
$$S_0 = S_1 = a^2, \quad S_2 = S_3 = 0.$$

Likewise, for linear polarization along Oz
$$S_0 = b^2, \quad S_1 = -b^2, \quad S_2 = S_3 = 0,$$

linear polarization at $45°$:
$$S_0 = 2a^2, \quad S_1 = 0, \quad S_2 = 2a^2, \quad S_3 = 0,$$

right circular:
$$S_0 = 2a^2, \quad S_1 = 0, \quad S_2 = 0, \quad S_3 = 2a^2;$$

left circular:
$$S_0 = 2a^2, \quad S_1 = 0, \quad S_2 = 0, \quad S_3 = -2a^2;$$

natural light:
$$S_0 = 2\langle a^2 \rangle, \quad S_1 = 0, \quad S_2 = 0, \quad S_3 = 0.$$

3. The equations require that one go back to the classical expressions (§ 8.2):
(a) $S_3 = S_0 \sin 2\beta, \quad 2ab \sin \phi = (a^2 + b^2) \sin 2\beta$
$$\sin 2\beta = \sin 2\alpha \sin \phi. \tag{5}$$

(b) $S_1 = S_0 \cos 2\beta \cos 2\theta,$ $a^2 - b^2 = (a^2 + b^2) \cos 2\beta \cos 2\theta$

$$\cos 2\beta \cos 2\theta = \cos 2\alpha. \tag{6}$$

(c) $S_2 = S_0 \cos 2\beta \sin 2\theta,$ $2ab \cos \phi = (a^2 + b^2) \cos 2\beta \sin 2\theta$ thus   using   (5),
$\sin 2\alpha \cos \phi = \cos 2\beta \sin 2\theta$:

$$\tan 2\alpha \cos \phi = \tan 2\theta. \tag{7}$$

II

1. Trace a great circle on the sphere passing through M and the origin of the longitudes A. In the spherical triangle ABM with a right angle at B (Fig. 24.2), one has

$$\cos AM = \cos AB \cdot \cos MB$$

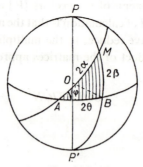

FIG. 24.2

so that, using (6)

$$AM = 2\alpha$$

and

$$\sin AM = \sin BM / \sin \widehat{MAB}$$

so that, with (5)

$$\widehat{MAB} = \phi.$$

2. For the linear vibrations, $\beta = 0$: the representative points are on the equator. For $\theta = 0$ the vibration has the direction Oy and it corresponds to point A, the origin of the longitudes. For $\theta = \pi/2$, the vibration is along Oz and it is represented by the point A' on the equator diametric with A ($2\theta = \pi$).

For circular light, $\beta = \pi/4$. The representative points are the poles P and P'. The point P represents right circular light ($0 < \phi < \pi$) and P' left circular light ($\pi < \phi < 2\pi$).

Right elliptical vibrations are in the northern hemisphere and left elliptical in the southern.

The points along a given parallel represent ellipses of the same form but different inclination.

The points on a given meridian represent vibrations of the same orientation and whose eccentricity varies from 0 on the equator to 1 at the poles. Figure 24.3 summarizes the statements above.

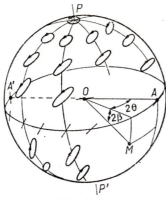

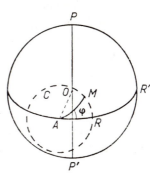

FIG. 24.3   FIG. 24.4

3. Let R be the point on the equator which represents the linear vibration and A the point which represents the vibration oriented along the optic axis Oy of this (uniaxial) birefringent system (Fig. 24.4). Curve $AR = 2\alpha$. With A as the centre, trace on the surface of the sphere, a circle, C, with radius AR. The intersection of this circle with the great circle passing through A and making angle ϕ with the equator is the required point.

For $\phi = \pi/2$, M is on the meridian of A and the ellipse has its axes along Oy and Oz. If, in addition, $\alpha = \pi/2$, R is at R', and the circle C is the meridian normal to OA and M passing through P or P', the vibration is circular. For $\phi = \pi$, the vibration is linear but with α changed in sign.

4. Let AA' be the diameter corresponding to the orientation of the birefringent system acting on the vibration represented by the point M. Trace the great circle passing through A and M (Fig. 24.5). As has been seen in 3, the angle ϕ_0 which this circle makes with the equator measures the phase difference between the vibrations entering along the neutral axis of the system and those entering along this one. The algebraic phase retardation ϕ is added to ϕ_0 and the point M' is then obtained from M by a rotation of ϕ about A.

If $\phi = \pi/2$, M passes into M'' on a great circle passing through A and making an angle ψ with the equator such that

$$\psi + \frac{\pi}{2} + \phi_0 = \pi.$$

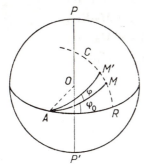

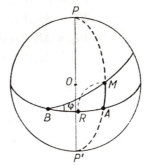

FIG. 24.5   FIG. 24.6

In the quarter-wave method, the linear vibration obtained from the polarizer is represented by point A (Fig. 24.6). The uniaxial lines of the birefringent system under study are represented by B and the diametric point. One adjusts $AB = \pi/2$. Consequently, the point M associated with the vibration leaving the system is on the meridian of A. The spherical triangle MBA is rectangular at M and A and from spherical trigonometry

$$\cos \phi = \cos MA = \cos 2\beta.$$

The neutral axes of the quarter-wave plate coincide by their adjustment with the ends of the diameter AA'. The quarter-wave plate rotates the point M by $\pi/2$ about A, and the vibration leaving is represented by the point R. It is rectilinear and $AR = AM = 2\beta$.

III

The results obtained in I.2 give the following expressions for the Stokes' vector taking the incident vibration to have unit intensity:

$$\begin{bmatrix} 1 \\ 0 \\ 0 \\ 0 \end{bmatrix} \qquad \begin{bmatrix} 1 \\ 1 \\ 0 \\ 0 \end{bmatrix} \qquad \begin{bmatrix} 1 \\ -1 \\ 0 \\ 0 \end{bmatrix} \qquad \begin{bmatrix} 1 \\ 0 \\ 1 \\ 0 \end{bmatrix} \qquad \begin{bmatrix} 1 \\ 0 \\ 0 \\ 1 \end{bmatrix}$$
$$(V_1) \qquad\quad (V_2) \qquad\quad (V_3) \qquad\quad (V_4) \qquad\quad (V_5)$$

V_1 = natural light, V_2 = linearly polarized along Oy, V_3 = linear along Oz, V_4 = linear at 45°, and V_5 = right circular.

We detail the symbolic operation $[V'] = [M][V]$. The multiplication rule for matrices gives for the component V'_{lc} of the vector $[V']$ related to row l and to column c:

$$V'_{lc} = \sum_i M_{li} V_{ic}$$

M_{li} is a term from row l of the matrix $[M]$ and V_{ic} is the term with the same index i in the column c of vector $[V]$.

For $[V]$ and $[V']$, c can only have the value one. Thus

$$V'_{11} = M_{11}V_{11} + M_{12}V_{21} + M_{13}V_{31} + M_{14}V_{41},$$
$$V'_{21} = M_{21}V_{11} + M_{22}V_{21} + M_{23}V_{31} + M_{24}V_{41},$$
$$V'_{31} = M_{31}V_{11} + M_{32}V_{21} + M_{33}V_{31} + M_{34}V_{41},$$
$$V'_{41} = M_{41}V_{11} + M_{42}V_{21} + M_{43}V_{31} + M_{44}V_{41}.$$

1. In this case $[V'] = [M_1][V_1]$, so that

$$V'_{11} = 1 \times 1 + 1 \times 0 + 0 \times 0 + 0 \times 0 = 1,$$
$$V'_{21} = 1 \times 1 + 1 \times 0 + 0 \times 0 + 0 \times 0 = 1,$$
$$V'_{31} = 0 \times 1 + 0 \times 0 + 0 \times 0 + 0 \times 0 = 0,$$
$$V'_{41} = 0 \times 1 + 0 \times 0 + 0 \times 0 + 0 \times 0 = 0,$$

then, taking into account the factor $\frac{1}{2}$ which acts on $[M_1]$, the vector $[V']$ is

$$[V'] = \begin{bmatrix} \frac{1}{2} \\ \frac{1}{2} \\ 0 \\ 0 \end{bmatrix}.$$

This can be seen to be identical to $[V_2]$ to within the factor $\frac{1}{2}$ which shows the reduction in intensity due to the polarization.

2. Likewise one finds $[M_2][V_1] = [V_4]$ to within the factor $\frac{1}{2}$ again due to polarization.

3. $[M_3][V_5] = [V_4]$ rectilinear at $45°$.

4. $[M_4][V_5] = [V_3]$ linear along Oz.

PROBLEM 25

Fresnel Formulas. Birefringent Prism

1. Recall the expressions for the reflection and transmission of a monochromatic, parallel beam of light incident from free space on the surface of a plane isotropic refracting medium with index n. Call r_n and r_p the amplitude reflection factors with the subscript p being with respect to Fresnel vibrations parallel to the plane of incidence and n normal to that plane. The corresponding transmission factors are designated by t_p and t_n.

2. Apply the results of 1 to the following two questions: A glass prism whose apex angle is $60°$ has an index of 1.52 for the radiation being studied. The face AB of the prism receives a parallel monochromatic beam of this radiation normal to the edge A and with an incident angle such that the deviation of the beam leaving the face AC is minimal. The incident beam has been polarized linearly so that its vibrations are at $45°$ to the plane of incidence. What is the angle that the emerging vibrations make with the plane of incidence?

3. In what way is it necessary to modify the angle A of the prism and the incident polarization so that one loses no light by reflection at the point of entry and at the exit point of the beam in the glass prism?

4. Assume now that the prism is of Icelandic spar which has been carefully cut so that the section ABC is an equilateral triangle, the face BC being planar and polished. The crystal axis is parallel to the face ABC. Show that a parallel beam of linearly polarized monochromatic light falling normally on the face AB and propagating with its vibrations at $45°$ to the face ABC, is totally reflected at BC, but that at its exit point on the face AC it is elliptically polarized. Neglect reflection losses in entering and leaving the prism since these are small near normal incidence and indicate how the emerging elliptical vibrations depend on the height h of the triangle ABC and on the two principal indices of spar $n_0 = 1.65$ and $n_e = 1.48$.

5. Assume finally that the crystal axis is normal to the entry face AB, that is, parallel to the incident beam. Show that it then has two distinct beams leaving the prism which have been reflected on BC. Find the direction of each of the emerging beams and indicate with what vibrations they propagate.

SOLUTION

2. One has (Fig. 25.1)

$$n \sin \frac{A}{2} = \sin \frac{D+A}{2} \quad \text{and} \quad D = 2i - A, \quad r = \frac{A}{2} = 30°.$$

$$\sin \frac{D+A}{2} = 1.52 \times 0.500 = 0.760, \quad \frac{D+A}{2} = 49° \, 28' = i.$$

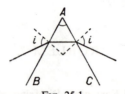

FIG. 25.1

Using the appropriate expression [(8.28) of § 8.4]

$$\tan \alpha_t = \cos (i-r) \tan \alpha_i, \quad \alpha_i = 45°, \quad \tan \alpha_i = 1.$$

The beam leaves at the same angle

$$\tan \alpha_t = \cos^2 (i-r),$$
$$i - r = 49° \, 28' - 30° = 19° \, 28', \quad \cos (i-r) = 0.94293,$$
$$\cos^2 (i-r) = 0.88912$$
$$\alpha_t = 41° \, 39'.$$

3. The reflection factor is zero for vibrations in the incident plane at the Brewster angle. It is therefore necessary to polarize the incident beam so that the vibration is normal to the plane of incidence and the incidence is such that $\tan i_B = n$, hence

$$i_B = 56° \, 40'.$$

For the emerging beam to be at the Brewster angle it is necessary for $\hat{A}' = 2\hat{r}$. Thus

$$\sin r = \cos i_B, \quad r = 90° - i_B = 33° \, 20'.$$
$$A' = 66° \, 40', \quad \text{hence} \quad A' - A = +6° \, 40'.$$

4. The incidence at I (Fig. 25.2) is at $60°$, hence $\sin i = 0.866$ is greater than $1/n_e = 0.675$ and $1/n_0 = 0.606$. Thus there is total reflection. The propagation of the e- and o-waves is normal to the axis so that the e- and o-rays which are reflected and emerge are mixed. However, they are polarized, the o-ray in the plane of the principal section and the e ray normal to this plane. Since the vibrations are coherent at the entry where the two principal axes are parallel and perpendicular to the edge, the birefringence is n_0-n_e. The distance travelled is $h \sin 30° = h/2$. One knows (§ 8.3.2) that for $\alpha = 45°$ the ellipse is oriented with its axes at 45° and the ratio of its axes is $\tan \beta$ such that $\tan 2\beta = \tan \phi$ (Fig. 25.2):

$$\phi = \frac{2\pi}{\lambda} \frac{h}{2} (n_0-n_e) = \frac{\pi h}{\lambda} \times 0.17.$$

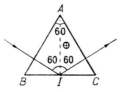

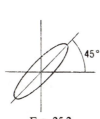

FIG. 25.2

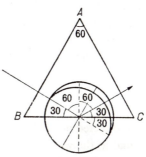

FIG. 25.3

5. The incident rays undergo no birefringence up to I (Fig. 25.3). They are totally reflected at an angle of 60°, that is, in a direction making an angle $\theta = 60°$ with the optic axis. The normals to the reflected waves remain mixed, but one wave has an index n_0 and the other n such that [(4.25) of § 4.4]

$$\frac{n^2 \sin^2 \theta}{n_e^2} + \frac{n^2 \cos^2 \theta}{n_0^2} = 1,$$

$$n^2 \left(\frac{3}{4 \times 2.19} + \frac{1}{4 \times 2.72} \right) = 1 = n^2 \frac{1}{0.434},$$

$$n^2 = 2.304, \qquad n = 1.517.$$

The corresponding rays are separated: the ordinary ray, coinciding with the wave normal, strikes the face AC at a normal and passes out without deviation. Its vibration is normal to the plane of the figure. The extraordinary ray makes an angle with the ordinary ray given by [(4.37) of § 4.9]

$$\tan \zeta = \frac{(n_e^2-n_0^2) \tan \theta}{n_e^2+n_0^2 \tan^2 \theta} = \frac{(2.19-2.72)1.732}{2.19+2.72 \times 3} = -0.088,$$

$$\zeta = -5°3'.$$

The minus sign shows that the extraordinary ray makes an angle of 54°58′ with the normal to BC. The incident angle at AC is ζ and the angle of refraction is

$$\sin i_e = 1.517 \times \sin 5°3' = 1.517 \times 0.0877 = 0.1335$$

$$i_e = 7°40'.$$

PROBLEM 26

The Field of Polarizing Prisms

A spar polarizing prism has the form of a parallelepiped whose face $ABCD$ (Figs. 26.1 and 26.2) is normal to the optic axis. The plane of the cut contains the optic axis and lies along the line AC. The two halves of the prism are separated by a layer with parallel faces formed either of a transparent cement with index $n = 1.540$ or of a layer of air. Determine in both cases the maximum angular range of the polarizer for rays normal to the optic axis, that is, the sum of the angles which these rays can make on either side of the normal to the face AB so that the emergent beam is polarized. Also find the size of this range when the angles are symmetric with respect to the normal. Calculate the ratio $R = L/h$ of the length AD of the prism to its height AB. The principal indices of spar are $n_0 = 1.658$ and $n_e = 1.486$.

SOLUTION

Propagation is directed normally to the optic axis and the ray direction is coincident with that of the normal to the waves.

1. It is the ordinary ray with the higher index which is eliminated by total reflection in the case where the two halves of the prism have cement between them. The range is limited on the upper side (Fig. 26.1) by the angle of incidence i corresponding to the direction AC of the refracted extraordinary ray whose angle of refraction is the angle α which defines the shape of the prism

$$\tan \alpha = \frac{h}{L} = \frac{1}{R}, \tag{1}$$

and one has, at A

$$\sin i = n_e \sin \alpha. \tag{2}$$

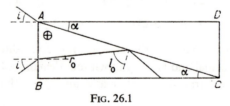

Fig. 26.1

On the lower side, the range is limited by the condition that the angle of refraction r of the ordinary ray be such that its angle of incidence on the cutting plane AC be at least equal to the limiting angle l_0 for spar-cement defined by

$$\sin l_0 = n/n_0.$$

Now

$$l_0 = \frac{\pi}{2} - (\alpha + r_0),$$

hence
$$\cos(\alpha+r_0) = n/n_0. \qquad (3)$$

The angle of incidence limiting the range on the lower side is such that
$$\sin i = n_0 \sin r_0. \qquad (4)$$

Combining equations (1), (2), (3), and (4), one gets
$$\sin i = \frac{n_e}{\sqrt{1+R^2}}$$

and
$$(n_0^2-n^2)R^4+[n_0^2-n_e^2-2n(n+n_e)]R^2-(n+n_e)^2 = 0.$$

The numerical factors give
$$R = L/h = 4.93$$
$$i = 17°10', \quad \alpha = 11°30'.$$

If one only wants to find the maximum symmetrical angular range, one must take $\alpha = r_0$ and (2) yields
$$\cos 2\alpha = n/n_0 = 0.9275$$
$$\alpha \approx 11°$$

and one gets from (1):
$$R = L/h = 5.14.$$

This prism is called a Glazebrook prism.

2. The condition for which the ordinary ray striking the face AB in the plane of the figure above the normal is totally reflected on AC is the same as above now with $n = 1$:
$$\sin i_0 = n \sin r_0, \quad \frac{\pi}{2}-\alpha-r_0 \geqslant l_0, \quad \sin l_0 = \frac{1}{n_0}.$$

But the extraordinary ray can also undergo total reflection and the range will be limited on the upper side by the condition that this ray can again exit at the face CD, which gives
$$\sin i_e = n_e \sin r_e, \quad \frac{\pi}{2}-\alpha+r_e \leqslant l_e, \quad \sin l_e = \frac{1}{n_e}.$$

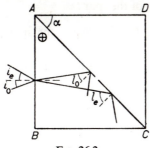

FIG. 26.2

For the range to be symmetrical it is necessary for $i_e = i_0$, so that

$$\sin i_0 = n_0 \sin r_0 = n_e \sin r_e. \tag{4}$$

Now the angles r_0 and r_e are small since

$$r_0 + r_e = l_e - l_0 = 5°11'.$$

Thus, one can write, in place of (4)

$$i_0 = n_0 r_0 = n_e r_e,$$

from which the range

$$2i_0 = \frac{2 n_e n_0}{n + n_0}(r_0 + r_e) = \frac{2 n_e n_0}{n_e + n_0}(l_e - l_0) = 8°10',$$

and the ratio

$$R = \cot \alpha = \cot (l_0 + r_0) = \cot \left(l_0 + \frac{i_0}{n_0} \right) = 0.825.$$

This is called a Glan polarizer. It is much shorter than the Glazebrook prism but the light wasted is much larger since the reflection factor is larger at the spar–air interface than at the spar–cement interface.

PROBLEM 27

Rotary Dispersion

A parallel beam of white light passes through a spar plate of thickness e placed between two nicols set at extinction. The optic axis of the plate is at $45°$ to those of the nicols. In addition, the beam passes through a grating with 500 lines per millimetre and a converging lens with a focal length of 1 metre. On a white screen placed in the focal plane one observes the first order spectra which are formed about the central fringe.

1. At what distances from the central fringe will the points P and Q be found where the radiation of wavelengths $\lambda_1 = 0.6 \ \mu$ and $\lambda_2 = 0.7 \ \mu$ converge?

2. There are dark bands at P and Q and one finds forty-one bands (fringes) between these two points. What is the thickness of the spar?

3. Consider now the white light which leaves the spar plate and which does not pass the second nicol.
  (a) What are the wavelengths λ for which the polarization is linear?
  (b) What are the wavelengths λ for which the polarization is circular?
Indices of spar: $n_0 = 1.658$, $n_e = 1.486$.

SOLUTION

1. A parallel beam of light which normally crosses a uniaxial birefringent plate with parallel faces cut parallel to the axis remains parallel and does not undergo doubling.

In effect, the normals to the ordinary and extraordinary waves are coincident within the crystal and are, in this case, both coincident with the rays since the normal to the extra-ordinary wave is in the equatorial plane of the indicial ellipsoid (Fig. 27.1). It thus strikes the grating normally. The principal maximum in first order for radiation with wavelength λ is in the direction i_1

$$\sin i_1 = \lambda/d,$$

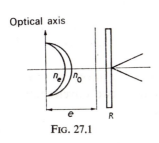

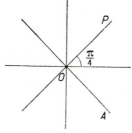

FIG. 27.1   FIG. 27.2

d being the grating step. One has $d = 2\mu$ and

$$\lambda = 0.6 \; \mu, \quad \sin i_1 = \frac{0.6}{2} = 0.30, \quad i_1 = 17°27'30'',$$

$$\lambda' = 0.7 \; \mu, \quad \sin i_1' = \frac{0.7}{2} = 0.35, \quad i_1' = 20°29'15''.$$

The distances from these two beams to the central image ($i = 0$) are respectively

$$r_1 = f \tan i_1 = 0.314 \; m, \quad r_1' = f \tan i_1' = 0.373 \; m.$$

There are two spectra in the first order symmetric about the central image.

2. The band spectra observed are due to the variation with wavelength of the phase difference between the principal vibrations introduced by the birefringent plate. The linear vibration OP from the first polarizer (Fig. 27.2) falls on the spar at 45° to the optic axes which are in the direction of the optic axis and normal to it, and has equal projections on these axes. After exiting from the plate, the vibration parallel to the axis, which is propagated with the index n_e less than the index n_0 of the perpendicular vibration, has taken a lead in optical path over the latter of:

$$\delta = e(n_0 - n_e).$$

This does not depend on the wavelength if one neglects the dispersion of the birefringence. This approximation, always good in the domain of radiation under consideration, is imposed by the statement of the problem which only gives a single value for the indices. However, the phase difference

$$\phi = \frac{2\pi\delta}{\lambda} = \frac{2\pi e}{\lambda}(n_0 - n_e)$$

varies with the wavelength. The intensity transmitted by the analyser (§ 6.19) is given by

$$I = \sin^2 \frac{\phi}{2}, \tag{1}$$

since the optic axes are at 45° to the direction of vibration given by the polarizer and which itself is crossed with the analyser. The intensity is zero for

$$\phi = 2K\pi \quad \text{or} \quad e(n_0 - n_e) = K\lambda, \quad (K \text{ integer}). \tag{2}$$

All wavelengths for which this relationship is satisfied—that is, for which the thickness of the plate is equal to K times that of a single wave plate—are quenched. They then correspond to dark bands in the spectrum.

To find the thickness of the plate, one needs to know the integer K (or K') relative to the λ band (or λ'). If one has forty-one bands between these two wavelengths

$$K = K' + 42.$$

Thus one has

$$e(n_0 - n_e) = K'\lambda' = (K' + 42)\lambda,$$

hence

$$K' = e \times \frac{0.172}{0.7}$$

and

$$e = \frac{42 \times 0.6}{0.172\left(1 - \dfrac{0.6}{0.7}\right)} = \frac{42 \times 0.6 \times 0.7}{0.172 \times 0.1} = 1026 \ \mu.$$

3. From one dark band to the next, the phase difference ϕ varies by 2π. For the wavelengths lying between these bands the vibration leaving the spar is generally elliptical. The axes of the ellipse are always oriented along the direction of the vibration OP defined by the polarizer and in the normal direction OA (§ 8.3).

(a) The vibration is linear when this ratio is zero or infinite, that is, if:

$$\frac{\phi}{2} = K\pi \quad \text{or} \quad \frac{\phi}{2} = (2K+1)\frac{\pi}{2} \quad (K \text{ integer}).$$

In the first case, one again gets equation (2) and the vibration parallel to OP is quenched by the analyser. In the second case, the thickness of the plate is equal to an odd number times that of a half-wave plate for the radiation under consideration and the linear vibration is parallel to OA and is passed by the analyser. Illumination of the spectrum is maximum for wavelengths defined by

$$e(n_0 - n_e) = (2K+1)\frac{\lambda}{2}.$$

(b) The vibration is circular for $\tan \phi/2 = 1$, so that

$$\frac{\phi}{2} = (2K+1)\frac{\pi}{4}.$$

The thickness of the plate is equal to an odd number times the thickness of a quarter-wave plate. This occurs for radiation whose wavelength satisfies

$$e(n_0 - n_e) = (2K+1)\frac{\lambda}{4}.$$

PROBLEM 28

Two Passes through a Quarter-wave Plate

A monochromatic light source emits a parallel beam which passes through a polarizing prism P. The emerging light rays fall at normal incidence on a crystalline quarter-wave plate (for the radiation utilized), L_1, and are then reflected normally at the surface of a perfectly reflecting metallic mirror M (Fig. 28.1).

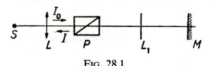

FIG. 28.1

Calling α the angle formed by the plane of the principal section of the polarizer P and the advancing optic axis of the plate L_1, and I_0 the luminous intensity of the incident beam before passing through polarizer P, one wants to know:

1. The value of the luminous intensity I of the returning beam emerging from the polarizer. Discuss the special cases of the angle α.

2. Rotating P uniformly about the incident light beam with a frequency $\nu (\alpha = 2\pi\nu t)$, what will be the modulation frequency ν' of the beam I?

In each case neglect the losses due to glassy reflection.

SOLUTION

The normal reflection on the mirror leads to a phase advance of π for all linear vibrations (§ 3.5). In addition, if one considers the most general polarization state of a light vibration—elliptical polarization—the reflection makes a phase advance of π both elliptical components thus the direction of the path is not modified.

However, the propagation direction of the light is inverted by reflection and the plate L_1 receives an elliptical vibration in an inverted sense to that which it originally produced.

On the whole the experiment is equivalent to that illustrated in Fig. 28.2:

FIG. 28.2

P and A: polarizer and analyser parallel; L_1: a quarter-wave plate; L_2: a half-wave plate having its optic axes parallel to the axes of the elliptical vibration exiting L_1, it inverts the sense of the ellipse as does the mirror but without altering the direction of propagation; L_1': a quarter-wave plate having its optic axes parallel to those of L_1; I: the direction of observation.

1. The linear vibration given by P is $a \cos \omega t$ with $a = \sqrt{I_0/2}$. Its components on the optic axes of L_1 can be written:

on the entry side $y_0 = a \cos \alpha \cos \omega t$, $z_0 = a \sin \alpha \cos \omega t$,

on the exit side $y_1 = a \cos \alpha \cos \omega t$, $z_1 = a \sin \alpha \sin \omega t$,

calling α the angle which the vibration a makes with the advancing optic axis y of the plate L_1. The vibration is left-handed in this case with its axes along y and z. The plate L_2 has its optic axes also directed along y and z. It transforms the components y_1 and z_1 into

$$y_2 = a \cos \alpha \cos \omega t, \quad z_2 = -a \sin \alpha \sin \omega t.$$

The plate L_1' gives

$$y' = a \cos \alpha \cos \omega t, \quad z' = -a \sin \alpha \sin \left(\omega t + \frac{\pi}{2} \right) = -a \sin \alpha \cos \omega t.$$

The analyser A permits passage of the components

$$y' \cos \alpha = a \cos^2 \alpha \cos \omega t \quad \text{and} \quad z' \sin \alpha = -a \sin^2 \alpha \cos \omega t,$$

so that

$$a(\cos^2 \alpha - \sin^2 \alpha) \cos \omega t = a \cos 2\alpha \cos \omega t.$$

The intensity is

$$I = \frac{I_0}{2} \cos^2 2\alpha. \tag{1}$$

It reaches its maximum value for $\alpha = 0$ and $\alpha = \pi/2$. In both cases a linear vibration directed along Oy or Oz passes through the entire apparatus.

The intensity is zero for $\alpha = \pi/4$. In this case, it exits from L_1 as a left-handed circular vibration (using the convention adopted). The plate L_2 transforms this into right-hand circular light. This is transformed into a linear vibration by L_1'. However, this latter vibration is perpendicular to the transmitting direction of the analyser which therefore quenches it.

2. The result is obtained immediately from (1) which gives

$$I = \frac{I_0}{2} \cos^2 4\pi\nu t,$$

where ν is the frequency of rotation of the polarizer-analyser. The modulation frequency is therefore $\nu' = 2\nu$.

PROBLEM 29

Birefringent Monochromator

A parallel beam of light from a sodium vapour lamp passes through a polarizer analyser pair which have their transmitting directions parallel and which are separated by a calcite plate with parallel faces whose optic axis is fixed in the plane of the faces. What must be the minimal thickness of this plate so that only one of the sodium D-lines, separated by 6 Å, leaves the analyser with maximum intensity? The following table gives the principal indices of calcite in the region of the D-lines:

$\lambda(\text{Å})$:	5876	5893
n_e :	1.486 47	1.486 41
n_0 :	1.658 46	1.658 36.

SOLUTION

The obtain a zero minimum, it is necessary to have the axis of the plate at 45° to the principal plane of the polarizers. The intensity is (§ 6.19)

$$I = \cos^2 \frac{\pi}{\lambda} e(n_0 - n_e)$$

since, for a plate parallel to the axis, the principal indices are n_0 and n_e. For the intensity at wavelength λ to be a maximum it is necessary that

$$e(n_0 - n_e) = K\lambda \qquad (K \text{ integer}). \tag{1}$$

For it to be zero at the same time for the wavelength $\lambda' = \lambda + \delta\lambda$, it is necessary that

$$e(n_0' - n_e') = (K + \tfrac{1}{2})\lambda' = (K + \tfrac{1}{2})(\lambda + \delta\lambda).$$

But

$$n_0' = n_0 - \frac{dn_0}{d\lambda} \delta\lambda, \quad n_e' = n_e - \frac{dn_e}{d\lambda} \delta\lambda,$$

hence

$$e\left\{ n_0 - n_e - \left(\frac{dn_0}{d\lambda} - \frac{dn_e}{d\lambda} \right) d\lambda \right\} = \left(K + \frac{1}{2} \right)(\lambda + \delta\lambda)$$

and, taking (1) into account,

$$\left\{ \frac{e(n_0 - n_e)}{\lambda} - e\left(\frac{dn_0}{d\lambda} - \frac{dn_e}{d\lambda} \right) \right\} \delta\lambda = \frac{\lambda + \delta\lambda}{2}$$

and finally, neglecting $\delta\lambda$ in the second term,

$$e = \frac{\lambda^2}{2\delta\lambda} \times \frac{1}{(n_0 - n_e) - \lambda\left(\dfrac{dn_0}{d\lambda} - \dfrac{dn_e}{d\lambda}\right)}.$$

With

$$\lambda = 5893, \quad \delta\lambda = 6, \quad \frac{dn_0}{d\lambda} = \frac{10^{-4}}{17}, \quad \frac{dn_e}{d\lambda} = \frac{6\times10^{-5}}{17}$$

one gets

$$e \approx 17\times10^6 \text{ Å} \approx 1.7 \text{ mm.}$$

This arrangement can effectively serve to separate the components of a doublet.

PROBLEM 30

Experiments of Fresnel and Arago

I

A collimator provided with a vertical slit and an astronomical telescope are situated in such a fashion that their optic axes are along the same horizontal. The two instruments are focused at infinity and the collimator is illuminated by a monochromatic radiation of wavelength 0.54 μ (the green line of mercury). Between the objectives of these two instruments an opaque screen is placed normal to their common optic axis and the screen is provided with two vertical windows F and F'. The windows have the same width $a = 1$ mm and their centres, situated on the same horizontal, are separated by a distance $d = 3$ mm. Sketch the appearance of the fringes as they appear in the telescope and, knowing that the magnification of the telescope is 20, find:
 (a) the apparent diameter of the central maximum;
 (b) the angular separation between two consecutive interference fringes.

II

A nicol polarizer whose principal section is vertical is positioned between the collimator and the screen containing the windows. In addition one has transparent, easily cleaved mica whose two principal indices for the green mercury line at normal incidence and for the two vibrations at right angles lying along the optic axes L' and L'' of the mica sheet are given by:

$$n' = 1.5998, \quad n'' = 1.5948.$$

A half-wave plate of this mica is required for the green mercury line. Determine its thickness and show how one can verify that the cleaved plate is precisely half-wave.

III

From this half-wave plate two pieces are cut having the form of elongated rectangles. One of these rectangles $ABCD$ has its long axis AB parallel to the optic axis L and the other, $A'B'C'D'$, has a long axis $A'B'$ which makes an angle of $\pi/4$ with the direction of L. These two plates are placed in front of and against the opaque screen with their sides AB and $A'B'$ vertical so that each one covers one of the windows. Explain why under these conditions the interference fringes previously seen disappear completely and show that in order to make them reappear again in the same place it is necessary to introduce an appropriately oriented nicol analyser behind the windows.

IV

Can one, by initially placing a thin plate of mica suitably oriented and covering the entire field, make the fringes continually visible when one turns the nicol analyser but displaced in a continuous fashion depending on its rotation?

SOLUTION

This problem is a variation on the experiments of Fresnel and Arago on interference in polarized light (§ 6.17).

Young's fringes are formed as with natural parallel monochromatic light. The angular distribution of the light passing through a system of windows is given by the usual expression (§ 6.1):

$$I = 4I_0 \left(\frac{\sin \frac{\pi a \sin i}{\lambda}}{\frac{\pi a \sin i}{\lambda}} \right) \cos^2 \left(\frac{\pi d \sin i}{\lambda} \right). \tag{1}$$

I_0 is the diffraction intensity along the axis of the system and i is the angle made with this central line in the plane of the diffraction. Since the magnification G of the telescope is, by definition, when focused at infinity equal to the ratio of the apparent diameters of the object when viewed with the instrument and the naked eye, it is only necessary to multiply the result given in (1) by G to obtain the required value.

(a) The first factor of (1), which corresponds to diffraction by each slit vanishes for $\sin i_1 = \lambda/a$. The angular diameter of the central maximum thus has the value

$$2G \sin i_1 = 2G \frac{\lambda}{a} = 40 \times \frac{0.546}{10^3} = 2.184 \times 10^{-2} \text{ rad} = 1°15'.$$

(b) The angular separation between two neighbouring fringes corresponds to the difference between two values of i which cause the second term of (1) to go to zero, namely $\sin i = \lambda/d$ and through the telescope

$$G \frac{\lambda}{d} = 20 \times \frac{0.546}{3 \times 10^3} = 0.364 \times 10^{-2} \text{ rad}.$$

Thus one has $2.184/0.364 = 6$ interference fringes in the central diffraction maximum. Since the central fringe is a maximum, the third bright fringe on either side of the central maximum falls in the direction i_1 where the intensity goes to zero. Therefore, five interference fringes are visible (Fig. 30.1).

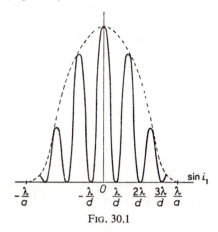

FIG. 30.1

The intensity due to the second factor of (1), namely

$$4I_0 \sin^2 \frac{\phi}{2} = 2I_0(1-\cos \phi) \tag{2}$$

is modulated by the variations in the first factor.

II

The thickness of a half-wave plate (§ 8.3) is

$$e = \frac{\lambda}{2(n'-n'')} = \frac{0.546}{2\times0.005} = 54.6 \ \mu.$$

Verification requires that one recall that a half-wave plate converts a linear vibration into a linear vibration. By placing it—in any orientation—between a crossed polarizer–analyser pair, one can find extinction by rotating the analyser. The procedure does not distinguish between a plate giving a phase retardation of $\lambda/2$ and one giving $k\lambda/2$ (k integral) but one knows that the plate is essentially a half-wave plate.

To increase the precision one uses the quarter-wave method (§ 8.6.4): the angle of rotation of the analyser which acts from the zero transmission region to restore the illumination is equal to $\phi/2 = 90°$.

III

The plate $ABCD$ does not affect the orientation of the linear vibration given by the polarizer which is parallel to the optic axis L'. The plate $A'B'C'D'$ rotates this same vibration through 90° since it is oriented at 45° to the optic axis L'. The diffracted beams from the two slits thus consist of coherent, linear vibrations of equal amplitude but oriented perpendicularly to one another. These recombine to give elliptical vibrations which vary according to the phase difference but which contains a uniform distribution of energy. Thus there are no longer any interference fringes.

The analyser prism allows passage of all linear vibrations parallel to its direction of transmission. These components which then have the same polarization can interfere. Calling OV and OV' (Fig. 30.2) the two linear components, OA the transmission direction of the analyser which makes an angle β with OV, and Ov and Ov' the respective projections of OV and OV' on OA, one has:

$$Ov = a \cos \beta \cos \omega t, \quad Ov' = a \sin \beta \cos (\omega t - \phi),$$

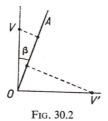

FIG. 30.2

with a the common amplitude of OV and OV' and their phase difference which depends on the angle of diffraction. Combining Ov and Ov' gives the intensity:

$$I = a^2 \sin^2 \beta + a^2 \cos^2 \beta - 2a^2 \sin \beta \cos \beta \cos \phi = a^2(1 + \sin 2\beta \cos \phi).$$

To recover the original system of fringes where the distribution of intensities is given by (1), it is necessary to have $\sin 2\beta = 1$ and $\beta = \pi/4$. The amplitudes Ov and Ov' are then equal and the fringe contrast is unity. The contrast is clearly zero for $\beta = 0$ or $\beta = \pi/2$ since only one of the vibrations OV or OV' is transmitted by the analyser and there can be no interference.

IV

The preceding discussion shows that the visibility of the fringes depends on the azimuth β of the analyser. To have them independent of this it is necessary that the projections Ov and Ov' do not depend on this, that is, that the vibrations OV and OV' are circular rather than linear. One can accomplish this by placing in front of the analyser a quarter-wave plate of mica with its optic axis at 45° to both OV and OV'. The two vibrations are transformed into circular vibrations with opposite senses (§ 8.3.3) and having the same amplitude. At any point in the field one of these has a phase difference ϕ with respect to the other. However, one knows (§ 8.5) that the resultant of such vibrations is linear with an azimuth equal to $\phi/2$. Between two points in the field where the phase difference varies by 2π, the linear resultant has rotated through π and at these points extinction will be obtained for the same orientation of the analyser. One has a system of fringes with the same separation as in natural light. Now, by turning the analyser through an angle α, one extinguishes all the vibrations inclined at an angle of α to those above and one sees a continuous displacement of the system of fringes.

This can be made more precise by a calculation. The components of OV on the optic axes OQ and OQ' of the quarter-wave plate (Fig. 30.3) are, at the exit side,

$$OQ_1 = \frac{a}{\sqrt{2}} \cos \omega t, \quad OQ_2 = \frac{a}{\sqrt{2}} \sin \omega t \quad \text{(right circular)},$$

while those of OV' are

$$OQ_1' = \frac{a}{\sqrt{2}} \cos(\omega t - \phi), \quad OQ_2' = \frac{-a}{\sqrt{2}} \sin(\omega t - \phi) \quad \text{(left circular)}.$$

The signs are those obtained by assuming that OQ_1 is the advanced optic axis (that with the smaller index).

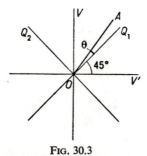

FIG. 30.3

Let θ be the azimuth of the analyser OA with respect to OQ_1. The projections of the vibrations on OA are

$$\frac{a}{\sqrt{2}} \cos \omega t \cos \theta + \frac{a}{\sqrt{2}} \sin \omega t \sin \theta = \frac{a}{\sqrt{2}} \cos(\omega t - \theta),$$

$$\frac{a}{\sqrt{2}} \cos(\omega t - \phi) \cos \theta - \frac{a}{\sqrt{2}} \sin(\omega t - \phi) \sin \theta = \frac{a}{\sqrt{2}} \cos(\omega t - \phi + \theta).$$

These are two vibrations in the same direction and with the same amplitude which have a phase difference $2\theta - \phi$ with a resultant

$$I = 2a^2 \cos^2(\theta - \phi/2).$$

If $\theta = 0$ or $K\pi$, one has the system of fringes represented by (1). If θ varies, the maxima are where $\theta = \phi/2$.

PROBLEM 31

Polarization Interferometer. Differential Method

I

A ray of monochromatic natural light falls at normal incidence on a quartz plate Q_1 having parallel faces and thickness e. The optic axis is in the plane of the figure and makes an angle of $45°$ with the normal to the plate (Fig. 31.1). One has $n_e = 1.533$ and $n_0 = 1.544$.

1. Construct the path of the rays in the crystal.

2. Calculate as a function of e, n_e, and n_0 the separation of the emerging rays. Give their states of polarization.

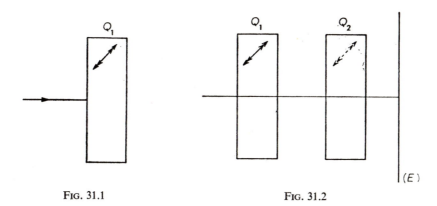

FIG. 31.1   FIG. 31.2

II

One adds a second quartz plate Q_2 identical to Q_1. The faces of these two plates are parallel. Let α be the angle between the plane of the principal sections of these two plates (Fig. 31.2). Show that in general four rays leave Q_2. Find for $\alpha = 0, 45°, 90°, 135°$ and $180°$:
the relative positions of the rays on the screen E (giving the polarization state of each);
the energy carried by each of them taking the source as unity.
In what follows assume that the axes Q_1 and Q_2 are parallel.

III

Place in front of Q_1 a polarizer P which only lets passage of vibrations oriented at $45°$ to the plane of the figure. Between Q_1 and Q_2 is placed a half-wave plate whose optic axes are at $45°$ to the plane of the figure. Sketch the path of the rays through the system and indicate the nature of the vibration transported by each one.

IV

Place behind the half-wave plate a transparent plate L to generate a phase variation (Fig. 31.3). Call ϕ the phase difference introduced between rays (1) and (2) displaced by Δ. What is the nature of the vibration leaving Q_2 in the case where

$$\phi = 0, \quad 0 < \phi < \frac{\pi}{2}, \quad \phi = \frac{\pi}{2}$$

$$\frac{\pi}{2} < \phi < \pi \quad \text{and} \quad \phi = \pi?$$

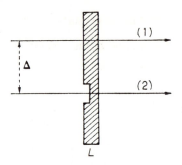

FIG. 31.3

V

1. One can cause the rays leaving Q_2 to interfere. What is the function of the polarizer P? How does it contribute to the production of good contrast?

2. One illuminates the system by a plane wave Σ parallel to the faces of the quartz. Show that the waves issuing from Q_2 are displaced laterally by an amount Δ. Assume that the plate L generates a phase constant variation throughout a groove of width a (Fig. 31.3). Find as a function of ϕ the illumination in the different regions of the image.

VI

In general one can detect objects with different phase and not measure the phase shift. Place behind Q_2 an objective O focused in the plane of the plate L. With this objective (with given characteristics) are associated the quartz plates Q_1 and Q_2 so that the separation Δ is less than the limit of resolution of the objective O.

Find e given that O has a numerical aperture $\sin u = 0.2$ and the wavelength used is $\lambda = 0.5 \ \mu$.

SOLUTION

1. *Huygen's construction* (§ 4.2).
The wave surface is made up of:

$$\left. \begin{array}{l} \text{a sphere of radius } 1/n_0 \\ \text{an ellipsoid of revolution with} \\ \quad \text{axes } 1/n_0 \text{ and } 1/n_e \end{array} \right\} \begin{array}{c} \text{tangent along the axis} \\ \text{(Fig. 31.4).} \end{array}$$

The incident plane wave breaks up within the crystal into an ordinary wave Σ_0 and an extraordinary wave Σ_e (Fig. 31.4). The ordinary wave is directly transmitted and the extra-ordinary ray is deviated toward the axis (in the case of a quartz crystal).

2. The intersection of the ellipsoid with the plane of the figure is an ellipse with equation

$$n_0^2 x^2 + n_e^2 y^2 = 1. \tag{1}$$

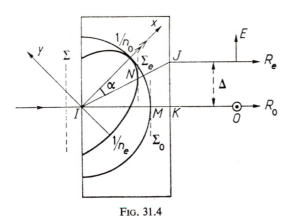

FIG. 31.4

At point N the tangent is parallel to the bisector of xOy and thus one has $dy/dx = 1$. Using equation (1) this becomes

$$1 = -\frac{n_0^2}{n_e^2} \frac{x}{y} \tag{2}$$

hence

$$\frac{y}{x} = \tan \alpha = -\frac{n_0^2}{n_e^2}. \tag{3}$$

The separation of the two emerging rays is

$$\Delta = e \tan \left(\frac{\pi}{4} + \alpha\right) = e \frac{1 + \tan \alpha}{1 - \tan \alpha} = e \frac{n_e^2 - n_0^2}{n_e^2 + n_0^2}. \tag{4}$$

Since the birefringence of the quartz is small, one can write

$$\Delta = e \frac{n_e - n_0}{n} = e \frac{9 \times 10^{-3}}{1.55}. \tag{5}$$

The extraordinary vibration is in the plane of the principal section (which contains the axis and the normal to the entry face). The ordinary vibration is perpendicular to the plane of the figure (Fig. 31.4).

II

One sees that the plate Q_1 splits the incident radiation into two rays situated in the plane of the principal section.

Take ray 1 incident on Q_2. If the vibration E does not coincide with the optic axis of Q_2 one has on the exit side two parallel rays in the plane of the principal section of Q_2 and carrying the vibrations EE and EO. Likewise, ray 2 is split and the two rays carry the OE and OO vibrations (Fig. 31.5).

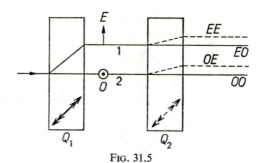

FIG. 31.5

$\alpha = 0$ OO and EE are in the plane of the figure separated by 2Δ. Each ray carries
          energy $\frac{1}{2}$.

$\alpha = \pi/4$ four rays each with energy $\frac{1}{4}$.

$\alpha = \pi/2$ two rays each with energy $\frac{1}{2}$.

$\alpha = 3\pi/4$ four rays each with energy $\frac{1}{4}$.

$\alpha = \pi$ two coincident rays with energy $\frac{1}{2}$.

The results are seen in Fig. 31.6.

Note. The assembly of two identical plates making an angle of $\alpha = \pi/2$ forms a "Savart plate".

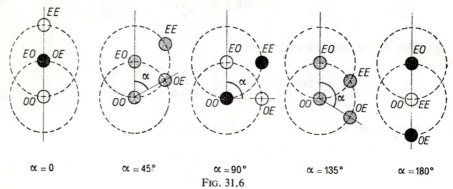

FIG. 31.6

III

The vibration transmitted by the polarizer P can decompose into two rectangular vibrations Px and Py having the same amplitude (Fig. 31.7). In the crystal Q_1, P_x becomes ordinary and P_y extraordinary. If one does not insert the half-wave plate, one has the situation shown in Fig. 31.6a. Knowing that a half-wave plate transforms a linear vibration into another linear vibration symmetric with respect to its optic axes, the vibration which is extraordinary in Q_1 behaves (after rotation by 90°) like an ordinary vibration in Q_2. Inversely, the ordinary ray in Q_1 becomes extraordinary in Q_2 (Fig. 31.7). The vibrations E and O are shifted out of phase in passing through Q_1; however, EO and OE are again in phase after passing through Q_2.

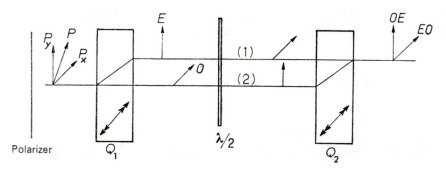

FIG. 31.7

IV

At the exit side of Q_2, one has, in general, an elliptical vibration (formed from two linear vibrations out of phase by ϕ). The ellipse can be inscribed in a square and its axes coincide with the diagonals of the square (§ 8.3.2).

$$\phi = 0 \qquad \text{linear}$$
$$0 < \phi < \pi/2 \quad \text{right elliptical}$$
$$\phi = \pi/2 \qquad \text{right circular}$$
$$\pi/2 < \phi < \pi \quad \text{right elliptical}$$
$$\phi = \pi \qquad \text{linear}$$

V

1. The polarizer makes the vibrations coherent. It is necessary to add an analyser after Q_2 to render the vibrations parallel. The contrast is maximum and equal to one when the vibrations have the same amplitude, that is, when the favoured direction of the analyser is at 45° to the plane of the figure.

2. The wave Σ is split into two waves Σ_O and Σ_E longitudinally and laterally out of phase. The two waves Σ_{OE} and Σ_{EO} are again found to be coincident and in phase as a result of the

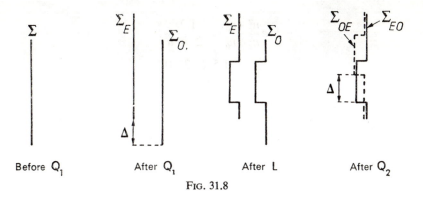

FIG. 31.8

9*

compensation by Q_2. However, if phase variations are introduced between Q_1 and Q_2, the field contains two phase-shifted objects separated by Δ (Fig. 31.8).

$$\Delta > a \quad \text{total splitting method} \quad \text{(Fig. 31.9a)},$$

$$\Delta < a \quad \text{differential method} \quad \text{(Fig. 31.9b)}.$$

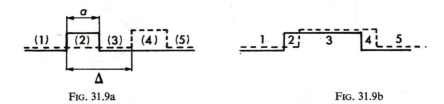

FIG. 31.9a   FIG. 31.9b

The phase shift and therefore the illumination varies in the different regions. The results are summarized in the following table:

Region	1	2	3	4	5
Phase difference	0	$+\phi$	0	$-\phi$	0
Illumination	1	$\cos^2 \phi/2$	1	$\cos^2 \phi/2$	1

This polarization interferometer leads to an interferometer with uniform brightness. If one uses white light, the phase variations show up in variations of colour.

VI

When one wishes to detect objects with different phase one uses the differential method and chooses a splitting less than the limit of resolution ϱ of the objective. This requires

$$\Delta < \varrho, \tag{6}$$

hence

$$e \frac{n_e^2 - n_0^2}{n_e^2 + n_0^2} < \frac{1.2\lambda}{2 \sin u}, \tag{7}$$

$$e(\mu) < \frac{1.5 \times 1.2 \times 0.5}{9 \times 10^{-3} \times 2 \times 0.2},$$

$$e < 250 \ \mu.$$

PROBLEM 32

Electrical Birefringence

A capacitor with rectangular plates A and B of length l and separated by a distance h (Fig. 32.1) is immersed in a cell containing carbon disulphide at 22°C. A parallel beam of light with wavelength λ in air is directed between the plates and parallel to their long axis. This beam is polarized by a polarizer P whose principal section makes an angle α with the plane of the plates. After ascertaining that the faces of the cell crossed by the incident light are isotropic, one applies to A and B the potentials V_1 and V_2. Determine the orientation

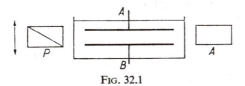

FIG. 32.1

and ellipticity of the light leaving the apparatus. For $\alpha = 45°$, show precisely on a figure the positions of a quarter-wave plate and of the principal section of an analyser A corresponding to complete extinction of the emergent ray.

For numerical purposes take $l = 20$ cm and $h = 4$ mm. For V_1 and V_2 take the pole potentials of a series of 5000 batteries of 2 volts each with the centre of the voltage source grounded.

Note. One knows that carbon disulphide placed in an electrostatic field E_0 behaves like a positive uniaxial crystal with axis parallel to the field and that the birefringence acquired by it at 22°C measured by the difference of the indices n_e and n_0 is such that:

$$n_e - n_0 = 3 \times 10^{-14} \times \lambda E_0^2,$$

the variables being in SI units. Neglect the effect of the edges of the capacitor and assume the field to be uniform.

SOLUTION

The cell acts like a crystalline plate with plane, parallel faces whose privileged directions are oriented along the direction Oz of the electric field (that is, normal to the plates of the capacitor) and Oy normal to Oz. Since the crystal is positive uniaxial, $n_e > n_0$. The direction of the extraordinary vibration Oz is retarded.

The linear vibration $OP = \sin \omega t$ given by the polarizer has the following components along the optic axes at the entrance to the cell:

$$y_0 = \cos \alpha \sin \omega t, \quad z_0 = \sin \alpha \sin \omega t;$$

at the exit from the cell

$$y = \cos \alpha \sin \omega t, \quad z = \sin \alpha \sin (\omega t - \phi),$$

with

$$\phi = \frac{2\pi}{\lambda}(n_e - n_0).$$

The expression given in the problem allows us to calculate the birefringence. The field

$$E_0 = \frac{V_1 - V_2}{h} = \frac{10^4}{4 \times 10^{-3}} = 25 \times 10^5 \text{ V/m}.$$

$$n_e - n_0 = 3 \times 10^{-14} \times \lambda \times (25 \times 10^5)^2$$

and

$$\phi = \frac{2\pi \times 0.20}{\lambda} \times 3 \times 10^{-14} \times \lambda \times 625 \times 10^{10} = 0.075\pi = 0.236 \text{ rad} = 13.2°.$$

The light which leaves the cell is elliptically polarized. For $\alpha = 45°$, expression (8.19) from § 8.6 shows that $\theta = \alpha$ whatever ϕ may be. The axes of the ellipse are respectively

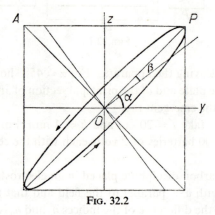

FIG. 32.2

parallel and perpendicular to the direction of the incident linear vibration OP (Fig. 32.2). Equation (8.16) of § 8.6 gives $\sin 2\beta = \sin \phi$. The ratio of the axes of the ellipse is

$$\tan \beta = \tan \frac{\phi}{2},$$

so that

$$\beta = 6.6°.$$

Since the direction Oz is retarded, the ellipse is traversed in the trigonometric sense (§ 8.3.1).

The emergent beam can be eliminated by the analyser if the vibration it carries is linear. The quarter-wave plate can convert to linear the elliptical vibration leaving the cell if its privileged directions are oriented parallel to the axes of the ellipse. One can adjust the quarter-wave plate before having applied the electric field by crossing the polarizer OP and the analyser OA then positioning the quarter-wave plate and rotating it until extinction occurs. The optic axes of the quarter-wave plate are parallel to OP and OA.

If the advancing optic axis of the quarter-wave plate is placed along OP, the OP component of the elliptical vibration, retarded by $\pi/2$ with respect to the component normal to it, is found to be advanced by $\pi/2$ after passage through the quarter-wave plate. The vibrations along OP and OA are then in phase and their resultant is a linear vibration OP which makes an angle β with OP. Extinction can be achieved then by adjusting the analyser, initially at OA, and bringing it to OA', that is, by turning it through an angle β in a sense inverse to that of the elliptical rotation (§ 8.6.4).

PROBLEM 33

Rotary Power. Circular Dichroism

Using the eye as a detector one wants to examine here the polarization states of a plane, monochromatic wave of sodium light, $\lambda = 0.589\ \mu$, which passes through a cell containing a liquid having an absorption and a natural rotary power.

I

(a) In the first experiment, the light is linearly polarized by a polarizer P (Fig. 33.1) and one determines the azimuth of the vibration before and after passing through the cell which has a length of 0.5 cm. Knowing that the vibration has turned through a clockwise angle

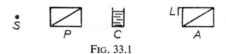

FIG. 33.1

$\alpha = 1.2°$ for the observer, and recalling that the rotary power is explained by circular birefringence, find the sign and magnitude of the difference in refractive indices for this liquid in right and left circular light of the wavelength given above.

(b) The inclination of the vibration is found using a half-shadow analyser which is made up of a half-wave plate L (Fig. 33.1) rigidly mounted on a Glazebrook prism A and covering half the light beam. The transmission direction of this prism makes a small angle ε with the optic axis of L. The assembly can be turned through a known angle. Explain the function of this apparatus.

II

(a) In a second series of experiments the light is circularly polarized. This is accomplished through the use of a linear polarizer (Glazebrook, for example) and a quarter-wave plate with known optic axes. Indicate briefly using a drawing, how one can produce right and left circular light with this apparatus.

(b) What is the thickness of the quarter-wave plate if it is cleaved from crystalline mica whose principal indices in the cleavage plane are $n = 1.5977$ and $n' = 1.5936$ in sodium light?

(c) Given this, one discovers, with the aid of a suitable flux detector, that a 1 mm cell full of the liquid transmits a fraction of 0.520 of the incident left-hand light and that a similar cell 2 mm in length transmits 0.320. Calculate the absorption coefficient of the liquid. Why is it necessary to measure it for two thicknesses?

Repeat this for right-hand light where the transmitted intensities are 0.503 and 0.301.

(Recall that for a homogeneous absorbing substance, the relative loss in intensity of a monochromatic light flux F in passing through a thickness dx of a substance is

$$-\frac{dF}{F} = K\,dx$$

where K is the absorption coefficient.)

III

Find the reduction in amplitude of right and left circular vibrations which pass through 0.5 cm of the liquid. If one illuminates the liquid by linearly polarized light, show that the emergent light is elliptically polarized and find the ratio of the axes.

IV

Repeat the measurement of rotary power made in part I. Now, without touching the analyser adjusted to extinction, place between the cell and the analyser a quarter-wave plate whose optic axes are parallel to those of the half-wave plate. What rotation of the analyser is required to re-establish extinction? Justify the result.

V

The absolute uncertainty in the orientation of a vibration as measured with an analyser at extinction and the eye varies inversely as the square root of the light flux received by the analyser. What thickness of this active, absorbing liquid will make the relative uncertainty minimal?

SOLUTION

I

(a) One has (§ 8.5):

$$\alpha = \frac{\pi l}{\lambda}(n_l - n_r),$$

α being given in radians. If the rotation occurs in the clockwise sense for the observer, it is the right circular vibrations which propagate most rapidly in the medium. $n_l > n_r$.

$$n_l - n_r = \frac{1.2 \times 0.589 \times 10^{-4}}{180 \times 0.5} = 7.85 \times 10^{-7}.$$

(b) Let Oy and Oz (Fig. 33.2) be the directions of the optic axes of the half-wave plate, OA be the direction of the vibration transmitted by the analyser, and OV_1 be the direction of the linear vibration carried by the beam falling on the plate. This vibration takes the direction OV_2, symmetric with respect to OV_1 about Oz in the half of the beam striking the half-wave plate. Let γ be the angle made by OV_1 with Oz. The projections of OV_1 and OV_2 on OA respectively are

$$Ov_1 = OV_1 \sin(\gamma + \varepsilon) \quad \text{and} \quad Ov_2 = OV_2 \sin(\gamma - \varepsilon).$$

The intensities of the corresponding two beams after passing through the analyser are

$$I_1 = I \sin^2(\gamma + \varepsilon) \quad \text{and} \quad I_2 = I \sin^2(\gamma - \varepsilon).$$

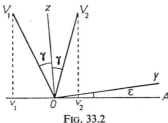

FIG. 33.2

They are equal when $\gamma = 0$, that is to say, when OV is parallel to Oz. They are faint since ε is a small angle; this is favorable for their comparison to be made by the eye (the solution $\gamma \approx \pi/2$ gives too large an intensity). One can then precisely locate the azimuth of the vibration OV before and after introduction of the active liquid.

II

(a) The transmission direction of the polarizer OP is oriented at $45°$ to the optic axes of the quarter-wave plate. To obtain a left-hand vibration for the observer receiving the light leaving the quarter-wave plate, the "advancing" optic axis, that is, the optic axis corresponding to the lower index of refraction, should be aligned along Oy ($n_y < n_z$). This is reversed for a right-hand vibration.

(b) For a quarter-wave plate of thickness e:

$$e(n - n') = \frac{\lambda}{4}$$

or:

$$e = \frac{0.589}{4(1.5977 - 1.5936)} = 36 \ \mu,$$

which can be easily obtained from mica because of the ease of cleaving.

(c) Expressing the absorption coefficient K in cm^{-1}, one has:

for the 1-mm cell $F_1/F_0 = A \exp(-0.1K)$,
for the 2-mm cell $F_2/F_0 = A \exp(-0.2K)$,

F_0 being the incident flux, A a constant coefficient which depends on the cell and especially which takes into account the reflection losses on its faces. Hence:

$$\frac{F_2}{F_1} = \exp(-0.1K).$$

For left-hand light:

$$\exp(-0.1K_l) = \frac{0.320}{0.520} = 0.615,$$

$$-0.1K_l = 2.3 \log 0.615 = 2.3(-0.21112) = -0.4856.$$
$$K_l = 4.86 \text{ cm}^{-1}.$$

For right-hand light:

$$\exp(-0.1K_r) = \frac{0.301}{0.503} = 0.599,$$

$$-0.1K_r = 2.3 \log 0.599 = 2.3(-0.22257) = -0.5119,$$
$$K_r = 5.12 \text{ cm}^{-1}.$$

It is to eliminate the effect of the cell accounted for by the coefficient A that requires the use of two measurements with varying thickness of cell.

III

(a) The intensity of a monochromatic light vibration is proportional to the square of the amplitude and the absorption coefficient for this latter is $K/2$. Hence, the values for the reduction factors resulting from passage through 0.5 cm are

Left-handed vibration: $\exp(-4.86 \times 0.5 \times 0.5) = \exp(-1.200) = 0.3012$.
Right-handed vibration: $\exp(-5.12 \times 0.5 \times 0.5) = \exp(-1.280) = 0.2791$.

The resultant of the two circular vibrations with the same amplitude and opposite sense is a linear vibration. For two vibrations occuring in the opposite sense with unequal amplitude the resultant is an elliptical vibration. To see this, one can refer the circular vibrations occurring after passage through the medium to two general rectangular axes, Oy and Oz. The left-handed one is, for example, given by the expressions

$$y_l = G \cos \omega t, \quad z_l = G \sin \omega t$$

and the right-handed by

$$y_r = D \cos(\omega t - \phi), \quad z_r = D \sin(\omega t - \phi)$$

with $\phi = 2\alpha$ where α is the angle of rotation. With respect to the axes OY and OZ with which these make the angle α, the equations take the form

$$Y_l = G \cos(\omega t - \alpha), \quad Z_l = G \sin(\omega t - \alpha),$$
$$Y_r = D \cos(\omega t - \alpha), \quad Z_r = -D \sin(\omega t - \alpha).$$

The resultant of circular vibrations is

$$Y = Y_l + Y_r = (G+D) \cos(\omega t - \alpha),$$
$$Z = Z_l + Z_r = (G-D) \sin(\omega t - \alpha).$$

These are the equations of an ellipse related to its axes. They are respectively equal to $2(G+D)$ and $2(G-D)$. The vibration is left-handed if $G-D > 0$ and right-handed if $G-D < 0$. The ellipse is described in the sense of a circular vibration less the absorption. The ratio of the ellipse axes is

$$\frac{G-D}{G+D} = \frac{0.3012-0.2791}{0.3012+0.2791} = 0.0381.$$

One can get this result also by a geometric reasoning. The two circular vibrations can be represented at each instant by the vectors $OD = D$ and $OG = G$ which rotate about O making equal angles about OY (Fig. 33.3). Construct the resultant OR of OG and OD then draw through R parallels to OY and OZ which strike OG at M and N. One finds $GN = GR = OD$. The point N describes a circle of radius $D+G$ and M of $D-G$. The locus of R is obtained

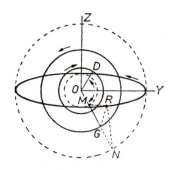

FIG. 33.3

by considering the two concentric circumferences, a moving radius ON and, at the points where it strikes the two circles, the parallels drawn respectively to OY and OZ. This is one definition of an ellipse.

Note that the elliptical vibration is obtained here by a mechanism quite different than that of linear birefringence. In the latter case, the ellipicity of the vibration varies with the angle of the incident vibration with respect to the optic axes of the birefringent system (§ 8.2). In the actual case, the ellipse does not depend on the angle of the initial vibration since optic axes in a liquid to which no field has been applied do not exist. The rotary power of a liquid turns the ellipse through an angle α independent of its orientation.

IV

The quarter-wave plate is situated with its optic axes parallel to the axes of the ellipse which leaves the cell, since the analyser at minimal transmission is set along the major axis of the highly flattened ellipse and since this axis is parallel to one of the optic axes of the half-wave plate. The quarter-wave plate transforms the elliptical light into linearly polarized light which makes an angle β with its optic axes such that

$$\tan \beta = \frac{G-D}{G+D} \approx \beta = 0.0330 \approx 2°.$$

The angle β is clockwise for the observer if the advancing optic axis of the quarter-wave plate coincides with the major axis of the ellipse.

V

The absolute uncertainty ε in the direction is

$$\varepsilon = C \sqrt{F} \qquad (C = \text{Const}).$$

The relative uncertainty is ε/α. However, $\alpha = Al$ ($A = \text{const.}$) and $F = F_0 \exp(-Kl)$. Hence

$$\frac{\varepsilon}{\alpha} = \frac{C \sqrt{F_0}}{A} \times \frac{\exp(-Kl/2)}{l}.$$

As a function of l this expression is minimal for

$$\frac{Kl}{2} = 1, \quad \text{from which} \quad l = \frac{2}{K} \approx \frac{2}{5} = 0.4 \text{ cm}.$$

PROBLEM 34

Faraday Effect

I

Between two polarizers P and P' (Fig. 34.1) set so that the direction of transmission of P' makes an angle of $+45°$ with that of P for an observer at O one places a column C of carbon disulfide CS_2, 0.5 m long, in a uniform magnetic field B parallel to the length of the column. What should be the direction and minimum value of B so that the maximum flux leaving S reaches O? What happens if one exchanges the positions of S and O without changing anything else in the apparatus?

The Verdet constant of CS_2 is 42×10^3 min/tesla-m.

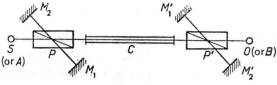

FIG. 34.1

II

S and O are now replaced respectively by two identical bodies A and B while maintaining the experimental set up described above such that one can operate in an adiabatic enclosure. The polarizers P and P' are birefringent prisms which eliminate the second beam by total internal reflection (Nicol prism, Glazebrook prism...). The rejected beams reflect normally on the perfect mirrors M_1, M_2, M_1', and M_2'. Show, by examining the polarization state of all of the beams, that the exchange of radiant energy between A and B does not alter the thermal equilibrium once it has been established.

SOLUTION

I

The Faraday effect must equal $+45°$ for the observer. The sense of this rotation is the same as that of the current which produces the field B. For the rotation to be right at O, the axial vector B must in the conventional sense be directed from O to S.

The magnitude of the rotation is given by Verdet's expression (§ 18.14)

$$\varrho = \varrho_0 Bl.$$

Hence, in SI units

$$B = \frac{\varrho}{\varrho_0 l} = \frac{45 \times 60}{42 \times 10^3 \times 0.5} = 0.1286 \text{ tesla.}$$

The Faraday effect always preserves the sense of the current generator of B and it changes the sense of the rotation for the observer when he exchanges positions with the source. The linear light formed by P' will then be found to be normal to the transmission direction of P and light will not pass.

This apparatus forms an optical valve; light passes freely in the sense SO but is stopped in the sense OS.

II

On the surface the reasoning involves taking the results of part I into account and saying that B receives half the flux Φ which A transmits to P (the other half being eliminated by total reflection) whereas A does not get half the flux Φ emitted by B and which passes P' (the other half being totally reflected). The thermal equilibrium is then broken down contrary to the second law of thermodynamics. This is Wein's paradox. In the complete picture, taking the mirrors into account, A recovers the flux $\Phi/2$ which it has emitted and which is reflected on M_1. Likewise, B recovers the flux $\Phi/2$ which it transmitted to M_1' after total reflection in P'. The other half, which contains the light vibration E_B crossing P' and C (Fig. 34.2) (the light vibration is E_B' at the exit side of C), is totally reflected at P, then in M_2,

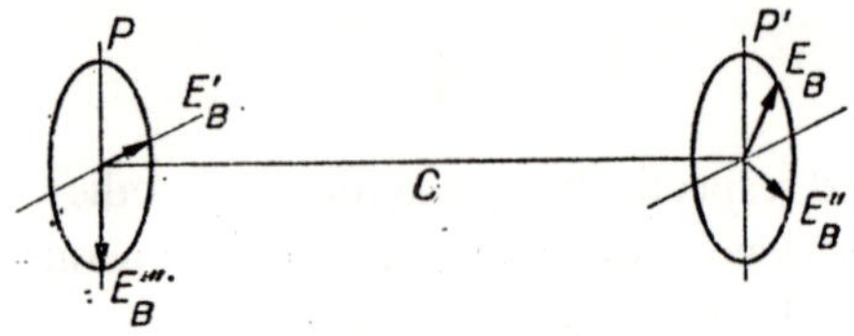

FIG. 34.2

then again in P and returns toward B. But, at the exit from the cell, the orientation of the vibration is E''_B normal to E_B. The beam E''_B is then reflected in P', then at M'_2, then in P', and returns toward P which allows it to pass since the vibration now has the orientation E'''_B. A then does, finally, receive the flux $\Phi/2$ coming from B and transmitted by P'.

DIFFRACTION

PROBLEM 35

Far-field Diffraction

Consider the apparatus shown in Fig. 35.1. It is made up of a centro-symmetric system of two lenses L_1 and L_2, both having the same focal length f. A luminous object placed at π_1 the focal plane of L_1 has its image formed at π_2 in the focal plane of L_2.

The source emits monochromatic radiation of wavelength λ.

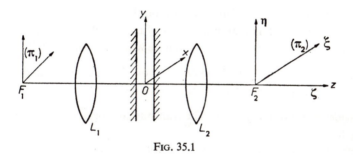

FIG. 35.1

I

Between L_1 and L_2 is placed a rectangular pupil of width a and length b ($b \gg a$). The centre of this slit coincides with the optic axis of the system. The coordinates of a point in the plane of the pupil are designated x and y.

1. Treat successively the cases dealing with the following type of object:
   (a) a dimensionless point placed at the focus of L_1;
   (b) a small, infinitely thin line segment passing through F_1 and parallel to the edges of the diffracting slit.

In both cases briefly describe the image. Find the distribution of the illumination in plane π_2 where the general point is represented by the coordinates ξ and η. Graphically represent this distribution along the axis $F_2\xi$. (Use as the ordinate the illumination and along the abscissa plot $u = (\sin i)/\lambda$ where i is the angle the diffracted ray makes with the normal to the plane of the pupil.)

2. The object now taken is a series of fine luminous lines parallel to one another and to the diffracting slit. These lines are equidistant from one another (period d) and the size of the object is taken to be very large.

What is the minimum value of d for which the image has a periodic structure:

(a) when the object emits in a totally incoherent fashion. Recall that

$$1 + \frac{1}{3^2} + \frac{1}{5^2} + \dots = \frac{\pi^2}{8};$$

(b) when the illumination is coherent?

Numerical application. $a = 5$ mm; $f = 1$ m; $\lambda = 0.5$ μ.

Question 2b is difficult to solve if one does not use the Fourier transformation.

II

Again a point source is placed at F_1 and in the plane xOy one places successively different gratings with step p.

1. The grating is made up of infinitely fine straight rulings parallel to Oy and separated by opaque intervals of width p.

(a) Determine the distribution of illumination on the plane π_2.

(b) Give the distribution of light graphically as a function of $u = (\sin i)/\lambda$.

Consider the cases where the grating has an infinite width and then a finite width L.

2. The grating rulings are all parallel to Oy but now the transparent and opaque intervals have the same width, namely $p/2$. Answer the questions above.

3. Now consider a sinusoidal grating. The transmitted amplitude at a point $P(x, y)$ in the pupil is of the form:

$$f(x) = \cos \frac{2\pi x}{p}.$$

The transmittance is constant along lines parallel to Oy. As in the questions above, determine the distribution of the illumination and present it graphically.

Numerical application: $p = 2$ μ; $\lambda = 0.5$ μ; $L = \infty$ then $L = 10$ cm.

SOLUTION

I. *Diffraction by a slit*

Since the slit is long and narrow, it gives diffraction only along planes parallel to xOz. Throughout this problem we are going to reduce to a one-dimensional problem. Only the variables x and ξ arise.

1. (a) *Point object.* If the pupil is very large, diffraction does not occur and one image, identical with the object is formed at F_2 (the conjugate point to F_1).

The insertion of the pupil causes a spreading of the image which does, however, remain centred on the geometric image F_2. Since the pupil is a slit parallel to Oy, the image spreads along the line $F_2\xi$ (Appendix A, III. 1).

A calculation will establish this result (§ 5.10).

In all of these problems we will normalize the results, that is, we will take the maximum intensity equal to 1. Thus

$$I(u) = \left(\frac{\sin \pi u a}{\pi u a}\right)^2, \quad \text{taking} \quad u = \frac{\sin i}{\lambda}. \tag{1}$$

The image is made up of a set of small luminous segments situated along $F_2\xi$ (Fig. 35.2a).

FIG. 35.2a

The variation in illumination is shown in Fig. 35.2b.

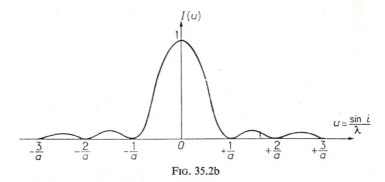

FIG. 35.2b

The central bright region is twice as wide $(2/a)$ as the other lateral fringes.

(b) *The object is a very fine bright line.* The problem does not specify the degree of spatial coherence of the light source. However, the result will be the same in all cases. The line segment can be thought of as being made up of a series of luminous points (Fig. 35.3) m_0, m_1, ..., m_n. Each of these points gives a line of diffraction centred on its geometric image and parallel to $O\xi$ (Fig. 35.4).

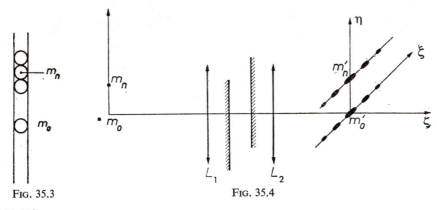

FIG. 35.3   FIG. 35.4

Since the locus of the diffraction lies in the planes parallel to $\xi F_2\zeta$, there is never any interference found along $F_2\eta$. When the source is made up of a short luminous line segment, the fringes formed are parallel to Oy (Fig. 35.5). The height of the fringes is equal to the height of the line source since in this case the magnification is unity.

The distribution of illumination along any line parallel to $O\xi$ is the same as previously calculated.

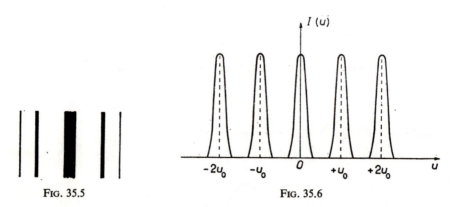

FIG. 35.5   FIG. 35.6

2. The object is a grating with step d.

(a) *Incoherent illumination.* Each slit gives a system of fringes identical to that found above and centred on its geometric image.

Since the line sources are incoherent, we have to consider the intensity given by each of these and then form the sum of these intensities.

For simplification we are going to assume that the diffraction figure given by each slit is limited to the central fringe centred on its geometric image. The distribution of the illumination is shown in Fig. 35.6 where u_0 is the ratio $d/f\lambda$.

One can clearly see that when d decreases eventually the situation will arise where the grating is no longer resolved.

If one adopts as the resolution criterion the case where the diffraction maximum coincides with the first minimum of the neighbouring image, one has (Fig. 35.7)

$$u_0 = \frac{1}{a} = \frac{\sin i}{\lambda} \approx \frac{i}{\lambda} = \frac{d}{f\lambda}. \tag{2}$$

Hence

$$d_{\min} = \lambda \frac{f}{a}. \tag{2 bis}$$

In reality, if one takes the secondary maxima into account, one finds

$$I_{\max} = I(0)$$

$$I_{\min} = 2\left\{ I\left(\frac{1}{2a}\right) + I\left(\frac{2}{3a}\right) + I\left(\frac{5}{2a}\right) + \cdots \right\}. \tag{3}$$

Equation (1) gives

$$I_{max} = 1,$$

$$I_{min} = 2\left[\frac{1}{\dfrac{\pi^2}{4}} + \frac{1}{\dfrac{9\pi^2}{4}} + \frac{1}{\dfrac{25\pi^2}{4}} + \cdots\right],$$

$$(4)$$

$$I_{min} = \frac{8}{\pi^2}\left[1 + \frac{1}{9} + \frac{1}{25} + \cdots\right] = \frac{8}{\pi^2}\times\frac{\pi^2}{8} = 1 = I_{max}.$$

For $u = u_0$, the image has no contrast.

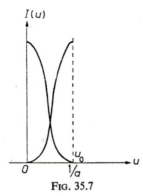

$$I(u)$$

$$0 \qquad \text{¹/a} \qquad u$$

Fig. 35.7

Periodic structure only appears for

$$u > u_0,$$
$$d > \lambda f/a.$$

$$(5)$$

Numerical application:

$$d(\mu) = 0.5\times\frac{10^3}{5}$$

$$d > 0.1 \text{ mm}.$$

(b) *Coherent illumination.* It is now necessary to sum the amplitudes rather than the intensities. There is no interference along $F_2\eta$, only along $F_2\xi$.

II. *Diffraction by a grating* (§ 7.7)

1. *Grating with fine slits*

(a) *Grating of width L.* Each slit, infinitely fine, diffracts uniformly into space.

As before, since the slits are parallel to Oy, there is only diffraction in planes parallel to xOz. In addition, as a result of the point source, the diffraction image is centred on the geometric image F_2 and spread out along $F_2\xi$.

The N diffraction slits produce an intensity

$$I(u) = A(u) A^*(u) = \left(\frac{\sin N\pi up}{\sin \pi up}\right)^2.$$

$$(6)$$

The examination of (6) allows us to determine the position of the spectra: they are equidistant with period $\Delta u = 1/p$.

Their width is $\delta u = 1/Np$. The number of spectra is limited by the condition $\sin i < 1$ which here appears as $|k| < 4$. Thus there are seven spectra visible all having the same intensity.

Figure 35.8a represents the intensity variations in the plane of the image. The variables $u = (\sin i)/\lambda$, $\sin i$, ϕ, and the interference order k are shown beneath each maximum on the figure.

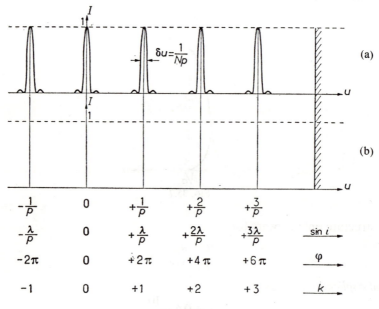

$-\frac{1}{p}$	0	$+\frac{1}{p}$	$+\frac{2}{p}$	$+\frac{3}{p}$	
$-\frac{\lambda}{p}$	0	$+\frac{\lambda}{p}$	$+\frac{2\lambda}{p}$	$+\frac{3\lambda}{p}$	$\sin i$
-2π	0	$+2\pi$	$+4\pi$	$+6\pi$	φ
-1	0	$+1$	$+2$	$+3$	k

FIG. 35.8

Note. The term spectrum as used here does not refer to a coloured spectrum since we are using monochromatic light, but rather to a maximum in the illumination in the diffraction figure.

(b) *Infinite grating.* The width of the spectrum $\delta u = 1/Np$ decreases when N increases. For infinite N, the image is made up of a series of bright points on the axis $F_2\xi$ (Fig. 35.8b).

2. *Foucault grating (step p, width of the slits $p/2$)*

(a) *Pupil with N slits.* The diffraction amplitude from one of the slits in direction u is given by

$$A_0(u) = \frac{p}{2} \frac{\sin \pi u \frac{p}{2}}{\pi u \frac{p}{2}}. \tag{7}$$

The grating diffraction intensity in direction u is

$$I(u) = \left(\frac{p}{2} \frac{\sin \frac{\pi u p}{2}}{\frac{\pi u p}{2}} \right)^2 \times \left(\frac{\sin N\pi u p}{\sin \pi u p} \right). \tag{8}$$

$\underbrace{\qquad\qquad\qquad}_{\substack{\text{diffraction} \\ \text{term}}}$ $\underbrace{\qquad\qquad}_{\substack{\text{interference} \\ \text{term}}}$

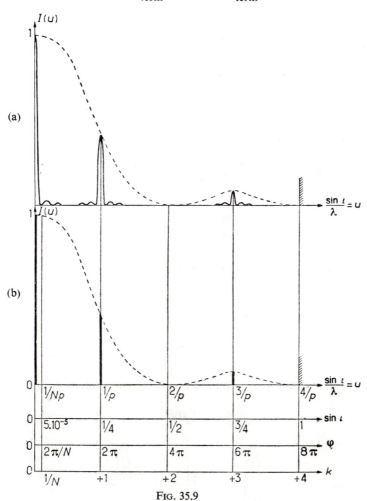

FIG. 35.9

The results are shown in Fig. 35.9a. The spectra of even order vanish and only spectra of orders 0, ± 1, ± 3 remain. The distribution of the diffraction figure is always along $F_2\xi$ since we are dealing with a point source and a one-dimensional pupil.

(b) *Infinite grating.* The discussion is the same as for the preceding case. Along the axis one finds five points of unequal intensity (Fig. 35.9b).

In summary, the characteristics of the grating lead to the following characteristics in the image:

The width of the slits determines the amount of modulation. When the slits have a finite width (as is always the case in practice) the spectra do not have the same intensity. Certain orders can even vanish.

The step of the grating determines the position of the spectra. They are equidistant in u space with period $\Delta u = 1/p$.

The width of the grating characterizes the width of the spectra ($\delta u = 1/Np = 1/L$).

3. Sinusoidal grating

(a) *Grating of width L.* The amplitude in the u direction is given by

$$A(u) = \int_{-L/2}^{+L/2} f(x)\, \mathrm{e}^{\mathrm{j}2\pi ux}\, \mathrm{d}x = \int_{-L/2}^{+L/2} \cos 2\pi \frac{x}{p}\, \mathrm{e}^{\mathrm{j}2\pi ux}\, \mathrm{d}x. \tag{9}$$

Taking

$$u_1 = 1/p. \tag{10}$$

The preceding integral is written

$$A(u) = \frac{1}{2} \int_{-L/2}^{+L/2} \mathrm{e}^{\mathrm{j}2\pi(u+u_1)x}\, \mathrm{d}x + \frac{1}{2} \int_{-L/2}^{+L/2} \mathrm{e}^{\mathrm{j}2\pi(u-u_1)x}\, \mathrm{d}x, \tag{11}$$

thus

$$A(u) = \frac{1}{2} L \left[\frac{\sin \pi(u+u_1)L}{\pi(u+u_1)L} + \frac{\sin \pi(u-u_1)L}{\pi(u-u_1)L} \right]. \tag{12}$$

Since L is very much larger than p, one finds

$$\frac{1}{L} \ll \frac{1}{p}$$

or

$$\frac{1}{L} \ll u_1.$$

The two spectra shown on Fig. 35.10a have practically no common point.

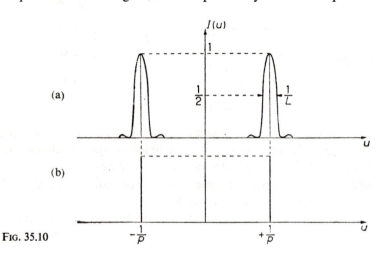

FIG. 35.10

Thus one can correctly write for the normalized intensity

$$I(u) = \left[\frac{\sin \pi(u+u_1) L}{\pi(u+u_1) L} \right]^2 + \left[\frac{\sin \pi(u-u_1) L}{\pi(u-u_1) L} \right]^2. \tag{13}$$

The variation in illumination is represented by two curves, analogous to the Fig. 35.2b and displaced by $2u_1$.

The image on axis ξ is made up of two spectra separated by $\Delta u = 2u_1 = 2/p$ and with width $\delta u = 1/L$.

One finds only spectra of orders $+1$ and -1 (Fig. 35.10a).

(b) *Infinite grating.* The image reduces to two bright points situated on the ξ-axis, symmetric with respect to F_2, and separated by $\Delta u = 2/p$ (Fig. 35.10b).

All of these problems can be treated more simply using the Fourier transformation.

I. *Pupil a slit*

1. (a) *Coherent illumination.* Since the source is a dimensionless point placed at the focus of L_1, a plane wave parallel to the plane of the pupil illuminates the slit.

The diffracted amplitude is the Fourier transform of the amplitude distribution on the pupil (see Appendix A, III. 1), namely

$$A(u) = F(u) = \frac{\sin \pi u a}{\pi u a}.$$

(b) *Illumination by a line.* $A(u)$ retains the same value as above in paragraph (a).

2. *The object a grating.* (a) *Incoherent illumination of the object. Notation.* Call M a point in the object plane and M' a point in the image plane. In addition, we will approach this question as if the planes of the object and the image were coincident (the separation of these two planes does not arise explicitly in the calculation). Therefore the vector $M - M'$ will be represented by the vector joining the point M' in the geometric image to point M. (M designates the vector $F_1 M$ or $F_2 M$ and P the vector OP.)

Each point M of the object gives a diffraction image centred on the geometric image (which we now call M also).

Since the diffraction images have no definite phase relationship, the intensities add in the image plane.

If $O(M)$ represents the intensity distribution in the object plane, the intensity at M' in the image is given by

$$I(M') = \int_{\text{object}} O(M) \, D(M' - M) \, dM \tag{14}$$

with $D(M' - M) = |A(M' - M)|^2 = $ the distribution of the intensity in the diffraction spot obtained with this pupil. Given that

$$\left. \begin{array}{l} I(M) \xrightarrow{\text{F.T.}} i(P) \\[4pt] O(M) \xrightarrow{\text{F.T.}} o(P) \\[4pt] D(M) \xrightarrow{\text{F.T.}} d(P) \end{array} \right\} \tag{15}$$

Parseval's theorem allows transformation of the convolution (14) into a product

$$i(P) = o(P) \times d(P). \tag{16}$$

We return to the special case in the text where the pupil, a fine slit, parallel to Oy, diffracts only along planes parallel to $\xi F_2 \zeta$. The only variables with which we have to deal are x, u, and u'.

Thus

$$I(u') = \int O(u) \, D(u'-u) \, du \tag{17}$$

and

$$i(x) = o(x) \times d(x). \tag{18}$$

We now will determine successively $d(x)$, $o(x)$, then $i(x)$.

(α) *Calculation of $o(x)$.* The intensity distribution in the object plane is

$$O(u) = \sum_{n=-\infty}^{+\infty} \delta(u - nu_0). \tag{19}$$

This is a "Dirac series" with step u_0.

The Fourier transform of a "Dirac series" with step u_0 is a "Dirac series" with step $1/u_0$. One has

$$o(x) = \sum_{n=-\infty}^{+\infty} \delta\left(x - \frac{n}{u_0}\right). \tag{20}$$

(β) *Calculation of $d(x)$.* We have taken

$$D(u) = A(u) \, A^*(u). \tag{21}$$

However, here $A(u)$ is real and

$$D(u) = [A(u)]^2. \tag{22}$$

Parseval's reciprocity theorem allows us to write

$$d(x) = \int_{-\infty}^{+\infty} f(X) f(x-X) \, dX \tag{23}$$

$d(x)$ is the autocorrelation function of the pupil transparency (in amplitude).

Since $f(X)$ is a "rectangular" function, the convolution is equal to the common area of two rectangles displaced by x (see Appendix A, B.III).

Several cases arise depending on the value of u_0.

$$\boxed{1/u_0 > a} \qquad \text{(Fig. 35.11a)}.$$

One has $i(x) = o(x) \, d(x) = \delta(x)$. Only the fundamental is passed by the slit.
One finds

$$I(u') = \text{F.T.}[\delta(x)], \quad \text{so that} \quad I(u') = 1 \quad \text{(Fig. 35.11b)}. \tag{24}$$

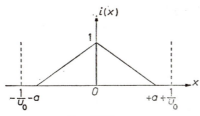

Fig. 35.11a

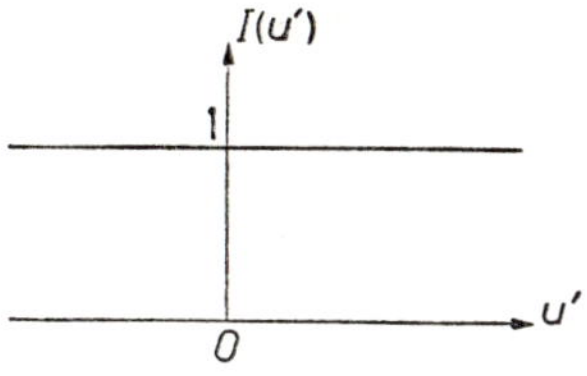

Fig. 35.11b

The image plane is uniformly illuminated.

$$\boxed{1/u_0 < a}$$ (Fig. 35.12a).

In addition to the fundamental a certain number of other spatial frequencies pass. These latter are always attenuated by the function $d(x)$. To establish these concepts take the example in Fig. 35.12. One finds

$$i(x) = \delta(x) + C\left[\delta\left(x+\frac{1}{u_0}\right) + \delta\left(x-\frac{1}{u_0}\right)\right],$$ (25)

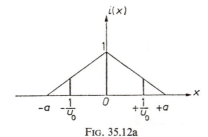

Fig. 35.12a

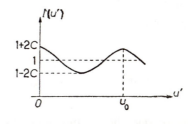

Fig. 35.12b

with

$$C = 1 - \frac{1}{au_0}.$$

One gets from this

$$I(u') = 1 + C[e^{j2\pi u'/u_0} + e^{-j2\pi u'/u_0}]$$ (26)

$$I(u') = 1 + 2C \cos 2\pi \frac{u'}{u_0}.$$

Since $1/u_0$ is less than a, the image has a periodic structure (Fig. 35.12b).

We conclude that the grating is resolved for

$$\frac{1}{u_0} < a, \quad \text{that is, for} \quad d > \lambda \frac{f}{a}.$$

Note. One sees that, even for values of u_0 greater than $1/a$, the image does not always conform to the object. In effect, even if all the spatial frequencies pass, their amplitudes are modified by $d(x)$. It is only the fundamental which is not affected.

(b) *Coherent illumination of the object.* Here it is necessary to take into account the phase relationships which exist between the various amplitudes transmitted to the image plane from points on the object. One needs then to evaluate the integral

$$E(M') = \int \Omega(M) \, A(M' - M) \, \mathrm{d}M,$$ (27)

$E(M')$ being the resultant amplitude at point M'.

Take

$$\begin{cases} \Omega(M) = \text{amplitude distribution in the object plane;} \\ A(M) = \text{amplitude distribution in the diffraction spot.} \end{cases}$$

If

$$\left. \begin{aligned} \Omega(M) &\xrightarrow{\text{F.T.}} \omega(P), \\ A(M) &\xrightarrow{\text{F.T.}} f(P), \\ E(M) &\xrightarrow{\text{F.T.}} e(P), \end{aligned} \right\}$$ (28)

Parseval's theorem leads to

$$e(P) = \omega(P) \times f(P).$$ (29)

The problem is always one-dimensional and thus one has
the convolution:

$$E(u') = \int \Omega(u) \, A(u' - u) \, \mathrm{d}u$$ (30)

and *the product:*

$$e(x) = \omega(x) \times f(x).$$ (31)

Take the product of the functions $\omega(x)$ and $f(x)$ and call this product $e(x)$ (Fig. 35.13a).

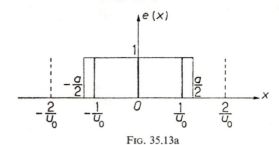

FIG. 35.13a

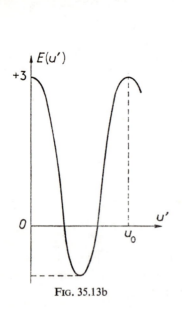

FIG. 35.13b

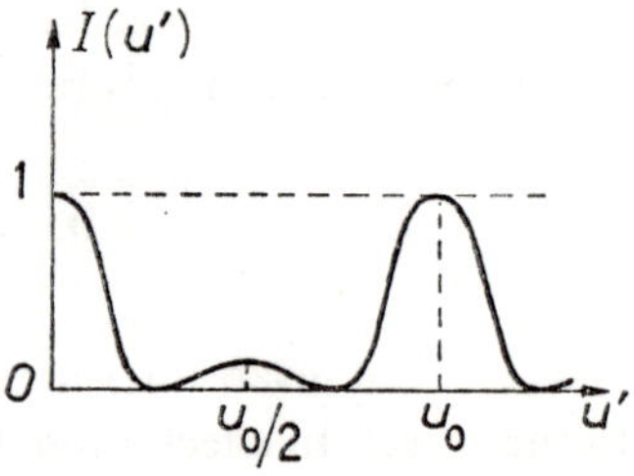

FIG. 35.13c

One gets the following results:
  for $1/u_0 > a/2$, only the fundamental is passed.
  $E(u') = 1$. The amplitude is uniform in the image plane.
  for $1/u_0 < a/2$, the fundamental and some of the harmonics pass.
In the example in Fig. 35.13a, one has

$$e(x) = \delta(x) + \delta\left(x - \frac{1}{u_0}\right) + \delta\left(x + \frac{1}{u_0}\right). \tag{32}$$

Hence

$$E(u') = 1 + e^{j2\pi u'/u_0} + e^{-j2\pi u'/u_0} = 1 + 2\cos 2\pi \frac{u'}{u_0}. \tag{33}$$

The curves 35.13b and 35.13c represent the amplitude and the corresponding intensity.

The image has the same period as the object, but secondary maxima appear between the principal maxima. In summary, when $1/u_0$ is less than $a/2$ the distribution of light in the image becomes periodic and the grating is resolved. The more spatial frequencies passed by the slit, the more the image resembles the object. In any case, the image will not be identical with the object since the width of the pupil is finite.

Conclusion. Comparing the results for coherent and incoherent illumination, one finds that in each case one "resolves" the object for the following limiting values of u_0:

incoherent object $1/u_0 < a,$

coherent object. $1/u_0 < a/2,$

In addition one finds:
  the resolution increases with the width of the aperture.
  the resolution is better in incoherent than in coherent light (the ratio being 2).

Note. The microscope is the optical instrument which allows one to vary to a great extent the coherence of the illumination. In effect, the operator can adjust the aperture of the condenser (condenser closed → coherent illumination, condenser open → incoherent illumination). Microscopists know that they can improve the resolution by using the most incoherent light possible.

II. *Grating*

Since the source is a fixed point it is no longer useful to introduce the variable u'. One needs only deal with the conjugate variables x and u.

A. *Grating with finite width* (see Appendix A).

The amplitude transparence of the pupil is characterized by

$$g(x) = h(x) \times f(x), \tag{34}$$

$h(x)$ is the amplitude transmitted by an infinite grating,
$f(x)$ is the amplitude transmitted by a slit of width L.

The diffracted amplitude in the direction u is given by the function $G(u)$ such that:

$$G(u) = H(u) \otimes F(u). \qquad (35)$$

1. The slits are infinitely fine.

It is easy to determine the convolution graphically:

$$F \otimes H$$

(see Figs. 35.14b, 35.15b, and 35.16).

    2. Foucault grating (see Figs. 35.14b, 35.17b, and 35.18).

    3. Sinusoidal grating (see Figs. 35.14b, 35.19b, and 35.20).

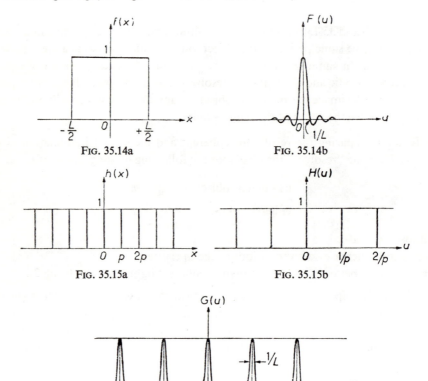

FIG. 35.14a   FIG. 35.14b

FIG. 35.15a   FIG. 35.15b

FIG. 35.16

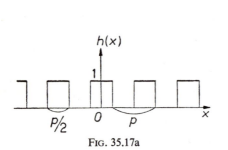

FIG. 35.17a

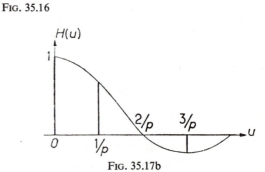

FIG. 35.17b

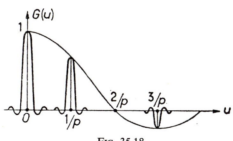

FIG. 35.18

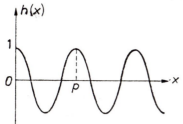

FIG. 35.19a

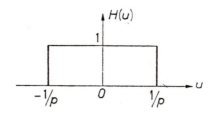

FIG. 35.19b

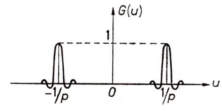

FIG. 35.20

B. *Infinitely wide grating*

The problem is simplified since there is no slit diaphragm of width L. The amplitude distribution in the image plane is given by $H(u)$ such that

$$H(u) = \text{F.T.}[h(x)]. \tag{36}$$

For the three gratings treated, the amplitude distribution is given by the curves 35.15b, 35.17b and 35.19b.

PROBLEM 36

Diffraction by Circular Pupils

Consider an objective O limited by a circular contour of radius $r_0 = 3$ cm.

This objective, assumed perfect, is illuminated by a point source at infinity along the axis of the objective O.

The source is monochromatic and radiates at wavelength $\lambda = 0.6\ \mu$.

Let α_0 be the maximum aperture of the objective (Fig. 36.1). Assume α_0 small.

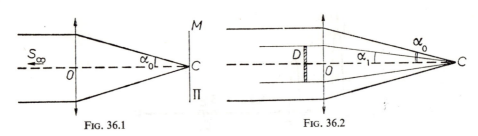

FIG. 36.1   FIG. 36.2

I. *Open pupil*

1. Give the amplitude distribution and the illumination in the plane π. Characterize a point M in the plane π by its distance ϱ from point C, the geometric image of the source. For simplification take $Z = (2\pi/\lambda)\,\alpha_0\varrho$. What is the illumination at the centre, C, of the diffraction figure?

2. Find the angular radius θ of the first dark ring in the diffraction figure (θ is the angle which one sees from the optic centre O of the lens to the radius of the first dark ring).

II. *Opaque disc*

In front of the objective O, an opaque circular screen D is placed normal to the incident light. The centre of the screen is on the optic axis of the objective. The screen D subtends the half-angle α_1 at point C (Fig. 36.2).

1. Give the amplitude and intensity distributions on plane π. What is the illumination intensity at the centre of the diffraction figure? Find the radius r_1 of D such that the intensity decrease not exceed 10% of that value found in question I.

2. Find the angular radius of the first dark ring in the case where $\alpha_0 = 2\alpha_1$.

3. Compare graphically the nature of the two diffraction spots:
    (a) without the screen D,
    (b) with the screen D.
Optical systems with a central screen occur in certain telescopes. What happens in these cases to the resolution of the components of double stars? Assume the component stars have the same intensity.

III. *Annular pupil*

Assume now that the screen D almost completely covers the objective O in such a way that light only passes through an infinitely narrow ring.
What is the structure of the diffraction pattern in plane π?
What is the angular radius of the central diffraction spot?

IV. *Identical screens distributed at random*

Replace the screen D by 1000 small opaque screens distributed at random in a plane in front of the objective O. Each screen subtends a very small angle at C equal to α_2 such that $\alpha_2/\alpha_0 = 10^{-2}$. Find the illumination in plane π at a distance $30/1.22$ times the radius of the diffraction spot formed by the free objective O (in the absence of the small screens). Show initially that the conditions are such that Babinet's theorem can be applied. In the remainder of this problem the illumination produced by an open pupil at C is taken as unity.

V. *Apodization*

The objective is now used with its full aperture (the screens are removed) and with a plate of glass, L, with parallel faces in front of it (Fig. 36.3). Deposited on one face of the plate L is a thin film with non-uniform absorption which does not introduce a phase variation. The

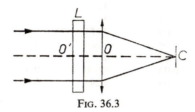

FIG. 36.3

absorbing film is deposited so that the absorption is the same for all points situated on a circumference whose centre is the intersection O' of the optic axis of the objective with the plate. The variation of the amplitude as a function of α is given by the expression $e^{-a\alpha^2}$ where a is a coefficient which fixes the maximum absorption.

Find the variation in illumination at the centre of the diffraction figure. *Numerical application:* $\alpha_0 = 1/5$ and $a = 1$. Can one form a diffraction figure if the absorption becomes very strong at the edge $(a \gg 1)$?

VI. *Focusing defects*

The plate used in the last section is now replaced by a perfectly transparent plate which has a uniformly varying thickness. The variation is, as in question V, cylindrically symmetric about the optic axis of the objective. The thickness variation of the plate introduces a phase variation (through a path difference) as a function of α given by $\varepsilon\alpha^2/2$ where ε is a coefficient which fixes the maximum path difference. Find the illumination at C. Examine the variations in illumination of the centre of the diffraction figure as a function of the phase difference $\Phi = \pi\varepsilon\alpha_0^2/\lambda$ (λ = wavelength of the light used). Plot the curve for values of Φ from 0 to 4π.

Show that by removing the plate L and by slowly displacing the focal plane parallel to itself to some point, the variation in illumination at the centre of the diffraction figure is given by the preceding curve.

Note. It is not necessary to know the properties of the Bessel functions to solve this problem. One simply requires several useful results:

$$\int_0^{2\pi} e^{-jK\alpha\varrho\cos\theta}\,\mathrm{d}\theta = 2\pi J_0(K\alpha\varrho)$$

$$\int_0^Z J_0(z)\,z\,\mathrm{d}z = Z\cdot J_1(Z).$$

TABLE OF NUMERICAL VALUES

Z	$J_0(Z)$	$2\dfrac{J_1(Z)}{Z}$
0.0	+1.000	+1.000
1.00	+0.765	+0.880
2.00	+0.224	+0.577
3.00	−0.260	+0.226
4.00	−0.397	−0.033
5.00	−0.178	−0.131
6.00	+0.151	−0.092
7.00	+0.300	−0.001
8.00	+0.172	+0.059
9.00	−0.090	+0.054
10.00	−0.246	+0.009

$J_0(Z)$ is zero for $Z = 2.405, 5.52, 8.65, \ldots$
$J_1(Z)/Z$ is zero for $Z = 3.83, 7.02, 10.17, \ldots$

SOLUTION

Throughout this problem, where the examples studied have rotational symmetry, one uses cylindrical coordinates (Fig. 36.4).

Determination of the amplitude at point M (§ 5.11).

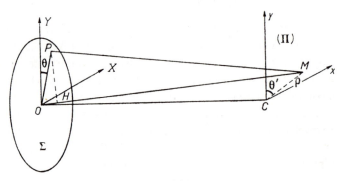

FIG. 36.4

The general expression for the amplitude at point M is given by

$$A(M) = \int_0^{\alpha_0} \int_0^{2\pi} \exp\left(-j\frac{2\pi}{\lambda}\alpha\varrho \cos(\theta-\theta')\right) \alpha \, d\alpha \, d\theta. \tag{1}$$

Changing the origin of the azimuths, one can write

$$A(M) = \int_0^{\alpha_0} \int_0^{2\pi} \exp\left(-j\frac{2\pi}{\lambda}\alpha\varrho \cos\theta\right) \alpha \, d\alpha \, d\theta. \tag{2}$$

I. Open pupil

1. One has

$$A(M) = 2\pi \int_0^{\alpha_0} J_0(K\varrho\alpha)\alpha \, d\alpha. \tag{3}$$

$$A(M) = \frac{2\pi}{K^2\varrho^2} \int J_0(K\varrho\alpha) \times (K\varrho\alpha) \, d(K\varrho\alpha) = \frac{2\pi}{K^2\varrho^2}(K\varrho\alpha_0) J_1(K\varrho\alpha_0),$$

$$A(M) = \pi\alpha_0^2 \frac{2J_1(Z)}{Z} \quad \text{with} \quad Z = \frac{2\pi}{\lambda}\varrho\alpha_0. \tag{4}$$

The diffracted intensity at M is then equal to

$$I(M) = (\pi\alpha_0^2)^2 \frac{4J_1^2(Z)}{Z^2}. \tag{5}$$

The intensity distribution is given by Fig. 36.5.

The diffraction "solid" has rotational symmetry about C. The first dark ring corresponds to $Z = 3.83$.

The diffraction image is always centred on the geometric image.

At point C one has

$$I(C) = (\pi\alpha_0^2)^2 \times 4\left(\frac{J_1(0)}{0}\right)^2 = (\pi\alpha_0^2)^2. \tag{6}$$

The intensity is always equal to the square of the area of the pupil S.

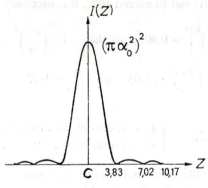

FIG. 36.5

This result seems paradoxical but is easily explained. In effect, the diffraction spot is spread out on a surface inversely proportional to the surface of the pupil. The total flux, equal to the volume of the "diffraction solid" is therefore proportional to S.

2. Since the first dark ring corresponds to $Z = 3.83$:

$$Z = \frac{2\pi}{\lambda}\alpha\varrho = \frac{2\pi}{\lambda}\frac{r_0}{f}\varrho = \frac{2\pi}{\lambda}r_0\frac{\varrho}{f} = \frac{2\pi}{\lambda}r_0\theta = 3.83.$$

Hence:

$$\theta = \frac{3.83\times\lambda}{2\times3.14\times r_0} = \frac{3.83\times0.6\times10^{-4}}{3.14\times6}$$

$$\theta = 1.2\times10^{-5}\,\text{rad.}$$

II. Opaque disc

1. Starting from the general expression for the amplitude, one can write

$$A(M) = 2\pi\int_0^{\alpha_0} J_0(K\alpha\varrho)\,\alpha\,\mathrm{d}\alpha - 2\pi\int_0^{\alpha_1} J_0(K\alpha\varrho)\,\alpha\,\mathrm{d}\alpha, \tag{7}$$

$$A(M) = \pi\alpha_0^2\frac{2J_1(Z_0)}{Z_0} - \pi\alpha_1^2\frac{2J_1(Z_1)}{Z_1} \tag{8}$$

taking $Z_0 = \pi\alpha_0\varrho$ and $Z_1 = \pi\alpha_1\varrho$.
The illumination then becomes

$$I(M) = \pi^2\left\{\alpha_0^4\left[\frac{2J_1(Z_0)}{Z_0}\right]^2 + \alpha_1^4\left[\frac{2J_1(Z_1)}{Z_1}\right]^2 - 2\alpha_0^2\alpha_1^2\frac{2J_1(Z_0)}{Z_0}\cdot\frac{2J_1(Z_1)}{Z_1}\right\}. \tag{9}$$

The illumination at C is equal to

$$I(C) = \pi^2\alpha_0^4\left[1 - \left(\frac{\alpha_1}{\alpha_0}\right)^2\right]^2 \tag{10}$$

whereas in the case of the open pupil one found $\pi^2\alpha_0^4$.
For the decrease in intensity not to exceed 10%, it is necessary to have

$$\left[1 - \left(\frac{\alpha_1}{\alpha_0}\right)^2\right]^2 \geqslant 0.90, \quad\text{so that}\quad 1 - \left(\frac{\alpha_1}{\alpha_0}\right)^2 \geqslant 0.95.$$

$$\left(\frac{\alpha_1}{\alpha_0}\right)^2 \leqslant 0.05 \quad\text{or}\quad \frac{\alpha_1}{\alpha_0} \leqslant 0.22.$$

Hence

$$\frac{r_1}{r_0} \leqslant 0.22 \qquad \begin{cases} r_0 = \text{radius of the objective,} \\ r_1 = \text{radius of the opaque disc.} \end{cases} \tag{11}$$

Note. The result given by (10) shows that one does not have the right to apply Babinet's theorem near the geometric image. In effect, an objective which has an open pupil of radius

r_1 gives at C an intensity

$$\pi^2\alpha_1^4 \neq \pi^2\alpha_0^4\left[1-\left(\frac{\alpha_1}{\alpha_0}\right)^2\right]^2. \tag{12}$$

2. *Radius of the first dark ring.* The amplitude $A(M)$ vanishes for values of ϱ such that

$$\alpha_0^2\frac{J_1(Z_0)}{Z_0}-\alpha_1^2\frac{J_1(Z_1)}{Z_1}=0. \tag{13}$$

Returning to the definition of Z_1 and Z_0, taking $m=\alpha_1/\alpha_0$ one can write

$$J_1(Z_0)=mJ_1(mZ_0). \tag{14}$$

In the special case where $m=0.5$, one finds

$$J_1(Z_0)=0.5J_1(0.5Z_0).$$

Using the table

$$Z_0=3.14 \quad J_1(3.14)-\tfrac{1}{2}J_1(1.57) \ = +0.00185$$
$$Z_0=3.15 \quad J_1(3.15)-\tfrac{1}{2}J_1(1.575) = -0.0023$$

Through linear interpolation

$$Z_0=3.144.$$

Thus

$$\theta=\frac{3.144\lambda}{2\times3.14\times r_0}\approx\frac{\lambda}{2r_0}=\frac{6\times10^{-4}}{60}$$

$$\theta=10^{-5}\,\text{rad}.$$

3. *Comparison of the diffraction images.* On Fig. 36.6 are plotted the distributions of illumination as a function of Z for the open pupil and for the opaque disc. One can see that the opaque centre leads to:

a decrease in the illumination at the central peak;

a slight improvement in the power of separation.

Note. In Fig. 36.6 only the central maximum is represented. If one examines the values of the intensity for $Z>3.14$, one discovers that the rings take a much more important role in the case of the Airy curve.

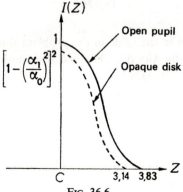

FIG. 36.6

One can approach this problem using the Heisenberg relation: if one decreases the dimensions of the pupil, there is a spreading of the diffraction figure.

In practice, this example arises in telescopes of the Cassegrainian form (see Fig. 36.7). Such an instrument is made up of two concentric mirrors. Covering of the pupil occurs because of the position of the small mirror.

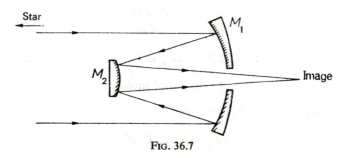

FIG. 36.7

III. *Annular pupil*

One can assume that the transparent ring, since it is taken to be infinitely thin, corresponds to a constant value of α_0. Under these conditions, the amplitude at M becomes:

$$A(M) = 2\pi J_0(K\alpha_0\varrho). \tag{15}$$

Figure 36.8 gives $J_0(Z)$ as a function of Z.

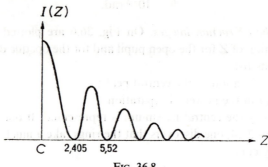

FIG. 36.8

The numerical values given in the table in the problem show that the decrease is much less rapid than in the case of the Airy disc. The first zero minimum in this diffraction figure is given by the first zero of $J_0(Z)$, namely

$$Z = 2.405$$

which gives $\theta = 0.77\lambda/2r_0$:

$$\theta = 0.77 \times 10^{-5} \text{ rad.}$$

Notes. 1. When one covers the pupil, the illumination of the rings increases at the expense of the illumination in the central peak.

2. A pupil of this type is used in the phase-contrast microscope. The condenser is provided with an annular pupil. With respect to ordinary microscopic observations such illumination leads to:

a loss of luminosity,

an improvement in the power of separation,

a considerable importance of the diffraction fringes which can make the interpretation of images difficult.

IV. *Identical screens distributed at random*

The amplitude at M is now given by:

$$A(M) = \underbrace{\int \int e^{-jK\alpha\varrho\cos\theta}\alpha\,d\alpha\,d\theta}_{\substack{\text{open}\\\text{pupil}}} - \underbrace{\int \int e^{-jK\alpha\varrho\cos\theta}\alpha\,d\alpha\,d\theta}_{\substack{\text{surface of the}\\\text{small screens}}}$$

$$A(M) = A_1(M) \qquad\qquad - A_2(M).$$

The first integral has already been evaluated in the first question of the problem. The second integral represents the amplitude diffracted by N openings, each of which has the same dimension as the small opaque screens (§ 5.16).

$$A_2(M) = a_0(M) \sum_{n=1}^{N} e^{jK\delta_n}. \tag{17}$$

(a_0 being the amplitude diffracted by a small hole on the axis of the instrument).

Application of Babinet's principle

Fix the distance $CM = \varrho_2$ set in the problem. If we call ϱ_1 the radius of the central diffraction spot given by the open pupil, we have

$$\varrho_2 = \frac{30}{1.22}\varrho_1, \tag{18}$$

$$\varrho_1 = \frac{1.22\lambda}{2\alpha_0}, \tag{19}$$

Then:

$$\varrho_2 = \frac{30\lambda}{2\alpha_0}. \tag{20}$$

Calculating the diffracted amplitude for the open pupil at this point

$$A_1(M) = \pi\alpha_0^2 \frac{2J_1(Z)}{Z}, \quad \text{with} \quad Z = \frac{2\pi}{\lambda}\alpha_0\varrho_2 = 30\pi.$$

For a large value of Z, the Bessel function of order 1 is practically zero. In this particular case we are able to state

$$|A(M)| = |A_2(M)| \tag{21}$$

hence:

$$I(M) = I_2(M).$$

$$I(M) = \underbrace{|a_0(M)|^2}_{\substack{\text{diffraction term for} \\ \text{1 small aperture}}} \times \underbrace{\left| \sum_{n=1}^{N} e^{jK\delta_n} \right|^2}_{\substack{\text{term due to} \\ \text{interference}}}.$$

So that finally

$$I(M) = |a_0(M)|^2 \sum_{n=1}^{N} \sum_{m=1}^{N} e^{jK\delta_n} \times e^{-jK\delta_m}.$$

Since the phase distribution is random one can assume that it has both positive and negative terms.

The preceding equation gives

$$I(M) = N|a_0(M)|^2. \tag{22}$$

Although the illumination is coherent, the random distribution of screens destroys the phase relationships; the intensities add as if the illumination were incoherent.

The point M is sufficiently removed from the geometric image that one can apply Babinet's principle.

Numerical application:

$$a_0(M) = \pi\alpha_2^2 \frac{2J_1(Z_2)}{Z_2},$$

with

$$Z_2 = \frac{2\pi}{\lambda} \varrho_2 \alpha_2 = \frac{2\pi}{\lambda} \frac{30\lambda}{2\alpha_0} \times 10^{-2}\alpha_0 \approx 1.$$

But

$$2J_1(1) \approx 1.$$

Thus one has

$$I(M) = N\pi^2\alpha_2^4 = N\pi^2\alpha_0^4 \left(\frac{\alpha_2}{\alpha_0}\right)^4.$$

Hence the normalized intensity

$$I(M) = N\left(\frac{\alpha_2}{\alpha_0}\right)^4 = 10^3 \times 10^{-8},$$

$$I(M) = 10^{-5}.$$

V. *Apodization: absorbing pupil*

Here the pupil is not uniformly transparent as was the case previously. The transmission is such that

$$\left. \begin{array}{ll} \tau(\alpha) = e^{-a\alpha^2} & \text{for} \quad 0 < \alpha < \alpha_0, \\ \tau(\alpha) = 0 & \text{for} \quad \alpha > \alpha_0. \end{array} \right\} \tag{23}$$

The general expression giving the amplitude at some point M is now

$$\int_0^{\alpha_0} \int_0^{2\pi} e^{-a\alpha^2} e^{-jK\alpha\varrho\cos\theta} \, d\alpha \, d\theta. \tag{24}$$

At point C, one finds

$$A(C) = \int_0^{\alpha_0} \int_0^{2\pi} e^{-a\alpha^2}\alpha \, d\alpha \, d\theta = \pi \int_0^{\alpha_0} e^{-a\alpha^2} \, d(\alpha^2),$$

$$A(C) = \frac{\pi}{a}(1-e^{-a\alpha_0^2}), \quad \text{hence} \quad I(C) = \frac{\pi^2}{a^2}[1-e^{-a\alpha_0^2}]^2. \qquad (25)$$

If $a\alpha_0^2$ is very small as is the case with the values given, one can write by a series expansion

$$A(C) = \frac{\pi}{a}[1-1+a\alpha_0^2-a^2\alpha_0^4/2] = \pi\alpha_0^2[1-a\alpha_0^2/2].$$

After introduction of the plate, one has $A(C) = \pi\alpha_0^2$.
 One gets the normalized intensity

$$I(C) = [1-a\alpha_0^2/2]^2. \qquad (26)$$

Numerical application:

$$\alpha_0 = 1/5, \quad a = 1.$$
$$I(C) = [1-0.02]^2 = 1-0.04 = 96/100.$$

Note. Returning to the general calculation of amplitude, one sees that the quantities α and ϱ are *conjugate variables* (by taking the wavelength as the unit of length).
 The results obtained in II and V can also be stated in the language of the Fourier transform.
 First example: a contraction in the α dimension leads to a dilation in the ϱ dimension.

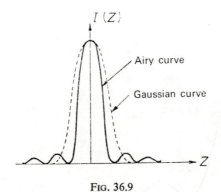

FIG. 36.9

Second example: the pupil's transmittance is governed by $\tau(\alpha) = e^{-a\alpha^2}$ which is a gaussian function. Knowing that the Fourier transform of a gaussian function is a gaussian function, one can immediately predict a gaussian distribution for the amplitude distribution in the image plane (Fig. 36.9). Without going into details, the figure shows the cross-section of the diffraction solid with and without the apodizing plate. The apodization causes the diffraction rings to vanish; however, it does diminish the resolving power of the instrument.

VI. *Phase plate, focusing defects*

The plate, L, with uniform transmittance does produce a variable phase-shift.

$$\left.\begin{array}{ll} \tau(\alpha) = e^{-jK\varepsilon\alpha/2} & \text{for} \quad 0 < \alpha < \alpha_0, \\ \tau(\alpha) = 0 & \text{for} \quad \alpha > \alpha_0. \end{array}\right\} \tag{27}$$

One has

$$A(C) = \int_0^{\alpha_0} \int_0^{2\pi} e^{-jK\varepsilon\alpha^2/2} \alpha \, d\alpha \, d\theta, \tag{28}$$

$$= 2\pi \int_0^{\alpha_0} e^{-jK\varepsilon\alpha^2/2} \, d\left(\frac{\alpha^2}{2}\right) = \frac{2\pi}{-jK\varepsilon} \, [e^{-jK\varepsilon\alpha^2/2}]_0^{\alpha_0}.$$

$$A(C) = \frac{2\pi}{jK\varepsilon} \, [1 - e^{-j\Phi}]. \tag{29}$$

The amplitude at C is now complex. The intensity becomes

$$I(C) = A(C) \times A^*(C),$$

so that:

$$I(C) = \frac{4\pi^2}{K^2\varepsilon^2} [1 - \cos\Phi] = \frac{16\pi^2}{K^2\varepsilon^2} \sin^2 \frac{\Phi}{2},$$

$$I(C) = \alpha_0^4 \left(\frac{\sin\Phi/2}{\Phi/2}\right)^2. \tag{30}$$

The variations of $I(C)$ as a function of Φ are shown on Fig. 36.10.

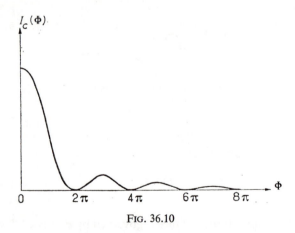

FIG. 36.10

Focusing errors

Displace the observation plane by a distance $CC' = \varepsilon$. By virtue of Malus' law and Fermat's principle, the path difference Δ between the ray passing through C and the ray passing through C' is equal to the separation between the aberrant wave centred on the point of observation C' and the sphere centred on the gaussian image (Fig. 36.11). One finds

$$\Delta = IJ = CJ - IK - KC = R - C'I - KC = R - (R - \varepsilon) - \varepsilon \cos \alpha.$$
$$\Delta = \varepsilon(1 - \cos \alpha) = \varepsilon \alpha^2 / 2.$$

One can see that the phase plate introduced above introduces the same phase-shift as a focusing defect.

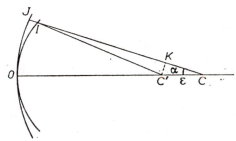

FIG. 36.11

Note. Using the results above, one sees that the centre of the diffraction image is alternatively bright and dark when one moves the focusing plane from one side to the other of the gaussian image. This method is used in industry to control objectives. In effect, the periodic succession of bright and dark central images is produced only if the objective is free of aberrations.

PROBLEM 37

Abbe's Experiment

A Fraunhofer grating consists of parallel opaque lines separated by transparent intervals. It has N lines. The collimator which is illuminated by monochromatic light is made up of an infinitely fine slit F placed in the focal plane of a lens L_1. A second lens L_2 is used behind the grating and a photographic plate is placed in the image focal plane. Use the following notations and values: grating step $p = 10\,\mu$, number of lines $N = 5000$, wavelength of the light $\lambda = 1.0\,\mu$, focal length of $L_1, f_1 = 50$ cm, and the focal length of $L_2 : f_2$, and finally, the angle which the diffracted rays make with the normal to the grating is i.

I

The transparent lines of the grating are infinitely fine.

1. Find $\sin i$ for the various images formed of F under the conditions of normal incidence.

2. Find the expression giving the angular width of a principal maximum in the diffraction. Find the theoretical resolving power in the spectra of various orders. Determine the focal length F_2 for which the photographic plate shows all the details which the resolving power allows one to distinguish. (Assume that the photographic plate separates precisely two images separated by a linear distance of $20\,\mu$.)

II

When the slit F is no longer narrow, what is the maximum value of its width which will allow one to make use of the resolving power of the grating?

III

Assume the width of the opaque rulings to be $2p/3$.
In the following consider the source slit to be very narrow and the light monochromatic.

1. Draw the curve representing the illumination in the focal plane of L_2 as a function of $\sin i$.

2. Replace the photographic plate by a lens L_3 whose focal length $f_3 = f_2/2$. What does one observe on a screen placed at a distance f_2 from L_3 in the following three cases: (a) one only allows the zero-order image to pass L_3, (b) one only allows the orders ± 1 to pass, and (c) one allows all the diffraction images to pass.

IV

Consider another grating having the same step and the same number of lines but with the opaque lines having the width $p/3$. Show that, for the same amplitude of incident plane wave, the amplitudes in the images of the first two orders are the same for both gratings. What is the ratio of the amplitudes in the zero order? Show that the results of part IV can be tied to Babinet's theorem.

SOLUTION

I. *Infinitely narrow linear source*

Characteristics of the grating:

$$\begin{cases} \text{infinitely narrow slits,} \\ \text{grating step } p, \\ \text{grating width } L = Np. \end{cases}$$

One finds diffraction only in planes normal to the slits in the grating.
There is no interference along lines parallel to the slits.

1. *Position of the diffraction maxima (normal incidence).* Between two homologous rays there is a path difference (Fig. 37.1):

$$\delta = p \sin i \tag{1}$$

hence a phase shift:

$$\phi = \frac{2\pi}{\lambda} p \sin i. \tag{2}$$

The interference is constructive when

$$p \sin i = k\lambda \qquad (k \text{ integer}).$$

From which one gets the values of i corresponding to diffraction maxima

$$\sin i = \frac{k\lambda}{p}. \qquad (3)$$

Numerical application: $\sin i = k\times\frac{1}{10}$.
The condition $\sin i \leqslant 1$ implies $k \leqslant 10$.

2. *Resolving power.* Each slit gives a highly spread out diffraction figure. The intensity in direction i is

$$I(i) = A(i)\,A^*(i) = \frac{\sin^2 N\phi/2}{\sin^2 \phi/2}. \qquad (4)$$

The principal maxima are given by: $\phi/2 = k\pi$.

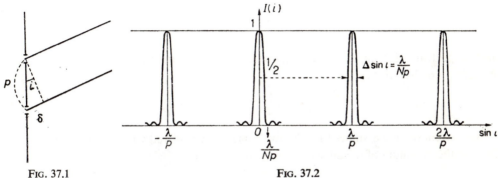

FIG. 37.1   FIG. 37.2

The zero minima correspond to $N\phi/2 = k'\pi$ $(k'/N \neq k)$.
The variations of I are given in Fig. 37.2.
By taking $\sin i$ as the variable, all the diffraction images are identical and they give:
the same illumination,
the same width $\sin i = \lambda/Np$,
the separation λ/p.
There are nineteen of them $(\sin i < 1)$.
Assume that the source emits two wavelengths λ and λ'. Since the deviation produced by the grating is proportional to the wavelength, the light distribution is similar to that shown in Fig. 37.3. For the wavelength λ, the positions of the maxima correspond to $\sin i = k\lambda/p$ and the width of a diffraction peak becomes

$$\Delta \sin i = \frac{\lambda}{Np},$$

so that

$$\Delta i = \frac{\lambda}{Np \cos i}, \qquad (5)$$

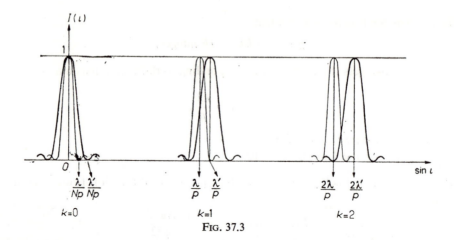

FIG. 37.3

For the wavelength λ' near λ such that $\lambda' = \lambda + d\lambda$, the position of the maxima is given by

$$\sin i' = k\,\frac{\lambda'}{p}.$$

(6)

In the spectrum of order k, two linear images corresponding to λ and λ' are separated by $\delta(\sin i) = \cos i \cdot \delta i$, hence

$$\delta i = \frac{k}{\cos i}\left[\frac{\lambda'}{p} - \frac{\lambda}{p}\right] = \frac{k}{\cos i}\,\frac{d\lambda}{p}.$$

(7)

By convention one assumes that the two images are resolved when the maximum of one falls on the minimum of the adjacent one, that is, when

$$\Delta i = \delta i$$

Equations (5) and (7) then yield

$$\frac{\lambda}{Np\cos i} = k\,\frac{d\lambda}{p}.$$

Hence, the resolving power of the grating is

$$R = \frac{\lambda}{d\lambda} = kN = k \times 5000.$$

(8)

Note. R is maximum for k maximum so that one should utilize the spectra of high order. R is minimal for $k = 1$ ($R = 5000$). In these spectra, one can separate two lines where

$$d\lambda = \frac{\lambda}{5000} = \frac{1000}{5000} = 2\text{Å}.$$

(9)

The angular separation of these two lines will be given by:

$$\Delta i = \frac{\lambda}{Np\cos i} \approx \frac{\lambda}{Np} = \frac{1}{10 \times 5000} = 2 \times 10^{-5}\text{ rad}$$

(10)

and their linear separation in the image plane will be

$$\Delta x = f_2 \, \Delta i. \tag{11}$$

It is now necessary to take into account the resolving power of the receiver $dx = 20 \, \mu$. The two lines are resolved on the photographic plate if

$$\Delta x > dx \quad \text{or} \quad f_2 \, \Delta i > dx. \tag{12}$$

The minimum focal length of L_2 is then

$$f_2 = \frac{dx}{\Delta i} = \frac{20 \times 10^{-6}}{2 \times 10^{-4}}.$$

The focal length for resolution $f_2 = 1$ m.

II. *Slit source of finite width (incoherent illumination)*

One can enlarge the slit source in such a way as to increase the luminance without in every case decreasing the resolving power of the grating. Let d be the limiting width of the source which can be attained without modifying the appearance of the image.

To treat this problem one can use two methods:

either generate an expression giving the distribution of illumination in the image as a function of the slit width and compare this with the results found in 1 and then derive the maximum width;

or directly determine the degree of coherence of the source.

First method

Determine $I(u)$ for a slit source of finite width. The problem is one-dimensional. Call x the abscissa of a general point in the pupil and u the conjugate variable such that $u = (\sin i)/\lambda$. Since the illumination is incoherent, the diffracted intensities of each point on the object add in the image plane. The resultant intensity at a point M' in the image plane will be:

$$I(u') = \int O(u) \, D(u'-u) \, du = O(u) \otimes D(u), \tag{13}$$

with: $O(u) = $ distribution of light in the object,

$D(u) = $ distribution of light in the diffraction image due to a point source (unit impulse).

Parseval's theorem allows us to transform the convolution (13) into a product (see Appendix A, B.II):

$$i(x) = o(x) \cdot d(x) \tag{14}$$

with

$$\left. \begin{array}{l} O(u) \xrightarrow{\text{F.T.}} o(x) \\[4pt] D(u) \xrightarrow{\text{F.T.}} d(x) \\[4pt] I(u) \xrightarrow{\text{F.T.}} i(x) \end{array} \right\} \tag{15}$$

Determination of the Fourier transforms:

$$\boxed{o(x)}$$

$O(u)$ is a slit function of width u_0.
One gets (see Appendix A)

$$o(x) = \frac{\sin \pi u_0 x}{\pi u_0 x}. \tag{16}$$

$$\boxed{d(x)}$$

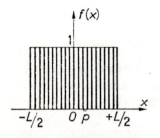

Transparency of the grating

FIG. 37.4a

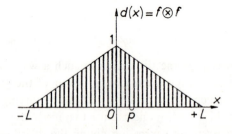

Transfer function

FIG. 37.4b

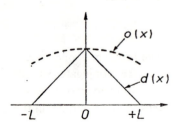

FIG. 37.5

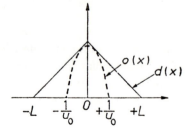

FIG. 37.6

1. $d(x)$ is the transfer function of the instrument (see p. 346). Here what is necessary is the determination of the autocorrelation function of a grating with infinitely narrow slits, step p, and width $L = Np$ (Fig. 37.4 a and b). In the determination of the product $o(x) \cdot d(x)$ many cases can arise:

when the source slit is highly narrowed, the curve $o(x)$ is very much spread out (Fig. 37.5) and one has

$$o(x) \times d(x) \approx d(x). \tag{17}$$

The distribution of light in the image is given by $D(u)$.

when the slit is widely opened the width of $o(x)$ can be less than that of $d(x)$ and the product $o(x) \, d(x)$ can be very different from $d(x)$ (Fig. 37.6).

Only the last case is of interest here. One has

$$i(x) = d(x)$$

that is, $I(u) = D(u)$ if $o(x)$ is about equal to 1 for $-L < x < +L$. Assume that this condition is fulfilled for

$$\frac{1}{4}\frac{1}{u_0} > L \quad \text{or} \quad \frac{1}{4}\frac{\lambda f_1}{d} > L, \tag{18}$$

hence

$$d < \frac{\lambda f_1}{4L}. \tag{19}$$

Numerical application:

$$d < \frac{1 \times 50 \times 10^4}{4 \times 10 \times 5000} = 2.5 \ \mu.$$

Second method

Use the Van Cittert–Zernike theorem which is stated and proved in Appendix B.

In the plane of the pupil (grating plane) place an artifical diffraction spot centred on P_0 (P_0 corresponds to an edge of the grating). The distribution of amplitude in this diffraction spot is equal to the F.T. of the energy distribution in the source plane.

Let

$$o(P) \xrightarrow{\text{F.T.}} O(M).$$

Of course this diffraction spot does not exist in the xOy plane which is in fact uniformly illuminated by the source. However, the amplitude distribution in the diffraction image is equal to the coherence distribution in the plane of the pupil.

The degree of coherence between a point P (a general point in the pupil plane) and the point P_0 is equal to the normalized amplitude $o(P)$, that is to say, $o(x)$ (Fig. 37.7).

On Fig. 37.7 is shown both $o(x)$ (dotted) and the pupil (amplitude) transparence $f(x)$.

The degree of coherence between P and P_0 will be 1 if one has $o(P) = 1$. The illumination of the grating will be considered coherent if $o(P_1) = 1$, that is, if $o(Np) = o(L) = 1$.

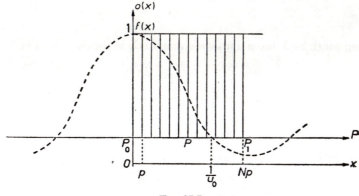

Fig. 37.7

One assumes that this condition is obtained if

$$\frac{1}{4u_0} > L.$$

One again finds (18).

III. *Abbe's experiment*

In the remainder of the problem assume that the illumination is coherent.

1. The transparent lines have a finite width namely $p/3$. The pupil amplitude transmittance is shown on Fig. 37.8.

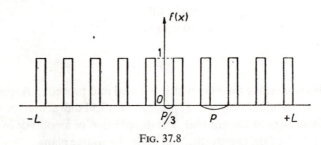

FIG. 37.8

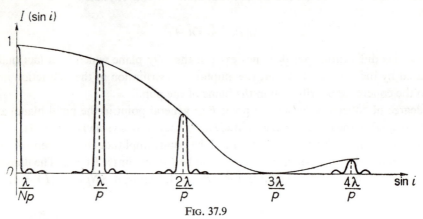

FIG. 37.9

Each slit, with width $2p/3$, has a diffraction amplitude in the direction i of

$$A(i) = \frac{p}{3} \frac{\sin \pi \left(\frac{p}{3} \frac{\sin i}{\lambda} \right)}{\pi \frac{p}{3} \frac{\sin i}{\lambda}}. \tag{20}$$

The (normalized) intensity transmitted by the N slits in this direction is given by

$$I(i) = \left(\frac{\sin \phi/6}{\phi/6} \right)^2 \left(\frac{\sin N\phi/2}{\sin \phi/2} \right)^2 \tag{21}$$

with

$$\phi = \frac{2\pi}{\lambda} p \sin i.$$

The variation of I as a function of $\sin i$ shown on Fig. 37.9. The spectra ± 3, ± 6, and ± 9 vanish.

2. The lens L_3 acts between its antiprincipal points. With the grating acting behind L_2, its image will be formed at O_2' with magnification 1 (Fig. 37.10). Let u and x be conjugate variables $(u = (\sin i)/\lambda)$. The amplitude distribution in the planes R, L_3, and O_2' are given successively by

$$\left. \begin{array}{l} f(x) \\ F(u) \text{ before the stop is inserted, and } \Phi(u) \text{ after} \\ \phi(x). \end{array} \right\} \tag{22}$$

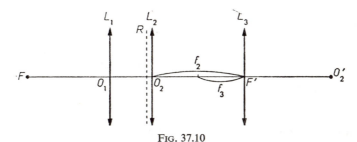

FIG. 37.10

(a) L_3 acts as a stop in such a way as to pass only the O order spectrum. Thus

$$\Phi(u) = \frac{\sin \pi u L}{\pi u L}. \tag{23}$$

$\Phi(u)$ is the diffracted amplitude from a slit of width L. The final image is thus made up of a band of uniform light having the width L and the height of the grating.

Using the language of the Fourier transformation

$$\phi(x) = \text{F.T.} [\Phi(u)] = \begin{cases} 1 & \text{for} \quad -L/2 < x < L/2 \\ 0 & \text{otherwise,} \end{cases} \tag{24}$$

These two functions are shown in Figs. 37.11a and 37.11b.

(b) Only the spectra of orders $+1$ and -1 are transmitted.

These two spectra constitute two secondary sources which are coherent and in phase and which produce Young type fringes. In effect, if one calls $\Phi'(u)$ the amplitude distribution after crossing L_3, one has

$$\Phi'(u) = \Phi\left(u + \frac{1}{p}\right) + \Phi\left(u - \frac{1}{p}\right) \qquad \text{(Fig. 37.12a).} \tag{25}$$

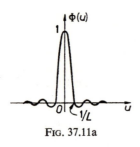

FIG. 37.11a

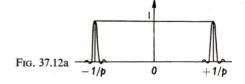

FIG. 37.12a

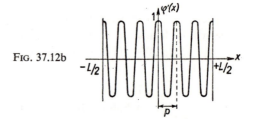

FIG. 37.12b

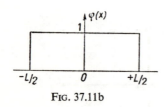

FIG. 37.11b

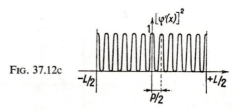

FIG. 37.12c

Hence an amplitude distribution in the image plane

$$\phi'(x) = \text{F.T.} \left[\Phi'(u) \right] = \phi(x)\, e^{j2\pi x/p} + \phi(x)\, e^{-j2\pi x/p}$$

$$\phi'(x) = 2\phi(x) \cos 2\pi \frac{x}{p}. \tag{26}$$

The image is made up of Young's fringes with step $p/2$ and width L (Figs. 37.12b and 37.12c).

(c) Lens L_3 passes all the spatial frequencies.

One has

$$\Phi(u) = F(u) \quad \text{hence} \quad \phi(x) = f(x) \qquad \text{The image and object are identical.}$$

Conclusion. If one suppresses or attenuates certain spatial frequencies, one can modify the image.

We see two examples of this.

(a) The fundamental passes and the image of the grating disappears completely.
(b) The first two harmonics pass and the illumination in the image plane is periodic. However, even though the grating step is p, the period of the image is $p/2$.
(c) It will be necessary for the lens to have infinite aperture for all spatial frequencies to pass so that the image will reproduce the object. In practice, objectives always have a finite aperture limited by the need to correct aberrations. One sees in Figs. 37.11 and 37.12 that it is often quite difficult to reconstruct the object knowing the image.

IV. *The complementary grating*

$f(x)$ and $I(\sin i)$ are given in Figs. 37.13 and 37.14. In the following table we summarize the performance of the grating treated above, (I), and this grating (II). Of course, since we wish to make comparisons of the intensities we no longer normalize the results.

	Grating I	Grating II
Width	L	L
Step	p	p
Slit width	$p/3$	$2p/3$
$A(u)$	$\dfrac{Np}{3}\dfrac{\sin \pi up/3}{\pi up/3}$	$N\dfrac{2p}{3}\dfrac{\sin \pi u2p/3}{\pi u\,2p/3}$
Spectra order O: $A(0)$	$\dfrac{Np}{3}$	$N\dfrac{2p}{3}$
Spectra order 1: $A(1/p)$	$\dfrac{Np}{3}\dfrac{\sin \pi/3}{\pi/3}$	$N\dfrac{2p}{3}\dfrac{\sin 2\pi/3}{2\pi/3}$

One sees that

$$A_{\mathrm{II}}(0) = 2A_{\mathrm{I}}(0)$$
$$A_{\mathrm{II}}(1/p) = A_{\mathrm{I}}(1/p).$$

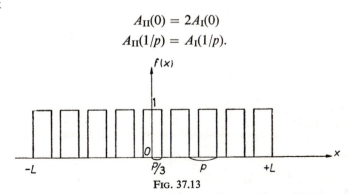

Fig. 37.13

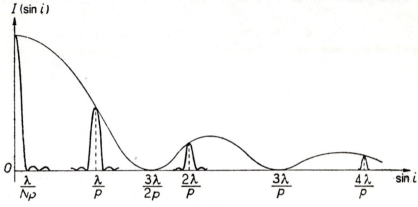

Fig. 37.14

12*

For the central image the amplitudes are in the ratio of two. For the spectra one finds the same amplitude.

This just verifies Babinet's theorem: two complementary screens give the same illumination at all points in space not illuminated in their absence (in their absence one has the central image given by a pupil of width L).

PROBLEM 38

Holography

I. *Recording of the hologram*

A coherent, monochromatic plane wave falls on the apparatus shown in Fig. 38.1. Pr: small angle prism, a, of index n. Ob: the object treated in transmission. At $P(x)$ one examines the interference of the wave Σ transmitted by the object with complex amplitude $A(x)e^{j\Phi(x)}$

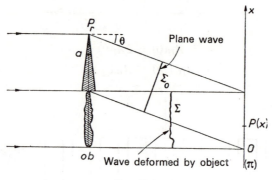

FIG. 38.1

and the wave Σ_0, known as the reference wave, which is deviated by the prism and which has the amplitude $A_0e^{j\Phi_0(x)}$. Assume that the object does not diffract. Determine the intensity function $I(x)$ by finding the relation which exists between the prism deviation θ and the phase of the reference wave.

II

The results are recorded by a photographic plate placed in the plane π. This is arranged so that one works in the linear part of its characteristic, that is, in the region where the density D is given by $D = \gamma \log E$. Find the relationship which ties $I(x)$ to the amplitude transmittance function $t(x)$ of the developed plate which is called a *hologram*. Show that $t(x)$ takes a simple form when $|A(x)|^2 \ll |A_0|^2$ (one looks for a series expansion of $t(x)$ where the significant terms are easily treated for values of γ near 3 or 4 for example).

III. *Restoration of the object*

In the expression for $t(x)$ which has been determined explain the role of each of the terms. The hologram behaves like a grating when it is placed in a parallel monochromatic beam. (One can find a "magnification" effect tied to the wavelength.) Show how the wave Σ is restored.

IV. *Holographic lens* (Fig. 38.2)

1. The reference beam remains unchanged. The object is replaced by an opaque screen containing a small hole T. The incident plane wave is transformed by diffraction into a spherical wave Σ centred on T. The distance from T to the plate is called f.

(a) Give expressions for $I(x)$ and $t(x)$.

(b) As before the hologram is illuminated by a coherent plane wave of wavelength λ.

Show that the hologram acts on this wave as:

   a converging lens of focal length f when observed in the $-\theta$ direction,
   a diverging lens when observed in the $+\theta$ direction.

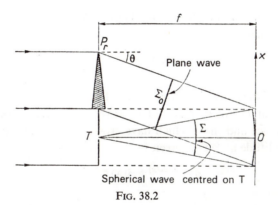

FIG. 38.2

2. The screen is now pierced with two identical holes T_1 and T_2 symmetric with respect to the OT axis and separated by $2d$. The waves Σ_0 and Σ have the wavelength λ.

(a) Give the expressions for $I(x)$ and $t(x)$.

(b) This new hologram is now illuminated by a point source S situated at a distance p from the centreline of the hologram and emitting the wavelength λ'.

Find the distance $2D$ which separates the images T_1' and T_2' from T_1 and T_2. Find the magnification of holographic lenses as a function of p, f, λ, and λ'.

V. *Sinusoidal grating*

The photographic recording of the interferogram obtained through the interference of two waves with greatly different amplitudes $A_1 \gg A_2$ (when the mirrors about $\delta_0 = 0$ are inclined at an angle θ with respect to one another) is nothing more than a hologram. Show that $I(x) = I(\delta)$ and that by illuminating such a photographic plate with coherent light one obtains directly the spectral distribution of the source. Show this by means of an elementary apparatus.

SOLUTION

Introduction

After crossing the object Ob, the electromagnetic wave carries an amplitude $\mathcal{A}$ in the form $\mathcal{A} = A(x)\,e^{j\Phi(x)}$. Both the real and complex amplitudes are required to characterize the structure of the wave. The majority of the detectors which are photographic plates are sensitive to variations in the illumination but do not provide information about the phase. The method shown here allows one to restore the wave Σ in its entirety.

I. *Recording the hologram*

The reference wave is deviated by an angle θ such that

$$\theta = (n-1)a.$$

This wave, Σ_0, has a constant amplitude A_0. Its phase $\Phi_0(x)$ varies linearly as a function of x in the plane π. One has

$$\Phi_0(x) = \frac{2\pi}{\lambda}\,\theta x. \tag{1}$$

In all of the problems here one takes the origin of the phase on Ox and assumes that the rays which interfere at O are in phase. The waves which have not crossed Ox have a positive phase difference and those which have crossed have a negative one.

The resultant amplitude at $P(x)$ is given by

$$A_0\,e^{-j\Phi_0(x)} + A(x)\,e^{j\Phi(x)}. \tag{2}$$

The resultant intensity is

$$I(x) = [A_0\,e^{-j\Phi_0(x)} + A(x)\,e^{j\Phi(x)}]\,[A_0\,e^{+j\Phi_0(x)} + A(x)\,e^{-j\Phi(x)}]$$
$$I(x) = A_0^2 + A^2(x) + 2A_0A(x)\cos[\Phi_0(x) + \Phi(x)]. \tag{3}$$

Notes. The phase $\Phi(x)$ occurs in the expression for $I(x)$. A variation in $\Phi(x)$ involves a modification in the step or the position of the fringes. A modification in $A(x)$ changes the contrast of the fringes.

II. *Amplitude transmitted by the photographic plate*

The plate is exposed to the illumination $E(x)$ such that:

$$E(x) = I(x) = \text{vibrational intensity.} \tag{4}$$

This plate is developed then illuminated by a parallel beam normal to the surface and with amplitude 1.

Recall that the density of the plate is given by

$$D = \gamma \log E. \tag{5}$$

On the other hand, one has

$$D = \log \frac{I_{\text{incident}}}{I_{\text{transmitted}}} = \log \frac{1}{T} = \log \frac{1}{t^2}. \tag{6}$$

T is the transmission factor and t the amplitude transmitted by the hologram.

Equating (5) and (6) one finds

$$t(x) = I(x)^{-\gamma/2}. \tag{7}$$

If the reference beam is much more intense than the beam which crossed the object, one has the condition

$$A_0 \gg A(x), \tag{8}$$

which allows us to transform the expression for $t(x)$:

$$t(x) = I(x)^{-\gamma/2} = \{A_0^2 + A^2(x) + 2A_0 A(x) \cos [\Phi_0(x) + \Phi(x)]\}^{-\gamma/2}$$

$$= A^{2-(\gamma/2)} - \frac{\gamma}{2} A_0^{-\gamma-2} A^2(x) - \gamma A_0^{-\gamma-1} A(x) \cos [\Phi_0(x) + \Phi(x)]$$

$$= A^{1-(\gamma/2)} \left\{ A_0^2 - \frac{\gamma}{2} A^2(x) - \gamma A_0 A(x) \cos [\Phi_0 + \Phi] \right\}.$$

Dividing by the constant factor $-2A_0^{-\gamma-2}$ one finds

$$t(x) \approx -2A_0^2 + \gamma A^2(x) + \gamma A_0 A(x) \, e^{j(\Phi_0 + \Phi)} + \gamma A_0 A(x) \, e^{-j(\Phi_0 + \Phi)}. \tag{9}$$

This equation can then be written

$$t(x) = -2A_0^2 + \gamma A^2(x) + \gamma A_0 \, e^{j\Phi_0} \times A(x) \, e^{j\Phi(x)} + \gamma A_0 \, e^{-j\Phi_0} \times A(x) \, e^{-j\Phi(x)}. \tag{10}$$

Notes. The γ of the plate occurs in the last three terms of (10). The amplitude of Σ, namely $A(x) \, e^{j\Phi(x)}$ appears directly in the three terms. The recording of the hologram can, in principle, be made with any coherent source, but the ratio $A_0/A(x)$ is very large and one must use a laser so that $A(x)$ will not be too small.

III. *Restoring the object*

The restoration is easy and may be done without an optical system.

The hologram is illuminated by a wave σ_0 plane-parallel and coherent at the plate (Fig. 38.3). The amplitude distribution in the plane of the plate is $t(x)$ and we need the amplitude distribution in Fourier space.

The four terms in $t(x)$ correspond to the following spatial frequencies:

$$\left.\begin{array}{l} -2A_0^2 + \gamma A^2(x) \rightarrow \text{frequency } 0, \\ \gamma A_0 A(x) \, e^{j\Phi(x)} \rightarrow \text{frequency } +u_0 = +\theta/\lambda, \\ \gamma A_0 A(x) \, e^{-j\Phi(x)} \rightarrow \text{frequency } -u_0 = -\theta/\lambda. \end{array}\right\} \tag{11}$$

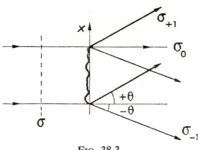

FIG. 38.3

The results are shown in Fig. 38.3. One has

a direct wave σ_0 with amplitude $-2A_0^2 + \gamma A^2(x)$ in the direction $\theta = 0$;

a wave σ_{+1} making an angle $+\theta$ with Ox reproducing to within the coefficient γA_0 the wave Σ;

a wave σ_{-1} having the same amplitude as σ_{+1} but opposite phase (in the direction $-\theta$).

To understand how the hologram acts on the plane wave σ, recall that a "apex downward" prism with angle a rotates the beam through the angle θ and introduces the phase-shift $+(2\pi/\lambda)\theta x$ (Fig. 38.4). The third term of (10) can then be interpreted as the amplitude of the object transmitted by the prism above.

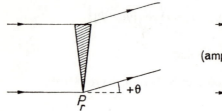

FIG. 38.4

The direct beam in the direction $\theta = 0$ is very bright while the beam diffracted in the direction $+\theta$ is highly attentuated and restores the object.

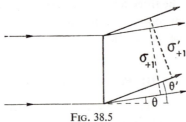

FIG. 38.5

Magnification. If the hologram is illuminated with a wavelength λ' different from the wavelength λ used to generate the hologram, the waves σ_{+1} and σ_{-1} propagate along the direction $+\theta'$ and $-\theta'$ such that

$$\frac{\theta}{\lambda} = \frac{\theta'}{\lambda'} \qquad \text{(from grating theory).} \tag{12}$$

One sees (Fig. 38.5) that the dimensions of the object vary directly with the wavelength.

IV. *Holographic lenses*

One causes the plane wave Σ_0 to interfere with the spherical wave Σ (Fig. 38.6).

1. (a) The resultant amplitude at point P is:

$$A_0 e^{-j\Phi_0(x)} + A e^{j\Phi(x)}. \tag{13}$$

One assumes that $A(x) = A = $ constant and does not vary when one passes from the centre O to the edge of the field. $\Phi(x)$ is the phase difference between the wave Σ centred at

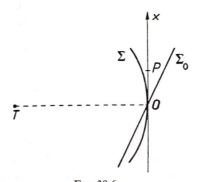

FIG. 38.6

T and the plane Ox (see, for example, the problem dealing with Newton's rings)

$$\Phi(x) = \frac{2\pi}{\lambda} \frac{x^2}{2f} = \frac{\pi}{\lambda f} x^2. \tag{14}$$

Putting this value of $\Phi(x)$ in the general equation (3) one finds

$$I(x) = A_0^2 + A^2 + 2A_0 A \cos \frac{2\pi}{\lambda} \left[\theta x + \frac{x^2}{2f} \right]. \tag{15}$$

(b) The amplitude transmitted by the hologram when illuminated by a coherent plane wave is

$$t(x) = -2A_0^2 + \gamma A^2 + \gamma A_0 \exp \left(j \frac{2\pi}{\lambda} \theta x \right) \times A \exp \left(j \frac{\pi}{\lambda f} x^2 \right)$$
$$+ \gamma A_0 \exp \left(-j \frac{2\pi}{\lambda} \theta x \right) \times A \exp \left(-j \frac{\pi}{\lambda f} x^2 \right). \tag{16}$$

Look at the expression for $t(x)$:

the first and second terms: the hologram transmits a plane wave σ_0 unperturbed having amplitude $-2A_0^2 + \gamma A^2$ in the direction $\theta = 0$;

the third term: the hologram behaves like a point down prism (deviating the wave in the direction $+\theta$) as if a divergent lens of focal length $-f$ (transforming the plane wave into a divergent wave (Fig. 38.7). σ_{+1} is a divergent wave restoring the object;

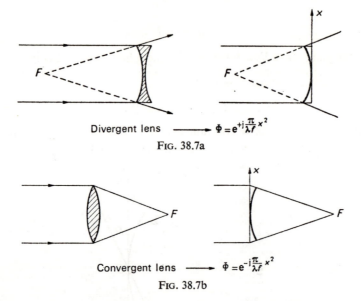

Divergent lens ⟶ $\Phi = e^{+j\frac{\pi}{\lambda f}x^2}$

FIG. 38.7a

Convergent lens ⟶ $\Phi = e^{-j\frac{\pi}{\lambda f}x^2}$

FIG. 38.7b

the fourth term: here one has the equivalent of a "point up" prism and of a converging lens of focal length f (Fig. 38.7b). σ_{-1} is a convergent wave.

The final images are shown in Fig. 38.8.

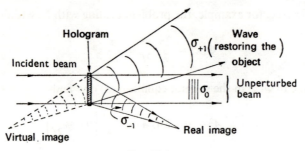

FIG. 38.8

2. (a) The resultant amplitude on the plate is

$$A_0 \exp\left(-j\frac{2\pi}{\lambda}\theta x\right) + A \exp\left(j\frac{\pi}{\lambda f}(x-d)^2\right) + A \exp\left(j\frac{\pi}{\lambda f}(x+d)^2\right). \tag{17}$$

(b) One derives for the amplitude transmitted by the hologram:

$$t(x) \approx 2A_0^2 - 2\gamma A^2\left[1 - \cos\frac{4\pi}{\lambda f}dx\right]$$

$$+ \gamma A_0 A\left\{\exp\left(j\frac{\pi}{\lambda f}(x-d)^2\right) + \exp\left(j\frac{\pi}{\lambda f}(x+d)^2\right)\right\} \exp\left(j\frac{2\pi}{\lambda}\theta x\right)$$

$$+ \gamma A_0 A\left\{\exp\left(-j\frac{\pi}{\lambda f}(x-d)^2\right) + \exp\left(-j\frac{\pi}{\lambda f}(x+d)^2\right)\right\} \exp\left(-j\frac{2\pi}{\lambda}\theta x\right). \tag{18}$$

One has

a wave in the direction $\theta = 0$,
two divergent waves in the direction $+\theta$,
two convergent waves in the $-\theta$ direction which give the real images T_1' and T_2' of T_1
and T_2.

For simplicity consider only these last two images. The last two terms of (18) show that T_1' and T_2' are obtained by comparing the hologram to

a point up prism,
two convergent lenses of focal length f for the wavelength λ whose centres are separated
by $2d$ (this distance is exaggerated in Fig. 38.9).

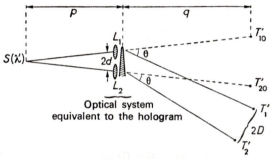

FIG. 38.9

The two lenses when acting on the wavelength λ' have a focal length f' such that (see
Fresnel Zones, § 5.4):

$$f\lambda = f'\lambda'. \tag{19}$$

The lens equation allows us to determine the positions of T_{10}' and T_{20}'. One finds:

$$\frac{1}{f'} = \frac{1}{p} + \frac{1}{q} = \frac{p+q}{pq}. \tag{20}$$

Using the triangles SL_1L_2 and $ST_{10}'T_{20}'$ we can write

$$g = \frac{2D}{2d} = \frac{p+q}{p} = \frac{q}{f'} = \frac{q}{f}\frac{\lambda'}{\lambda}. \tag{21}$$

If the plate is exposed with X-rays and the hologram illuminated with laser light at
6328 Å, one gets a very significant magnification. It is necessary, however, that one does not
forget that the photographic plate has a finite resolution.

V. *Sinusoidal grating*

One has

$$I(x) = A_1^2 + A_2^2 + 2A_1A_2 \cos\frac{2\pi}{\lambda}\theta x. \tag{22}$$

If $A_1 \gg A_2$, the developed plate transmits the amplitude

$$t(x) = -2A_1^2 + \gamma A_2^2 + \gamma A_1 A_2 \exp\left(j\frac{2\pi}{\lambda}\theta x\right) + \gamma A_1 A_2 \exp\left(-j\frac{2\pi}{\lambda}\theta x\right). \tag{23}$$

This sinusoidal grating allows only the spatial frequencies 0, $+u_0$, and $-u_0$ to pass. Apart from the direct image, one has the spectra of orders $+1$ and -1 where the dispersion is proportional to the wavelength (Fig. 38.10). This spectrogram does not contain spectra of order higher than 1. Unfortunately, the grain of the plate impairs the resolution.

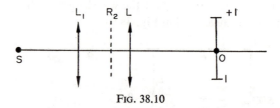

FIG. 38.10

PROBLEM 39

Reflection Gratings

I

Consider a reflecting and diffracting pupil of width a and height $h \gg a$ illuminated by a parallel monochromatic beam of light.

The incident rays are normal to the large dimension h of the pupil. Let i be their angle of incidence and i' the angle of the diffracted rays relative to the normal to h which will form the basis of the problem. Find the expression for the far-field diffraction intensity in the direction i and specify the sign of i and i'.

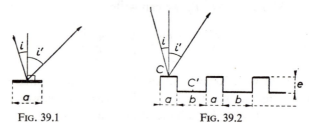

FIG. 39.1   FIG. 39.2

II

Consider a set of N pupils of width a and N pupils of width b illuminated as in part I but displaced by e as indicated in Fig. 39.2. Their second dimension is always $h \gg b$ and $\gg a$.

1. What is the delay of the vibration diffracted by the centre C' of (b) with respect to that diffracted by the centre C of (a)?

2. Derive the diffracted intensity for the set of pupils (a) and (b).

Assume that the angles i and i' are sufficiently small that the incident and diffracted beams will not be noticeably separated.

3. This set of N pupils (N large) forms a reflection grating. Show that the proper choice of e allows one to cancel out some orders. Take $a = b$ and $i = 0$.

III. *Echelle grating*

Littrow's mounting (Fig. 39.3) uses an echelle grating.
The ruling of the grating is as shown in Fig. 39.4.

FIG. 39.3

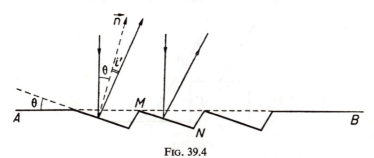

FIG. 39.4

On the metallic surface AB one rules grooves whose profile is saw-toothed. The surface MN of each tooth makes an angle θ with AB and has a width $MN = a$.

1. Explain the diagram.

2. The incident rays are normal to the surface AB.

(a) Give as a function of $\sin i'$ the diffracted intensity in the direction i' for a tooth of width a.

(b) Find the intensity of diffraction in this direction for N teeth.

3. What minimum value must θ have so that the diffracted energy from the grating will be concentrated in a particular spectrum near $\lambda = 1\ \mu$. What will be the order of this spectrum? $MN = a = 4\ \mu$.

SOLUTION

I

Amplitude diffracted by a pupil of width a (§ 7.8).

$$F_0(u) = a\frac{\sin \pi u a}{\pi u a},$$ (1)

by taking

$$u = \frac{\sin i + \sin i'}{\lambda}.$$ (2)

It is convenient to take the origin of the angles on the normal to the pupil and to take the usual trigonometric sense for positive and negative.

II. *Amplitude diffracted by a step*

1. Between the incident rays which reflect at C and C' (Fig. 39.5) one finds a path difference

$$\delta_i = C'H = C'L + LH = C'K \sin i + e \cos i.$$

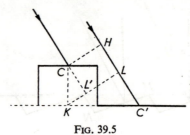

Fig. 39.5

Since i is small

$$\delta_i = \frac{p}{2} \sin i + e, \quad \text{with} \quad p = a + b.$$

Between the rays reflected at C and C' one has a path difference

$$\delta = \frac{p}{2}[\sin i + \sin i'] + 2e,$$

hence

$$\phi = \pi p u + \frac{4\pi}{\lambda}e.$$ (3)

2. Diffracted amplitude for a step of width p:

$$F_1(u) = a\frac{\sin \pi u a}{\pi u a} + b\frac{\sin \pi u b}{\pi u b} e^{j\phi},$$ (4)

hence the intensity is

$$I_1(u) = |F_1(u)|^2$$

3. Diffracted amplitude for a grating with N identical steps (§ 7.8):

$$F(u) = F_1(u)[1 + e^{j\Phi} + \ldots + e^{j(N-1)\Phi}],$$

from which

$$I(u) = [F_1(u)]^2 \frac{\sin^2 N\Phi/2}{\sin^2 \Phi/2}, \tag{5}$$

with

$$\Phi = 2\pi up \tag{6}$$

If $a = b = p/2$, equation (4) simplifies to

$$F_1(u) = a \frac{\sin \pi ua}{\pi ua} [1 + e^{j\phi}] = 2a\, e^{j\phi/2} \frac{\sin \pi ua}{\pi ua} \cos \phi/2, \tag{6'}$$

from which

$$I(u) = p^2 \left(\frac{\sin \pi ua}{\pi ua}\right)^2 \cos^2 \frac{\phi}{2} \frac{\sin^2 N\Phi/2}{\sin^2 \Phi/2},$$

$$I(u) = 4a^2 \left(\frac{\sin \pi ua}{\pi ua}\right)^2 \times \cos^2 \left[\pi ua + \frac{2\pi e}{\lambda}\right] \times \left(\frac{\sin N2\pi ua}{\sin 2\pi ua}\right)^2. \tag{7}$$

One can write finally:

$$I(u) = 4a^2 \underbrace{\left(\frac{\sin \pi ua}{\pi ua}\right)^2 \times \left(\frac{\sin N2\pi ua}{\sin 2\pi ua}\right)^2}_{\text{I}} \underbrace{\cos^2 \left[\pi ua + \frac{2\pi e}{\lambda}\right]}_{\text{II}}.$$

The product I represents the diffracted intensity for a Foucault grating of step $p = 2a$ having N lines (see Problem 35). The spectra of even order (except $k = 0$) vanish. Only the spectra of orders $k = 0, \pm 1, \pm 3, \pm 5, \ldots$ remain. The variations of I as a function of u are shown in Fig. 39.6 by the solid line.

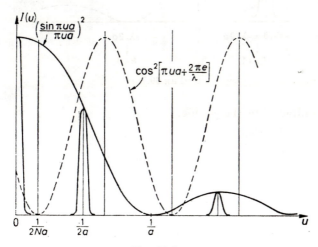

FIG. 39.6

I is modulated by II. The function $\cos^2(\pi ua + 2\pi e/\lambda)$ has period $1/a$ and the positions of the maxima and minima depend on the value of e. The variations of II are shown in Fig. 39.6 as dotted curves. Depending on the position of this function certain spectra can vanish.

(a) If $e = K\lambda/2$ (K integer) $\rightarrow II = \cos^2(\pi ua)$. A maximum of the dotted curve coincides with O. Only the zero order spectrum is reflected. The odd order spectra vanish.

(b) If $e = K\pi/4$ (K odd) $\rightarrow II = \sin^2(\pi ua)$. A minimum of the function occurs at the origin. Only the remaining odd-order spectra have maximal intensity.

III. *Echelle grating* (§ 7.9)

1. Littrow mounting (§ 20.5).

2. (a) Amplitude diffracted by one tooth:

$$F_1(u) = a\,\frac{\sin \pi ua}{\pi ua},\tag{8}$$

with:

$$u = \frac{1}{\lambda}(\sin \theta + \sin i')\tag{8'}$$

$I = |F_1(u)|^2$ (see the dotted curve in Fig. 39.7).

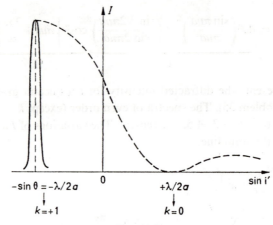

FIG. 39.7

(b) Amplitude diffracted by the grating.
One only requires (5) taking

$$\Phi = 2\pi\frac{\delta}{\lambda} = \frac{2\pi}{\lambda}\overline{CH} = \frac{2\pi}{\lambda}\overline{CD}\sin\widehat{CDH}\qquad\text{(see Fig. 39.8)}$$

$$\Phi = \frac{2\pi}{\lambda}\frac{a}{\cos\theta}\sin(\theta - i').\tag{9}$$

3. Using equation (8) one can see that the modulation due to the diffraction by one tooth does not arise if $u = 0$ or $i' = -\theta$. This corresponds to the ordinary law of reflection.

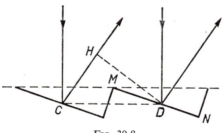

FIG. 39.8

For that value of i' corresponding to a principal maximum

$$\Phi = 2k\pi$$

or finally, using (9)

$$\frac{a}{\cos \theta} \sin (\theta - i') = k\lambda = \frac{a}{\cos \theta} \sin 2\theta = 2a \sin \theta.$$

The smallest value allowed for θ corresponds to $k = 1$, namely

$$\sin \theta = \frac{\lambda}{2a} = \frac{1}{8} \rightarrow \theta = 7°24'. \qquad (10)$$

With the echelle ruled at this angle θ the most intense spectrum corresponds to the order $k = +1$ and the diffracted rays lie in the direction i' such that:

$$\sin i' = -\frac{\lambda}{2a}.$$

The position of the other spectra is given by (9) where θ is replaced by its value but one can see that this coincides with the zero minima of (8). All the energy is concentrated in the $+1$-order spectrum (Fig. 39.8).

PROBLEM 40

Irregular Grating

A Fraunhofer diffraction grating has $3N+1$ lines which are assumed infinitely narrow with the result that the diffraction factor of each of these can be thought of as constant. One covers one slit in three (including the first and last). Find the conditions for maximum intensity and the expression for the relative intensities when N becomes infinite.

SOLUTION

I

Number the slits from 0 to $3N$ (Fig. 40.1). The diffracted amplitudes are all the same. From one slit to the next the phase difference varies by

$$\phi = \frac{2\pi d \sin i}{\lambda}$$

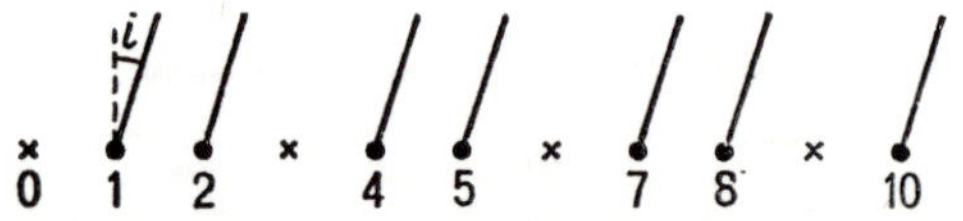

FIG. 40.1

by taking the origin at slit 0. One must find the sum

$$\exp(-j\phi) + \exp(-2\phi) + \exp(-4j\phi) + \exp(-5j\phi) + \ldots$$

which can be written as the difference between two other sums. The first is obtained by considering the entire grating

$$\sum_0^{3N} \exp(-jn\phi) = \frac{1 - \exp[-j(3N+1)\phi]}{1 - \exp(-j\phi)}$$

$$= \frac{\exp[-j(3N+1)\,\phi/2]}{\exp(-j\phi/2)} \times \frac{\exp[j(3N+1)\,\phi/2] - \exp[-j(3N+1)\,\phi/2]}{\exp(j\phi/2) - \exp(-j\phi/2)}$$

$$= \exp\left(-j3N\right)\frac{\phi}{2} \times \frac{\sin(3N+1)\,\phi/2}{\sin\phi/2}.$$

The second sum is due to the covered slits

$$\sum_0^{3N} \exp(-j3n\phi) = \frac{1 - \exp[-j3(N+1)\,\phi]}{1 - \exp(-j3\phi)} = \exp\left(-j3N\,\frac{\phi}{2}\right) \times \frac{\sin 3(N+1)\phi/2}{\sin 3\phi/2}.$$

The resultant vibration is given by

$$\exp\left(-j3N\,\frac{\phi}{2}\right) \left[\frac{\sin(3N+1)\,\phi/2}{\sin\phi/2} - \frac{\sin 3(N+1)\,\phi/2}{\sin 3\phi/2}\right]$$

and the intensity is proportional to the square of the expression between the brackets which we will call A.

The maxima are generated for:

(a) $\phi/2 = K\pi$ (K integer). The phases of 1 and 2 differ by 2π. All the emitted waves are in phase:

$$A = 3N + 1 - (N+1) = 2N,$$
$$I \propto 4N^2.$$

(b) $\phi/2 = \{(3K+1)/3\}\pi$. The phases of 1 and 2 differ by $2\pi/3$ and those of 1 and 4 by 2π:

$$A = \frac{\sin(3N+1)\pi/3}{\sin \pi/3} - \frac{\sin(N+1)\pi}{\sin \pi}.$$

In the second term replace π by $\pi - \varepsilon$:

$$B = \frac{\sin[(N+1)(\pi-\varepsilon)]}{\sin(\pi-\varepsilon)} = \frac{\sin[(N+1)\pi-(N+1)\varepsilon]}{\sin \varepsilon}.$$

Make ε go to zero:

if N is even, $B \to N+1$, $A = 1-(N+1) = -N$,

if N is odd, $B \to -(N+1)$, $A = -1+N+1 = N$.

In the second case $I \propto N^2$.

(c) $\phi/2 = \{(3K+2)/3\}\pi$. The phases of 1 and 2 differ by $4\pi/3$ and those of 1 and 4 by 4π. Reasoning similar to that of (b) shows that

$$I \propto N^2.$$

Figure 40.2 shows the resultant amplitude values for the vibrations emitted by slits 1 and 2 in cases (a), (b), and (c).

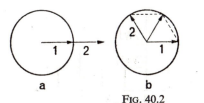

a   b   c

FIG. 40.2

The set of slits, 1 and 2, form the "basis" of a periodic grating which in its entirety is generated by a set of translations through $3d$. This problem gives a simple model useful in the analysis of crystalline structures whose sites contain a basis composed of several atoms.

II

These results are obtained immediately using the Fourier transformation.

One knows that the Fourier transform of a Dirac series of step p is a Dirac series of step $1/p$ (Appendix A).

Consider successively a grating of step p and a grating of step $3p$. The diffracted amplitudes are represented respectively on Fig. 40.3 and 40.4.

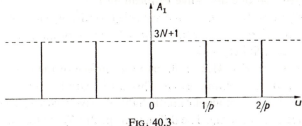

FIG. 40.3

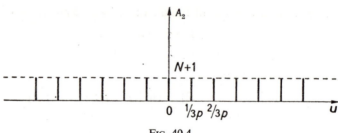

FIG. 40.4

The amplitude is proportional to the surface of the pupil, thus, here, to the number of slits. Let A be the amplitude diffracted by the covered grating:

$$A = A_1 - A_2.$$

The values of I are shown as ordinates in Fig. 40.5.

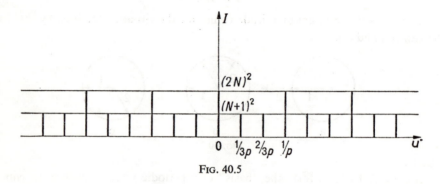

FIG. 40.5

One finds:

intensity of the principal maxima $\propto [(3N+1)-(N+1)]^2 = 4N^2;$
intensity of the secondary maxima $\propto [N+1]^2 \approx N^2.$

PROBLEM 41

Phase Grating

(It is recommended that one first solve Problem 39.)

Consider a metallic surface on which one has ruled parallel grooves of width a and height e. This is to form a reflection grating with step $p = 2a$ whose profile is shown in Figure 41.1. The grating is assumed infinitely wide.

Use the set-up shown in Fig. 41.2 to observe the surface of the grating. A slit source is placed at the focus F of the collimator, L_1. This slit is parallel to the grating rulings and normal to the figure. It emits monochromatic light of wavelength λ. The incident rays are normal to the surface of the grating. The diffraction spectra S_p are formed in the focal plane

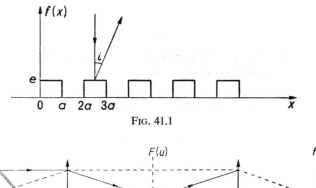

FIG. 41.1

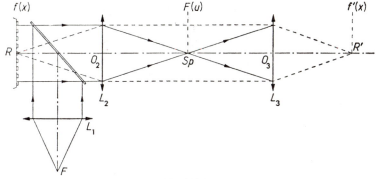

FIG. 41.2

of L_2. Lenses L_2 and L_3 are identical. The distances RO_2, O_2S_p, S_pO_3, and O_3R' are equal to the focal lengths of L_2 or L_3. As a result, the planes R and R' are conjugate with a magnification of 1.

Assume that the lens openings are very large in order that there be no stop. Assume the Fourier transform applicable.

I. Ordinary observation

Determine the amplitude distribution in the plane S_p. What are the amplitude and illumination distributions in the image plane R'? Find the appearance and the image contrast. Take as the definition of the contrast

$$\Gamma = \frac{I_{max} - I_{min}}{I_{max}}.$$

Treat the special case where $e = \lambda/2$, $\lambda/4$, and e small.

II. Observation by phase contrast

The heights e of the rectangles are small with the result that the phase shift introduced by them is very small.

Place, in the plane S_p, a phase plate which retards the direct wave by $\pi/2$. Compare these results with those which are given by a Schlieren experiment.

III

To improve the sensitivity of the phase contrast method, one can reduce as desired the amplitude of the direct wave. To do this, one adds a phase plate in front of a birefringent plane (Fig. 41.3). This latter plate is made up in the following way: from a half-wave plate one cuts three surfaces $S_1, S_2,$ and S_3 and inserts the element S_1 between S_2 and S_3 (Fig. 41.4)

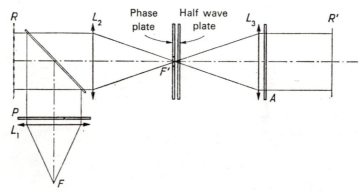

FIG. 41.3

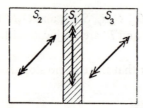

FIG. 41.4. Half-wave plate shown from the front. The slow axes are represented by two arrows.

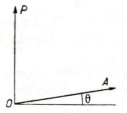

FIG. 41.5

The strip S_1 covers the image F' of the slit F. The retarding optic axes are oriented as indicated on Fig. 41.4. A polarizer P which only passes vibrations OP parallel to the lines of the grating and an analyser A are set in position. The vibrations transmitted by A make the angle $\pi/2 - \theta$ with OP (θ variable) (Figs. 41.3 and 41.5). Find the extremes of the intensity. Find the value of θ as a function of e for which the image is made up of bright bands of width a interspaced with dark bands of the same width.

SOLUTION

I. Direct observation

1. *Amplitude distribution in the spectra.* If one recalls (6′) in Problem 39 after normalization and a change of the origin of phases it becomes:

$$F(u) = \frac{\sin \pi u a}{\pi u a} \times \frac{\sin 2N\pi u a}{\sin 2\pi u} \cos \left(\pi u a + \frac{2\pi e}{\lambda} \right). \tag{1}$$

Since $N \to \infty$, the spectra are very narrow. They are equally spaced at a distance

$$\Delta u = \frac{1}{p} = \frac{1}{2a}. \tag{2}$$

The positions of certain spectra coincide with the zeros of the function

$$\frac{\sin \pi u a}{\pi u a}, \quad \text{with} \quad u = \frac{1}{a}, \frac{2}{a}, \frac{3}{a} \cdots$$

Finally, one only has the spectra of orders $0, \pm 1, \pm 3, \pm 5$, etc.

By taking

$$\phi = \frac{4\pi e}{\lambda}, \tag{3}$$

one gets for the amplitudes of these spectra

$$F(0) = \cos \frac{\phi}{2}$$

$$F\left(\pm \frac{1}{2a}\right) = \mp \frac{2}{\pi} \sin \frac{\phi}{2}$$

$$F\left(\pm \frac{3}{2a}\right) = \mp \frac{1}{3} \frac{2}{\pi} \sin \frac{\phi}{2}$$

$$F\left(\pm \frac{5}{2a}\right) = \mp \frac{1}{5} \frac{2}{\pi} \sin \frac{\phi}{2}$$

$$\vdots \qquad \vdots$$

or, more generally,

$$F(u) = F\left(\pm \frac{K}{2a}\right) \begin{cases} K = 0, \quad F(0) = \cos \frac{\phi}{2} \\ \\ K \text{ odd}, \quad F\left(\frac{K}{2a}\right) = \mp \frac{1}{K} \frac{2}{\pi} \sin \frac{\phi}{2}. \end{cases} \tag{4}$$

$F(u)$ is shown on Fig. 41.6.

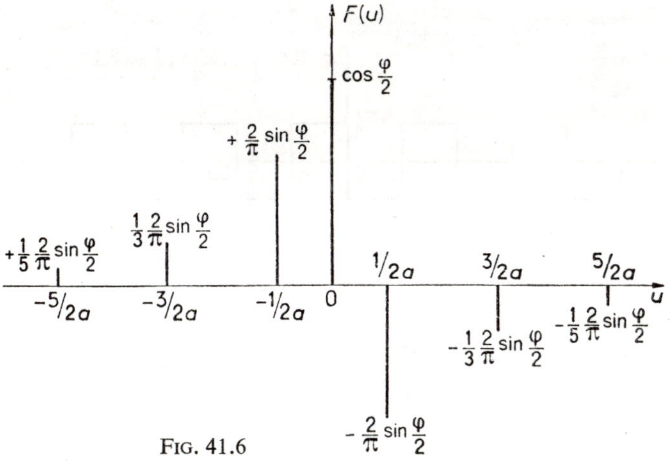

FIG. 41.6

2. *Amplitude distribution in the image plane.* This is precisely the same as in the object plane since the spectrum is transmitted in its entirety. One gets from the inverse of the Fourier transform

$$f'(x) = f(x) = \text{F.T. } [F(u)]. \tag{5}$$

Knowing that

$$\delta\left(u - \frac{K}{2a}\right) \xrightarrow{\text{F.T.}} e^{j2\pi(K/2a)x} = e^{jK\pi x/a} \tag{6}$$

one finds

$$f'(x) = f(x) = \cos\frac{\phi}{2} - \frac{2}{\pi}\sin\frac{\phi}{2}\left[(e^{j\pi x/a} - e^{-j\pi x/a}) + \frac{1}{3}(e^{j3\pi x/a} - e^{-j3\pi x/a})\right.$$

$$\left. + \frac{1}{5}(e^{j5\pi x/a} - e^{-j5\pi x/a}) + \cdots\right] \tag{7}$$

$$f'(x) = f(x) = \cos\frac{\phi}{2} - \frac{4j}{\pi}\sin\frac{\phi}{2}\left[\sin\pi\frac{x}{a} + \frac{1}{3}\sin 3\pi\frac{x}{a} + \frac{1}{5}\sin 5\pi\frac{x}{a} + \cdots\right]. \tag{8}$$

3. Distribution of the illumination in the image plane (Fig. 41.8):

$$I(x) = |f'(x)|^2 = |f(x)|^2 = \cos^2\frac{\phi}{2} + \frac{16}{\pi^2}\sin^2\frac{\phi}{2}\left[\sin\frac{\pi x}{a} + \frac{1}{3}\sin\frac{3\pi x}{a} + \frac{1}{5}\sin\frac{5\pi x}{a} + \cdots\right]^2. \tag{9}$$

Since the periodic function between the brackets oscillates between $+\pi/4$ and $-\pi/4$ (Fig. 41.7), one finds

$$I_{\text{max}} = \cos^2\frac{\phi}{2} + \sin^2\frac{\phi}{2} = 1,$$

$$I_{\text{min}} = \cos^2\frac{\phi}{2}.$$

Thus:

$$\Gamma = \sin^2\frac{\phi}{2}. \tag{10}$$

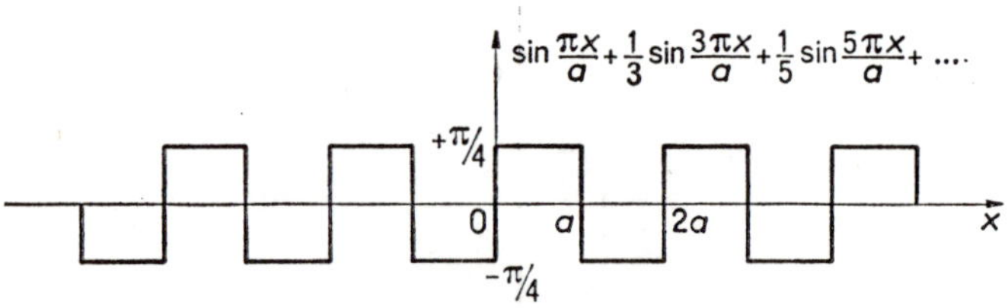

Fig. 41.7

Special cases:

$e = \lambda/2$, $\phi = 2\pi$. In the focal plane of L_2 one finds only the zero order spectrum. The image plane is uniformly illuminated ($\Gamma = 0$). The periodic structure of the object disappears completely.

$e = \lambda/4$, $\phi = \pi$. One only finds odd-order spectra:

$$f'(x) = -\frac{4j}{\pi}\left[\sin\frac{\pi x}{a} + \frac{1}{3}\sin\frac{3\pi x}{a} + \dots\right].$$

The image plane, uniformly illuminated, has black striations equally spaced at a distance a ($\Gamma = 1$). It is not possible to distinguish the relief of the grating.

e small, ϕ small. The image is similar to that above but with poor contrast.

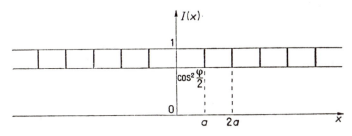

FIG. 41.8

II. *Observation by phase contrast* (§ 5.15)

The diffracted wave is advanced by $\pi/2$.

The amplitude of the odd spectrum is multiplied by $e^{j\pi/2} = +j$.

Taking account of the fact that ϕ is small,

$$F(u) = F\left(\frac{K}{2a}\right) \begin{cases} K = 0, & F(0) = \cos\dfrac{\phi}{2} = 1 \\[2mm] K \text{ odd}, & F\left(\dfrac{K}{2a}\right) = -j\dfrac{2}{K\pi}\sin\dfrac{\phi}{2} = -\dfrac{j}{K\pi}\phi. \end{cases} \quad (11)$$

Thus the amplitude in the image plane

$$f'(x) = \text{F.T.}\,[F(u)] = 1 + \frac{2\phi}{\pi}\left[\sin\frac{\pi x}{a} + \frac{1}{3}\sin\frac{3\pi x}{a} + \dots\right] \quad (12)$$

and the intensity

$$I(x) = [f'(x)]^2 = \left\{1 + \frac{2\phi}{\pi}\left[\sin\frac{\pi x}{a} + \frac{1}{3}\sin\frac{3\pi x}{a} + \dots\right]\right\}^2.$$

Neglecting powers of ϕ greater than 2, one can write

$$I(x) = 1 + \frac{4}{\pi}\phi\left[\sin\frac{\pi x}{a} + \frac{1}{3}\sin\frac{3\pi x}{a} + \dots\right], \quad (13)$$

$$I_{max} = 1 + \phi, \quad I_{min} = 1 - \phi, \quad (14)$$

$$\Gamma = \frac{2\phi}{1+\phi}, \quad \text{hence} \quad \Gamma = 2\phi. \quad (15)$$

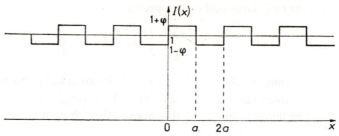

Fig. 41.9 Phase contrast.

Notes. (a) One has transformed the phase variations into amplitude variations to which the detector is sensitive. The grating relief appears as variations in the illumination, but the image has a poor contrast (proportional to ϕ) (Fig. 41.9).

(b) In the case of Schlieren photography, one shifts the direct wave. The amplitude in the plane R' is then

$$f'(x) = +\frac{2}{\pi}\,\phi\left[\sin\frac{\pi x}{a}+\sin\frac{3\pi x}{a}+\ \ldots\right].\tag{16}$$

Thus

$$I(x) = \frac{4}{\pi^2}\,\phi^2\left[\sin\frac{\pi x}{a}+\sin\frac{3\pi x}{a}+\ \ldots\right]^2\tag{17}$$

$$I_{max} = \frac{\phi^2}{4},\quad I_{min} = 0 \to \Gamma = 1.\tag{18}$$

The image formed has bright bands of width a separated by fine black lines (Fig. 41.10).

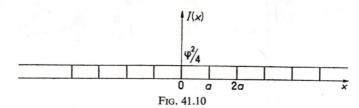

Fig. 41.10

III. *Variable phase contrast (Kastler–Montarnal apparatus)*

Equation (11) gives the amplitude of the spectra after crossing the phase plate.

The polarizer P polarizes the incident light.

The strip S_1 does not modify the orientation of the vibration carried by the direct wave.

The half-wave plates S_2 and S_3 turn the vibrations of the odd spectra through 90° (Fig. 41.11) (see the properties of half-wave plates, § 8.3).

The direct and diffracted waves are always at right angles, and their vibrations are normal.

$$F(0) = 1 \qquad \text{parallel to } OP$$

$$F\left(\frac{K}{2a}\right) = -\frac{j}{K\pi}\,\phi \quad \text{perpendicular to } OP.\tag{19}$$

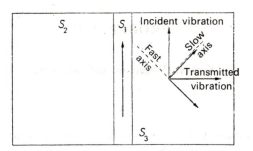

FIG. 41.11. Direction of vibrations after crossing the half-wave plate.

The analyser only passes the components of the vibrations along OA so that

$$\left.\begin{array}{ll} F(0) = \sin \theta & \text{parallel to } OA, \\ F\left(\dfrac{K}{2a}\right) = -\dfrac{j}{K\pi} \phi \cos \theta & \text{parallel to } OA. \end{array}\right\} \tag{20}$$

Taking the Fourier transform of the amplitude distribution in the plane S_p

$$f'(x) = \sin \theta + \frac{2\phi}{\pi} \cos \theta \left[\sin \frac{\pi x}{a} + \frac{1}{3} \sin \frac{3\pi x}{a} + \dots \right]. \tag{21}$$

Hence, in the image plane R',

$$I(x) = \sin^2 \theta + \sin \theta \cos \theta \frac{4\phi}{\pi} \left[\sin \frac{\pi x}{a} + \frac{1}{3} \sin \frac{3\pi x}{a} + \dots \right]. \tag{22}$$

The extremes of the intensity are

$$\left.\begin{array}{l} I_1 = \sin \theta [\sin \theta + \phi \cos \theta], \\ I_2 = \sin \theta [\sin \theta - \phi \cos \theta]. \end{array}\right\} \tag{23}$$

$\theta = 0$. The direct wave is totally stopped. $I(x) = \phi^2/4$.

This is ordinary Schlieren photography.

$\theta > 0$. If $\phi > 0$, one has $I_1 > I_2$. The phase advance is accompanied by an intensity increase. The phase contrast is positive. The intensity minimum is 0 if $I_2 = 0$ so that:

$$\phi = \tan \theta \approx \theta = \frac{2\pi}{\lambda} \times 2e.$$

Thus

$$e = \frac{\lambda \theta}{4\pi}. \tag{24}$$

$\theta < 0$. The effect is inverted and the phase contrast is negative.

PROBLEM 42

X-rays. Production and Diffraction

I

Describe briefly, using a drawing, the construction of a tube to produce X-rays.

The radiation emitted from the tube consists of a continuous spectrum upon which is superimposed intense lines which characterize the anode. Briefly indicate the origins of the line spectra and the continuous spectrum (without taking into account their fine structure).

1. Knowing the potential difference applied to the tube to be $V = 40$ kV, find the minimum wavelength of the continuous spectrum.

2. Given a copper anode whose ionization potential for the deepest level electrons (*K*-shell) is 8.98 kV, find the wavelength of the *K*-absorption limit of copper. What should the minimum potential difference applied to the tube be for one to observe the line corresponding to the transition between the *L*- and *K*-shells?

II

One wants to isolate the line (K_α) with wavelength $\lambda = 1.54$ Å by means of a crystal monochromator. A beam of X-rays from the tube T is collimated by fine slits F and F' (Fig. 42.1) and allowed to fall on the face of a cube of sodium chloride. What relationship

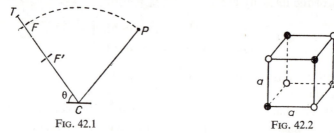

FIG. 42.1   FIG. 42.2

exists between the wavelength λ, the distance d of the crystal planes parallel to the face of the cube, and the angle θ which the beam makes with the face of the crystal, so that the beam with wavelength λ will be selectively reflected? Find the smallest value of θ, given that the ions of Cl^- and Na^+ alternate in the sodium chloride structure at a distance of $a = 2.81$ Å (Fig. 42.2).

III

Replace the sodium chloride crystal by another crystal whose crystal planes are separated by an unknown distance d' which we wish to measure by measuring the angle θ' of selective reflection of order K for $\lambda = 1.54$ Å. The angle θ' is determined with an uncertainty of 6′.

For what values of θ' will one get the best precision for d'? From what angle forward will the relative precision reach $1/1000$? Near 3.1 Å, what is the smallest value of K which corresponds to the greatest precision on d'?

IV

Show that if one removes the slit F' in Fig. 42.1 and if then one rotates the crystal about a point passing through C normal to the figure, the rays with wavelength λ which diverge at F after reflection will pass through a point P in the plane such that $CF = CP$.

One is given the values of c, h, e, and m_e.

SOLUTION

I

1. The continuous radiation from an X-ray tube is due to the transformation of the kinetic energy W_k of the electrons into radiant energy through interaction with the anode (§ 11.3). A single electron can undergo numerous decelerations in passing through the atoms of the metal each time producing a photon with energy $h\nu = -\Delta W_k$. The maximum frequency ν_m of the photon which one can get through this process corresponds to the transformation of the entire kinetic energy of the electron into radiant energy at one time. With $W_k = eV$, one finds:

$$\lambda_m = \frac{c}{\nu_m} = \frac{hc}{eV} = \frac{6.62\times10^{-34}\times3\times10^8}{1.60\times10^{-19}\times4\times10^4} = 0.31\times10^{-10} \text{ m (0.31 Å)}.$$

2. The K absorption limit separates the radiations of very short wavelength which can produce ionization of the atomic K-shell and the longer wavelengths. For copper:

$$\lambda_K = \frac{hc}{eV_k} = \frac{6.62\times10^{-34}\times3\times10^8}{1.60\times10^{-19}\times8980} = 1.38\times10^{-10} \text{ m}.$$

The emission of an X-ray corresponding to a transition of an electron between the various L, M, N, ... levels and the K level can only occur if the K-shell, filled in all atoms from helium onward, has lost through ionization one of its two electrons. This is true in particular for the K_α line ($L \rightarrow K$, Fig. 42.3). It is necessary to have, therefore, the potential difference $V_K = 8.98$ kV available.

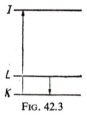

FIG. 42.3

II

The expression required is the Bragg formula (§ 7.14):

$$2d \sin \theta = K\lambda. \quad (K \text{ integer}) \tag{1}$$

The smallest value of θ is obtained for $K = 1$ and for d equal to the minimum separation of two parallel crystal planes in sodium chloride parallel to the cube face* so that $d = a$, and

$$\sin \theta = \frac{1.54}{5.63} = 0.273, \quad \theta = 15°50'.$$

III

From (1), one gets

$$2d' \cos \theta' \cdot \Delta\theta' + 2 \sin \theta' \cdot \Delta d' = 0$$

$$\left| \frac{\Delta d'}{|d'} \right| = \frac{\Delta\theta'}{\tan \theta'} .$$

The greatest precision will be obtained for $\theta' = \pi/2$. If one wants $|\Delta d'/d'| < 10^{-3}$, it is necessary for

$$\tan \theta' = \frac{6 \times 3 \times 10^{-4}}{10^{-3}} = 1.8, \quad \theta' = 58°.$$

The minimal value of θ' is 58°. For all the integral values of K, only the value $K = 4$ gives $\theta' > 58°$:

$$\sin \theta' = \frac{4 \times 1.54}{2 \times 3.1} = 0.993.$$

IV

Let α be the angle through which the reflecting surface of the crystal has been rotated (Fig. 42.4) so that the ray FC' leaving F strikes this plane at the same angle θ as the ray FC before rotation. FC' makes an angle α with FC. Let P be the point of intersection of the

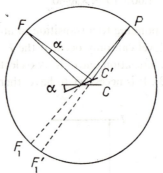

Fig. 42.4

* Note that the cube in Fig. 42.2 is not basic for NaCl; a is half the translation which will map the crystal lattice onto itself.

reflected ray corresponding to FC with the circumference of a circle with centre C and radius CF. Find the symmetric positions F_1 and F_1' of F with respect to the two successive positions of the reflecting plane. The arc $F_1 F_1' = 2\alpha$ and the angle $\widehat{F_1 P F_1'} = \alpha$. The reflected ray in the second orientation is then $C'P$.

PROBLEM 43

X-ray Spectrometer with a Curved Crystal

Cut a thin plate parallelepiped of sodium chloride whose thickness along OZ is small; then subject this to an external force so that it takes the form of a cylinder in which the planes initially parallel to XOY become coaxial cylinders whose axis is $R = OC = 1$ m and with the curved surface parallel to OY. With this deformation one can assume that the crystal lattice is not deformed but only its orientation varies.

1. Show that the rays obtained by reflection from the crystal planes which remain parallel to YOZ before deformation are tangent to the same curve for a given order of diffraction K. Assume the plate infinitely thin.

2. What happens to the above result if the plate has an appreciable thickness?

3. Show that the rays corresponding to a diffraction order K are practically brought to a focus and find its position.
   Where are the various foci F_k obtained with $\lambda = 1$ Å located? The distance between crystal planes parallel to YOZ is $d = 2.8$ Å.

SOLUTION

1. The crystal planes, initially parallel to the plane YOZ and thereby parallel to the faces of the plate, reflect among the various incident X-rays those which, in the plane of the figure, make with their normal CO in this same plane, the angle θ_k given by the Bragg formula (§ 7.14):

$$2d \sin \theta_k = K\lambda. \tag{1}$$

K is the diffraction order and d the crystal spacing.
   Assume initially that the plate is infinitely thin. After it is distorted into an arc SS' about the centre C (Fig. 43.1) the selective diffraction of order K at any point P all occur giving the reflected ray with the same angle θ_k with the normal CP to the crystal plane passing through P, or, in other words, with the radius to the centre C. When P is displaced across the plate, the distance $CP = R$ and the angle θ_k remains fixed. From C draw the normal CF to the ray reflected at P. One has

$$CF = R \sin \theta_k = \text{const.} \tag{2}$$

   In addition, the rays reflected from various points on SS' corresponding to the same diffraction order K are all tangent to the circle with radius $R \sin \theta_k$ and centre C.

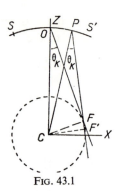

FIG. 43.1   FIG. 43.2

2. Consider a plate of finite thickness and a ray PF reflected on the crystal plane CP at P through angle θ_k. This ray passes through a point P' on the plate and makes an angle θ'_k with CP' to the crystal plane passing through P' (Fig. 43.2). One has

$$CP \sin \theta_k = CP' \sin \theta'_k = CF. \tag{3}$$

from the geometry:

$$\frac{d}{d'} = \frac{CP}{CP'}$$

and expression (3) becomes

$$d \sin \theta_k = d' \sin \theta'_k.$$

so that the direction PF is the same for the rays reflected at each point such that P satisfies (1). The thickness of the plate thus makes no change in the conclusions of the first part of the problem.

3. Since the plate has small dimensions, Fig. 43.1 shows that near F all of the reflected rays are concurrent to within the second order. The coordinates of F are

$$x = CF \cos \theta_k = R \sin \theta_k \cos \theta_k,$$
$$z = CF \sin \theta_k = R \sin^2 \theta_k. \tag{4}$$

For the various values of K, equation (1) lets us find the various values of $\sin \theta_k$ and (4) gives the values of the corresponding focus F_k. One finds

k	$=$	1	2	3	4
$\sin \theta_k$	$= 0.1786$		0.3572	0.5358	0.7144
θ_k	$= 10°17'$		20°56'	32°24'	45°33'
$\cos \theta_k$	$= 0.9839$		0.9340	0.8443	0.7003
x_k (in m)	$= 0.175$		0.333	0.452	0.500
z_k (in m)	$= 0.032$		0.127	0.287	0.510

When θ_k varies, the geometric locus of F_k is a circle with the equation

$$x^2 + \left(z - \frac{R}{2}\right)^2 = \frac{R^2}{4}.$$

Its centre is at $x_0 = 0$ and $z_0 = R/2$. It is then situated about the centre of CO (Fig. 43.1). The circle passes through C and is tangent to the plate at O.

REFRACTION AND DISPERSION

PROBLEM 44

Group and Phase Velocity

Calculate the group velocity U of waves whose phase velocity V has the following functional variation with wavelength:

1. Acoustic waves in air: $V = A$, A a constant.
2. Transverse elastic waves in a bar: $V = A/\lambda$.
3. Deep water waves: $V = A\sqrt{\lambda}$.
4. Capillary ripples: $V = A/\sqrt{\lambda}$.
5. Ionospheric electromagnetic waves: $V = \sqrt{c^2 + A^2\lambda^2}$ (c is the free-space velocity of light).

SOLUTION

The group velocity of sinusoidal waves, each characterized by its angular frequency ω and by its wave vector $\sigma = 2\pi/\lambda$, whose angular frequencies are distributed about a mean value ω, are given by (§ B.3)

$$U = \frac{d\omega}{d\sigma} = \frac{d\omega}{d\left(\dfrac{\omega}{V}\right)} = \frac{dv}{d\left(\dfrac{1}{\lambda}\right)} = V - \lambda\frac{dV}{d\lambda}.$$

1. If V is constant the medium is not dispersive, $dV/d\lambda = 0$, and $U = V$.

2. $U = \dfrac{A}{\lambda} + \dfrac{\lambda A}{\lambda^2} = 2V.$

3. $U = A\sqrt{\lambda} - \dfrac{\lambda A}{2\sqrt{\lambda}} = \dfrac{A}{2}\sqrt{\lambda} = \dfrac{V}{2},$

4. $U = \dfrac{A}{\sqrt{\lambda}} + \dfrac{\lambda}{2}\dfrac{A}{\sqrt{\lambda^3}} = \dfrac{3}{2}V.$

5. $U = \sqrt{c^2 + A^2\lambda^2} - \dfrac{\lambda 2 A^2\lambda}{2\sqrt{c^2 + A^2\lambda^2}} = \dfrac{c^2}{\sqrt{c^2 + A^2\lambda^2}} = \dfrac{c^2}{V}.$

Also see Problems 54 and 56.

PROBLEM 45

Foucault Method for the Velocity of Light

I

The measurement of the velocity of electromagnetic waves using stationary waves gives the phase velocity v_ϕ. The toothed-wheel method of Fizeau where one uses very short-wave trains gives the group velocity v_g. Show that the Foucault rotating mirror method gives the group velocity. To do this take into account the frequency variation which arises from a moving mirror as a result of reflection. Show that after reflection on the rotating mirror M

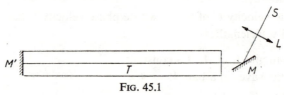

FIG. 45.1

(Fig. 45.1), the wave front propagating in the dispersive medium contained in the tube T with length l undergoes an additional rotation after exiting into the air and this rotation adds to that previously produced by the mirror.

II

Michelson, using this method, measured the velocity of light in carbon disulphide contained in the tube T and found it to be $v = c/1.77$. The source S emits white light. The index of refraction of carbon disulphide has the following values for the wavelengths in air measured in microns:

λ_0	0.589	0.550	0.486
n	1.628	1.640	1.652

Show that these values are in agreement with the conclusions of the first part.

SOLUTION

I

After passing through the lens L, the plane wave, obliquely reflected on the rotating mirror M, enters the tube T. If one neglects the typically small relativistic effects (§ 9.10), displacement of a mirror with a velocity u in a direction normal to its plane produces, for radiation with frequency v falling on the mirror with angle of incidence i, the same effect that one would get by displacing the source with a velocity $2u \cos i$. Thus, the frequency variation due to the Doppler effect is

$$\frac{\Delta v}{v} = \pm \frac{2u}{c} \cos i.$$

One takes the $+$ or $-$ signs depending upon whether the mirror is moving toward or away from the ray which strikes it.

With this in mind, let MM' (Fig. 45.2) be the face of the mirror rotating about the axis O with the angular velocity ω. The reflected wave is no longer monochromatic. Its frequency is altered from M to M' by the amount

$$\Delta v = v \, \frac{4\omega R \cos i}{c} \tag{1}$$

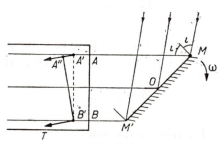

FIG. 45.2

by taking $MM' = 2R$, since $u = \omega R$ at M and at M'. The wave front which enters the tube at AB at normal incidence propagates in the dispersive medium contained in the tube with a phase velocity v_ϕ as a function of v

$$\Delta v_\phi = \frac{dv_\phi}{dv} \times \Delta v, \tag{2}$$

thus, using (1),

$$\Delta v_\phi = \frac{dv_\phi}{dv} \times \frac{4\omega vR \cos i}{c}. \tag{3}$$

For carbon disulphide, as for all transparent media, $dv_\phi/d\lambda$ is negative in the visible. Therefore, with the rotation ω in the direction indicated in the figure, one can easily see that the velocity of the wave front is greater at A than at B. The front propagates with a mean velocity, that of the phase velocity v_ϕ, of the radiation incident at O where the frequency is not changed, but it undergoes a rotation which per unit time is given by

$$\omega' = \frac{A'A''}{A'B'} = \frac{\Delta v_\phi}{2R \cos i} = -\frac{dv_\phi}{dv} \times \frac{2\omega v}{c}, \tag{4}$$

the minus sign being due to the fact that the direction of ω' is opposite that of ω. The expression relating the group velocity v_g to the phase velocity (§ B.3)

$$\frac{1}{v_g} = \frac{d}{dv}\left(\frac{v}{v_\phi}\right) = \frac{1}{v_\phi} - \frac{v}{v_\phi^2}\frac{dv_\phi}{dv}$$

allows one to write (4) in the form

$$\omega' = \frac{2\omega}{c}\, v_\phi^2\!\left(\frac{1}{v_g} - \frac{1}{v_\phi}\right).$$

After the time $\Delta t = 2l/v_\phi$ necessary for up and back travel through the tube, the angle through which the front of the wave has turned is

$$\beta = \omega'\,\Delta t = \frac{4\omega l}{n}\left(\frac{1}{v_g} - \frac{1}{v_\phi}\right),$$

$n = c/v_\phi$ being the index for radiation of frequency ν. When it enters the air, the wave front, which is inclined at a small angle β to the interface, undergoes refraction which transforms β into

$$\alpha = n\beta = 4\omega l\left(\frac{1}{v_g} - \frac{1}{v_\phi}\right).$$

This rotation resulting from the reflection turns out to be in the same direction as the rotation of the mirror. The latter, neglecting the Doppler effect, is given by (§ 1.5)

$$\theta = \frac{4\omega l}{v_\phi}. \tag{5}$$

for radiation of frequency ν. The effective rotation is

$$\alpha + \theta = \frac{4\omega l}{v_g},$$

rather than (5). Thus one measures the group velocity.

II

Using the results above, the velocity $c/1.77$ is the group velocity. To relate this to the data given in the statement of the problem, it is necessary to express it in terms of the dispersion $dn/d\lambda_0$ of carbon disulphide. One has (§ B.3)

$$v_g = v_\phi - \lambda\,\frac{dv_\phi}{d\lambda}$$

where λ is the wavelength in the medium. Since

$$n = \frac{c}{v_\phi}, \qquad \frac{dn}{n} = -\frac{dv_\phi}{v_\phi},$$

$$v_g = v_\phi\!\left(1 + \frac{\lambda}{n}\frac{dn}{d\lambda}\right).$$

Using $\lambda_0 = \lambda n$

$$\frac{d\lambda}{\lambda} = \frac{d\lambda_0}{\lambda_0} - \frac{dn}{n},$$

hence

$$\frac{\lambda}{d\lambda} \times \frac{dn}{n} = \frac{\lambda_0}{n} \frac{dn}{d\lambda_0} : \left(1 - \frac{\lambda_0}{n} \frac{dn}{d\lambda_0}\right).$$

The term between the parentheses represents a small correction. Hence, finally,

$$v_g = v_\phi\left(1 + \frac{\lambda_0}{n} \frac{dn}{d\lambda_0}\right).$$

The data from the problem statement gives

$$\frac{dn}{n} = \frac{0.024}{1.640}, \qquad \frac{\lambda_0}{d\lambda_0} = \frac{0.55}{0.1031}.$$

For the mean wavelength $v_\phi = c/1.64$, hence,

$$v_g - v_\phi = \frac{c}{1.64}(-0.077)$$

(the negative sign because $dn/d\lambda < 0$). One gets

$$v_g - v_\phi = c\left(\frac{1}{1.77} - \frac{1}{1.64}\right) = \frac{c}{1.64}(-0.073).$$

The agreement is satisfactory.

PROBLEM 46

Velocity of Light in Moving Water

The Fizeau experiment on the velocity of light in moving water has been repeated with improved precision by Zeeman.

The length l of each tube is 6 m and the velocity of the water is 5.50 m/s. The index of refraction for the green line of mercury ($\lambda_0 = 546$ nm) in water is 1.3345 at the temperature at which the experiment was conducted. One registers the position of the central fringe for a given direction of flow and then reverses the direction of flow and finds the new position of the same fringe.

1. Find the observed displacement as a function of the fringe spacing.

2. In this experiment, the water through which the light is transmitted is in motion with respect to the source. One has, therefore, a Doppler effect. Calculate the new fringe positions taking this effect into account. Can one neglect this effect if the fringe position can be found within an uncertainty of 0.01 fringe separations? The index of water for $\lambda_0 = 589$ nm is 1.3330.

SOLUTION

1. The difference in time of passage of the light in the two tubes is (§ 9.5)

$$\Delta t = \frac{l}{v_{\phi_1}} - \frac{l}{v_{\phi_2}} \tag{1}$$

and if one neglects the Doppler effect, one has

$$v_\phi = \frac{c}{n} \pm u\left(1 - \frac{1}{n^2}\right), \tag{2}$$

hence

$$\Delta t = \frac{l}{\dfrac{c}{n} - u\left(1 - \dfrac{1}{n^2}\right)} - \frac{l}{\dfrac{c}{n} + u\left(1 - \dfrac{1}{n^2}\right)} \approx \frac{2lu}{c^2}(n^2 - 1). \tag{3}$$

The displacement in terms of the fringe spacing is

$$\Delta p' = \frac{\Delta t}{T} = \frac{c\Delta t}{\lambda_0} = \frac{2lu}{c\lambda_0}(n^2 - 1),$$

$$\Delta p' = \frac{4 \times 6 \times 5.5}{3 \times 10^8 \times 546 \times 10^{-9}}(1.7817 - 1) = 0.64.$$

2. The radiation with wavelength λ_0 emitted by the source has a wavelength $\lambda_0' = \lambda_0 + \Delta\lambda$ in a system bound to the water. The index of the water must be found for λ_0'. Now

$$n' = n + \frac{dn}{d\lambda_0}\Delta\lambda_0$$

and (§ 9.6)

$$\frac{\Delta\lambda_0}{\lambda_0} = \pm\frac{u}{nc},$$

the + sign relates to the tube in which the water moves away from the source. Hence

$$n' = n \pm \lambda_0 \frac{dn}{d\lambda} \times \frac{un}{c} = n\left(1 \pm \lambda_0 \frac{dn}{d\lambda}\frac{u}{c}\right)$$

The phase velocity given by (2) becomes

$$v_\phi = \frac{c}{n'} \pm u\left(1 - \frac{1}{n'^2}\right) = \frac{c}{n}\left(1 \pm \lambda_0 \frac{dn}{d\lambda}\frac{u}{c}\right) \pm u\left(1 - \frac{1}{n'^2}\right),$$

$$v_\phi \approx \frac{c}{n} \pm u\left(1 - \frac{1}{n^2} - \frac{\lambda_0}{n}\frac{dn}{d\lambda_0}\right),$$

by replacing n'^2 with n^2 which is equivalent to neglecting the terms in c/u. Equation (3) becomes

$$\Delta t \approx \frac{2lu}{c^2}\left(n^2 - 1 - n\lambda_0 \frac{dn}{d\lambda_0}\right)$$

and the displacement of the fringes is

$$\Delta p'' = \frac{2lu}{c\lambda_0}\left(n^2-1-n\lambda_0\frac{dn}{d\lambda_0}\right) = \Delta p'\left(1-\frac{n\lambda_0}{n^2-1}\frac{dn}{d\lambda_0}\right).$$

Numerical application:

$$\frac{dn}{d\lambda_0} = -\frac{0.0015}{43\times10^{-9}} \qquad n^2-1 = 0.78$$

$$\Delta p'' = 0.64\left(1+\frac{1.3345\times456\times10^{-9}\times0.0015}{0.78\times43\times10^{-9}}\right)$$

$$= 0.64(1+0.034) = 0.66 \text{ fringes.}$$

This effect should be taken into account as Zeeman has done in his experiment.

PROBLEM 47

Propagation of Waves in a Periodic Discontinuous Medium

Consider the longitudinal elastic waves which propagate along an infinite lattice of point particles each having mass m and distributed along the Ox axis with spacing d. Each of these particles exerts a repulsive force on its neighbours proportional to the change in their spacing from the equilibrium position (Fig. 47.1).

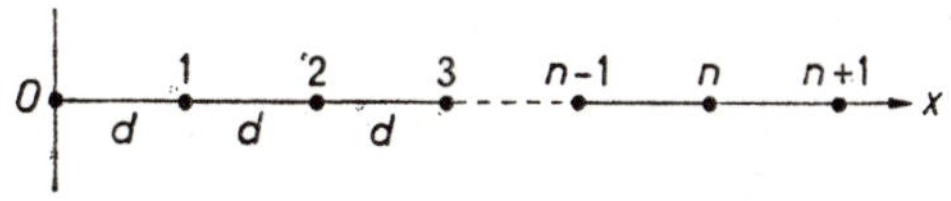

FIG. 47.1

1. Find the equation of motion of the particle indexed by n, calling the displacement from the equilibrium position s_n and show that the simultaneous equations for the motions of the particles yields a solution of the type

$$s_n = S\cos(\omega t - \sigma nd).$$

Show that the angular frequency ω varies with the wave vector σ and always remains less than some value ω_m which should be determined.

2. Find the phase velocity of the sinusoidal waves which can propagate along this lattice of particles and examine its variation as a function of σ. Find the group velocity of the waves whose angular frequency is near a given value.

3. Given that the refractive indices of crystalline NaCl and KCl have the following values at the two extremities of the visible spectrum

NaCl	1.537	1.568
KCl	1.483	1.510

what must be the number of particles per wavelength to account for the dispersion due to the mechanism studied above? Is the preceding mechanism applicable to light waves?

SOLUTION

1. The equation of motion of the particle n is

$$m \frac{d^2 s_n}{dt^2} = -k(s_n - s_{n-1}) - k(s_n - s_{n+1}) = k(s_{n-1} + s_{n+1} - 2s_n). \qquad (1)$$

k being the Hooke's law coefficient.

One wants to satisfy this equation with a progressive wave

$$s(t) = S \exp [j(\omega t - \sigma x)].$$

The displacement is only defined about the abscissas nd ($n = 1, 2, \ldots$) where one finds the particles, hence

$$s_n(t) = S \exp [j(\omega t - \sigma n d)]. \qquad (2)$$

By substituting the solution (2) in equation (1), one finds, following division by s_n,

$$m\omega^2 = -k[\exp(-j\sigma d) + \exp(j\sigma d) - 2] = 4k \sin^2 \frac{\sigma d}{2},$$

hence

$$\omega = \omega_m \sin \frac{\sigma d}{2} = \omega_m \sin \frac{\pi d}{\lambda}, \qquad (3)$$

with

$$\omega_m^2 = \frac{4k}{m}. \qquad (4)$$

The angular velocity ω varies periodically with σ. The period of the variation is $\sigma = 2\pi/\lambda$ ($\lambda = 2d$). For the values of the wave vector

$$\sigma_K = \sigma + K \frac{2\pi}{d} \qquad (K \text{ integer})$$

ω has the same value as for σ. The maximum value $\omega = \omega_m$ is obtained for $\sigma = \pi/d$. Figure 47.2 respresents these results. Only the angular frequencies lying between 0 and ω_m can be transmitted along this lattice.

When $\exp(j2K\pi) = 1$, the solutions of (2) for σ and σ_K are identical. If one wants to get a unique relation between the vibrational state and the modulus of the wave vector, it is

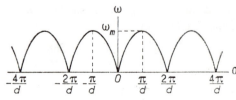

FIG. 47.2

necessary to limit the latter to an interval $2\pi/d$. One typically selects the following interval:

$$-\frac{\pi}{d} \leqslant \sigma \leqslant \frac{\pi}{d}.$$

The positive values of σ correspond to the waves which propagate along the direction of the lattice and the negative values to those which propagate in the inverse direction.

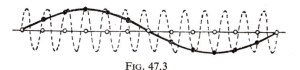

FIG. 47.3

One can find the conditions imposed on the wave vector by another method by considering the wavelength. Figure 47.3 represents a wave of length $\lambda = 12d$ by the solid line so that $\sigma = 2/12d$ and the dotted curve represents a wave

$$\sigma' = 2\pi\left(\frac{1}{12d} + \frac{1}{d}\right),$$

sothat $\lambda' = 12d/13$. The displacements s are the same.

2. Using (3), the phase velocity of the waves (2) is given by

$$v_\phi = \frac{\omega}{\sigma} = \frac{\omega_m d}{2}\frac{\sin \sigma d/2}{\sigma d/2} = v_0 \frac{\sin \sigma d/2}{\sigma d/2}. \tag{5}$$

In this form the function $(\sin x)/x$ has an absolute value varying as a function of σ as shown in Fig. 47.4. The phase velocity is a maximum for $\sigma = 0$ $(\lambda = \infty)$ and $v_\phi = v_0$. It has fallen to $2/\pi = 0.695$ of this value for $\sigma = \pi/d$ and it goes to zero with ω for $\sigma = 2\pi/d$.

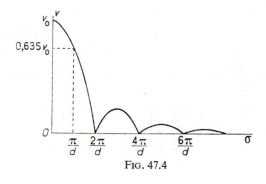

FIG. 47.4

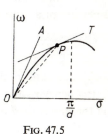

FIG. 47.5

In Fig. 47.2 the phase velocity at a point P is represented by the slope of segment OP (Fig. 47.5). The group velocity $v_g = d\omega/d\sigma$ is represented by the slope of the tangent PT at this point. One has, using (3),

$$v_g = \frac{\omega_m d}{2}\cos\frac{\sigma d}{2} = v_0 \cos\frac{\sigma d}{2}.$$

For very small values of σ, the curve $\omega(\sigma)$ can be taken to coincide with its tangent OA at the origin. As a result v_ϕ is constant and the lattice of particles, which can be thought of as a continuous medium for the propagation of waves only when $\lambda \gg d$, is not dispersive. One knows that this is the case for elastic waves propagating along strings at an audio frequency.

3. Expression (5) can be written

$$v_\phi = v_0 \frac{\sin \pi/N}{\pi/N},$$

calling $N = \lambda/d$ the number of particles contained in a wavelength. The following table gives the values of v_ϕ as a function of N:

N	2	4	8	12	16	20	∞
v_ϕ/v_0	0.636	0.900	0.974	0.989	0.994	0.996	1

Now the relative variations of the index and hence of the phase velocity approach 0.02 for NaCl and 0.018 for KCl. The number of particles contained in a wavelength can never exceed ten, if one uses this model for a dispersive medium. But one know that the distances d are of the order of several angstroms and thus one has at least a thousand atoms in a wavelength of visible light. The theory of dispersion in this region of the spectrum must involve intramolecular mechanisms.

PROBLEM 48

Sellmeier Dispersion Equation

I

The quantum mechanical expression giving as a function of the frequency ν the refractive index dispersion of a pure material removed from regions of spectral absorption is written

$$n^2 = 1 + \frac{2N}{3h\varepsilon_0} \sum_i \frac{\nu_i d_i^2}{\nu_i^2 - \nu^2}. \tag{1}$$

h is Planck's constant, ε_0 the permittivity of free space, d_i the dipole moment of the transition with frequency ν_i, and N the number of molecules per unit volume.

Taking the moments d_i as being produced by harmonic oscillators with charge q_i, mass m_i and energy $W = h\nu_i$, show that expression (1) is equivalent to the classical expression:

$$n^2 = 1 + \frac{1}{4\pi^2\varepsilon_0} \sum_i \frac{N_i f_i q_i^2}{m_i(\nu_i^2 - \nu^2)}. \tag{2}$$

In this latter expression, each oscillator has only one eigenfrequency and there are as many oscillators as eigenfrequencies. f_i is a numerical coefficient called the oscillator strength. Find an expression for f_i in the case where only a single transition is considered. One sees then that f_i is of the order of unity and in that which follows take $f_i = 1$.

II

The dispersion of hydrogen under normal conditions of temperature and pressure can be represented between 0.4 μ and 9 μ by

$$n^2 = 1 + 2.721 \times 10^{-4} + \frac{2.11}{\lambda^2} \times 10^{-18} \qquad (\lambda \text{ in metres}). \qquad (3)$$

To connect this expression with the theoretical expression (2), one writes it in the form

$$n^2 = 1 + A\left(1 + \frac{B}{\lambda^2}\right). \qquad (4)$$

Find the values of A and B and verify that the eigenfrequency lies in the ultraviolet. Find the ratio q/m. To what particle does it apply? The density of hydrogen is $\mu = 9.0 \times 10^{-2}$ kg-m^{-3}, the Faraday $\mathscr{N}e = 9.65 \times 10^7$ coulombs, where $\mathscr{N}$ is the number of molecules per kilomole, and e is the elementary charge.

III

Between 0.3 μ and 10 μ the index of refraction of CaF$_2$ is given by

$$n^2 = 6.09 + \frac{6.12 \times 10^{-15}}{\lambda^2 - 8.88 \times 10^{-15}} + \frac{5.10 \times 10^{-9}}{\lambda^2 - 1.26 \times 10^{-9}} \qquad (\lambda \text{ in metres}). \qquad (5)$$

Put equation (2) in the form

$$n^2 = A + \frac{C_1 \lambda_1^4}{\lambda^2 - \lambda_1^2} + \frac{C_3 \lambda_2^4}{\lambda^2 - \lambda_2^2}. \qquad (6)$$

By comparison with (5), give the expressions for A, C_1, and C_2. Find λ_1 and λ_2 as well as the ratio C_2/C_1. By assuming that the oscillator responsible for the infrared absorption is made up of a set of two F$^-$ ions displaced with respect to a fixed Ca^{++} ion, find the ratio m_H/m_e of the mass of the proton to that of the electron and compare it with the theoretical value 1830. $m_F = 19 m_H$.

IV

In the X-ray region, show that one can neglect all the electron eigenfrequencies and not consider the ions. Put (2) in the form

$$n^2 = 1 - K\lambda^2. \qquad (7)$$

Give the expression for K and its numerical value for copper, $A = 63$, density $\mu = 8 \times 10^3$ kg-m^{-3} and atomic number $Z = 29$.

Calculate the phase velocity v_ϕ and the group velocity v_g in the X-ray region.

What must the dispersive law be for a substance which satisfies $v_\phi v_g = c^2$ where c is the free-space velocity of light?

SOLUTION

I

The energy of a harmonic oscillator with frequency v_i is given by

$$W_i = hv_i = 4\pi^2 m_i v_i^2 s_i^2,$$

s_i being the displacement. The dipole moment $d_i = q_i s_i$. For the oscillators of type i, one has

$$\frac{2N v_i d_i^2}{3\varepsilon_0 h} = \frac{2N v_i^2 s_i^2 q_i^2}{3\varepsilon_0 h v_i} = \frac{2N q_i^2}{3\varepsilon_0 4\pi^2 m_i}.$$

Therefore, comparing (1) and (2) one sees that

$$f_i = 2/3.$$

II

For $v \ll v_i$, equation (2) with a single term can be written

$$n^2 = 1 + \frac{Nq^2}{4\pi^2 \varepsilon_0 m v_i^2} \frac{1}{1 - \dfrac{v^2}{v_i^2}} = 1 + \frac{Nq^2}{4\pi^2 \varepsilon_0 m v_i^2} \left(1 + \frac{v^2}{v_i^2}\right)$$

and with $\lambda = c/v$ and $\lambda_i = c/v_i$:

$$n^2 = 1 + A\left(1 + \frac{B}{\lambda^2}\right),$$

with

$$A = \frac{Nq^2}{4\pi^2 \varepsilon_0 m v_i^2}, \qquad B = \frac{v^2}{v_i^2}\lambda^2 = \lambda_i^2, \qquad \frac{A}{B} = \frac{Nq^2}{4\pi^2 \varepsilon_0 c^2 m}.$$

Comparing the coefficients of (3) and (4)

$$A = 2.712 \times 10^{-4}, \qquad B = \frac{2.11 \times 10^{-14}}{2.721} = 0.78 \times 10^{-14} = \lambda_i^2$$

hence

$$\lambda_i = \sqrt{0.78} \times 10^{-7} \approx 0.9 \times 10^{-7} \text{ m} \approx 900 \text{ Å}$$

λ_i is in the far ultraviolet.

To determine the charge to mass ratio of the oscillators, q/m, one writes

$$\frac{A}{B} = k^2 \frac{e}{m} \frac{Ne}{4\pi^2 \varepsilon_0 c^2},$$

taking $q = ke$ where e is the elementary charge and k an integer. One has $N = \mathcal{N}\mu/M$, (M is the molecular mass in this case 1) hence

$$Ne = \mathcal{N}e\mu = 9.65 \times 10^7 \times 9 \times 10^{-2} = 0.87 \times 10^7 \text{ coulombs}$$

and

$$4\pi\varepsilon_0 c^2 = \frac{4\pi\times9\times10^{16}}{4\pi\times9\times10^9} = 10^7,$$

hence

$$k^2\frac{e}{m} = \frac{2.721\times10^{-4}}{0.78\times10^{-14}}\times\frac{3.14\times10^7}{0.87\times10^7} = 1.3\times10^{11} \text{ coulombs/kg.}$$

Thus the value of e/m is 1.76×10^{11} C-kg^{-1}. Now $k = 1$ and the oscillators which determine the ultraviolet absorption are the electrons. One knows that the ultraviolet transitions cause the electronic energy of the molecule to vary.

III

When one only retains the first two terms in (2), one has

$$n^2 = 1+\frac{N_1 q_1^2}{4\pi^2\varepsilon_0 m_1}\times\frac{1}{v_1^2-v^2}+\frac{N_2 q_2^2}{4\pi^2\varepsilon_0 m_2}\times\frac{1}{v_2^2-v^2}. \tag{8}$$

Replacing v by c/λ this becomes

$$n^2 = 1+\frac{N_1 q_1^2}{4\pi^2\varepsilon_0 c^2 m_1}\times\frac{\lambda^2\lambda_1^2}{\lambda^2-\lambda_1^2}+\frac{N_2 q_2^2}{4\pi^2\varepsilon_0 c^2 m_2}\times\frac{\lambda^2\lambda_2^2}{\lambda^2-\lambda_2^2},$$

$$n^2 = 1+C_1\frac{\lambda^2\lambda_1^2}{\lambda^2-\lambda_1^2}+C_2\frac{\lambda^2\lambda_2^2}{\lambda^2-\lambda_2^2} = 1+C_1\frac{\lambda_1^2(\lambda^2-\lambda_1^2+\lambda_1^2)}{\lambda^2-\lambda_1^2}+C_2\frac{\lambda_2^2(\lambda^2-\lambda_2^2+\lambda_2^2)}{\lambda^2-\lambda_2^2},$$

$$n^2 = 1+C_1\lambda_1^2+C_2\lambda_2^2+\frac{C_1\lambda_1^4}{\lambda^2-\lambda_1^2}+\frac{C_2\lambda_2^4}{\lambda^2-\lambda_2^2}.$$

Thus

$$A = 1+C_1\lambda_1^2+C_2\lambda_2^2, \quad C_1 = \frac{N_1 q_1^2}{4\pi^2\varepsilon_0 c^2 m_1}, \quad C_2 = \frac{N_2 q_2^2}{4\pi^2\varepsilon_0 c^2 m_2}.$$

Comparison with the empirical expression (5) gives

$$\lambda_1 = \sqrt{88.8\times10^{-16}} = 9.42\times10^{-8} \text{ m} = 942 \text{ Å (ultraviolet),}$$

$$\lambda_2 = \sqrt{12.6\times10^{-10}} = 3.55\times10^{-5} \text{ m} = 35 \ \mu \text{ (infrared),}$$

$$\frac{C_2}{C_1} = \frac{C_2\lambda_2^2}{C_1\lambda_1^2}\times\frac{\lambda_1^4}{\lambda_2^4} = \frac{5.10\times10^{-9}}{6.12\times10^{-15}}\left(\frac{8.88\times10^{-15}}{1.26\times10^{-9}}\right)^2 = 4.15\times10^{-5}\frac{N_2 q_2^2}{N_1 q_1^2}\times\frac{m_1}{m_2}$$

The first term of the dispersion expression is the electron term and $q_1 = e$ and $m_1 = m_e$. On the other hand, $q_2 = e$ (the F$^-$ ion is monovalent) and $m_2 = 19m_H$. There are two valence electrons which are both optical electrons and two F$^-$ ions. One takes $N_1 = N_2$. Hence

$$\frac{m_H}{m_e} = \frac{1}{19\times4.15\times10^{-5}} = 1270.$$

This value is of the order of magnitude of the ratio 1830. This kind of calculation with many approximations played an important role at the turn of the century in suggesting the atomic origin of the oscillators active in the infrared.

IV

(a) In the X-ray domain, ν is very large (of the order of 10^{18}) and λ is very small (of the order of several Å). Thus $\nu \gg \nu_1 \gg \nu_2$. On the other hand, the masses m_2 of the ions are about 2000 times greater than those of the electrons while in (5) N_1 is of the same order as N_2 and q_1 as q_2. Therefore, the latter term in this expression is negligible. One has then

$$n^2 \approx 1 - \frac{Ne^2}{4\pi^2\varepsilon_0 m_e c^2} = 1 - \frac{Ne^2\lambda^2}{4\pi^2\varepsilon_0 c^2 m_e} = 1 - K\lambda^2,$$

taking

$$K = \frac{Ne^2}{4\pi^2\varepsilon_0 c^2 m_e}.$$

N is the total number of electrons per unit volume since here ν is greater than all the electron eigenfrequencies. Each atom has Z electrons and one has

$$Ne = \mathcal{N}eZ\frac{\mu}{A}.$$

For copper:

$$Ne = 9.65 \times 10^7 \times 29 \times \frac{8 \times 10^3}{63} = 3.5 \times 10^{11},$$

hence

$$K = 3.5 \times 10^{11} \times 1.76 \times 10^{11} \times \frac{1}{3.14 \times 10^7} = 2 \times 10^{15}.$$

(b) By definition:

$$v_\phi = \frac{c}{n} = \frac{c}{\sqrt{1 - K\lambda^2}},$$

$$v_g = \frac{\partial \nu}{\partial\left(\dfrac{\nu}{v_\phi}\right)} = \frac{\partial \nu}{\partial\left(\dfrac{n\nu}{c}\right)} = \frac{\partial\left(\dfrac{c}{\lambda}\right)}{\partial\left(\dfrac{n}{\lambda}\right)}. \qquad (9)$$

One finds from (7)

$$\left(\frac{n}{\lambda}\right)^2 + K = \left(\frac{1}{\lambda}\right)^2.$$

Hence

$$v_g = nc.$$

Then one has

$$v_\phi \times v_g = c^2. \qquad (10)$$

In general for (10) to be satisfied it is necessary for

$$v_g = nc, \quad \text{since} \quad v_\phi = c/n.$$

But, using (9),

$$v_g = \frac{d\left(\dfrac{c}{\lambda}\right)}{d\left(\dfrac{n}{\lambda}\right)} = \frac{-\dfrac{c}{\lambda^2}}{\dfrac{1}{\lambda}\dfrac{dn}{d\lambda} - \dfrac{n}{\lambda^2}} = \frac{c}{n - \lambda\dfrac{dn}{d\lambda}}$$

thus

$$n - \lambda\frac{dn}{d\lambda} = \frac{c}{nc} = \frac{1}{n}$$

where

$$\frac{2n\,dn}{n^2 - 1} = \frac{2\,d\lambda}{\lambda}.$$

One finds then

$$n^2 - 1 = \text{const.} \times \lambda^2.$$

which is the dispersion law (7).

PROBLEM 49

Dispersion in a Region of Weak Absorption

1. Starting with equation (7) from Problem 51 giving the complex amplitude of the electric dipole moment induced by an electromagnetic wave on a classical model of the atom, find the expression giving the complex index $\mathbf{n} = n - jk$ in a gaseous medium containing N atoms per unit volume.

2. From the preceding expression derive an expression for the index of refraction n and the absorption index k in the case where the following assumptions can be made: the absorption is very weak so that one can neglect k in finding n; the region with absorption is very narrow so that one can take

$$\omega_0^2 - \omega^2 = 2\omega_0(\omega_0 - \omega),$$

relating ω and ω_0 when their difference is not a major factor; take $n = 1$ in the product nk.

3. Graph the variations in n and k in this case using the ratio

$$u = \frac{2}{g}(\omega_0 - \omega).$$

as the variable.

4. The gas above is crossed by a parallel beam of polychromatic light carrying the energy flux per unit area $\Phi_0 = \int \Phi_\omega \, d\omega$ where the interval $d\omega$ entirely covers the region of spectral absorption. Assuming that the monochromatic flux Φ_ω has the same value for all the radiation, find the flux absorbed $-d\Phi_0$ in passing through a thickness dx of the gas as a function of the absorption index k found in question 3 and the radiant energy density $w(\omega)$. Recall that

$$\int_{-\infty}^{+\infty} \frac{du}{u^2+1} = \pi.$$

What is the expression giving the absorbed flux if the energy density is expressed as a function of the frequency ν rather than the angular frequency ω?

SOLUTION

1. The starting expression for an oscillator with charge q and mass m is

$$p = \frac{q^2 E}{m(\omega_0^2-\omega^2+jg\omega)}. \tag{1}$$

The polarization p of a gas formed of such atoms is

$$P = Np$$

and the complex index

$$n^2-1 = \varepsilon_r-1 = \frac{P}{\varepsilon_0 E} = \frac{Nq^2}{\varepsilon_0 m(\omega_0^2-\omega^2+jg\omega)}. \tag{2}$$

Separating the real and imaginary parts of $n = n-jk$, this becomes

$$n^2-k^2 = 1+\frac{Nq^2}{\varepsilon_0 m}\frac{\omega_0^2-\omega^2}{(\omega_0^2-\omega^2)^2+g^2\omega^2} \tag{3}$$

and

$$2nk = \frac{Nq^2}{\varepsilon_0 m}\frac{g\omega}{(\omega_0^2-\omega^2)^2+g^2\omega^2}. \tag{4}$$

2. The approximations indicated in the problem statement give

$$n-1 = \frac{Nq^2}{2\varepsilon_0 m}\times\frac{2\omega_0(\omega_0-\omega)}{4\omega_0^2(\omega_0-\omega)^2+g^2\omega_0^2} = \frac{Nq^2}{2\varepsilon_0 m\omega_0 g}\times\frac{u}{u^2+1} = \frac{Cu}{u^2+1}, \tag{5}$$

$$k = \frac{Nq^2}{2\varepsilon_0 m}\times\frac{g\omega}{4\omega_0^2(\omega_0-\omega)^2+g^2\omega_0^2} = \frac{Nq^2}{2\varepsilon_0 m\omega_0 g}\times\frac{1}{u^2+1} = \frac{C}{u^2+1}, \tag{6}$$

calling C the constant

$$\frac{Nq^2}{2\varepsilon_0 m\omega_0 g}.$$

3. The curves representing $n-1$ and k as a function of u are given in Fig. 49.1. For $u = 0$ ($\omega = \omega_0$), $n-1$ vanishes and k passes through a maximum equal to C. For $u = \pm 1$ ($|\omega_0-\omega|$

$= g/2$), $n-1$ passes through maxima (in absolute value) equal to $C/2$ at the same time as k has the value $C/2$. The interval $2\,|\omega_0-\omega| = g$ is the width of the absorption line.

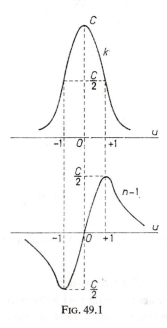

FIG. 49.1

4. The absorption law for a beam of monochromatic light is:

$$-\frac{\mathrm{d}\Phi_m}{\mathrm{d}x} = 2K\Phi_\omega,$$

where K is the absorption coefficient related to the absorption index by $K = k\omega/c$ (§ 2.5). Throughout the absorption region the flux loss is

$$-\frac{\mathrm{d}\Phi_0}{\mathrm{d}x} = \frac{2\Phi_0}{c}\int k\omega\,\mathrm{d}\omega. \tag{7}$$

Replacing k in this expression by (6), ω by ω_0, and $\mathrm{d}\omega$ by $-g/2\mathrm{d}u$, it becomes

$$\frac{\mathrm{d}\Phi_0}{\mathrm{d}x} = \frac{\Phi_0 Nq^2}{2\varepsilon_0 mc}\int \frac{\mathrm{d}u}{u^2+1}.$$

The integral measuring the area lying between the curve $k(u)$ and the abscissa in Fig. 49.1, is found to be arctan u. It has about the same value in the spectral interval where the absorption is significant and in the interval $-\infty$ to $+\infty$. With the latter limits it has the value π, hence

$$\frac{\mathrm{d}\Phi_0}{\mathrm{d}x} = \frac{\Phi_0 N\pi q^2}{2\varepsilon_0 mc}. \tag{8}$$

The flux carried by a parallel beam is related to the energy density by $\Phi_0 = cw$ (c being the velocity of the electromagnetic waves in the medium where n is presumed close to unity):

15 R & M: PIA

Now w can be expressed as a function of ω, with Φ_0 from (7), (8) now becomes

$$\frac{\mathrm{d}\Phi_0}{\mathrm{d}x} = \frac{N\pi q^2}{2\varepsilon_0 m}\, w(\omega).$$

If one takes $v = \omega/2\pi$ as the variable with $\Phi_0(\omega)\,\mathrm{d}\omega = \Phi_0(v)\,\mathrm{d}v$

$$\Phi_0(\omega) = \frac{1}{2\pi}\Phi_0(v) = \frac{c}{2\pi}w(v)$$

and

$$\frac{\mathrm{d}\Phi_0}{\mathrm{d}x} = \frac{Nq^2}{4\varepsilon_0 m}\, w(v). \tag{9}$$

PROBLEM 50

Band Spectra. Anomalous Dispersion of a Vapour

A Michelson interferometer is adjusted in such a way that the image M_2' of mirror M_2 given by the beam splitter Sp makes an angle α with the mirror M_1 and the distance between M_1 and M_2 (image of M_2 in Sp) measured along AIA' has a given value $AB = e_0$. A lens L produces on P an image of M with unit magnification (Fig. 50.1).

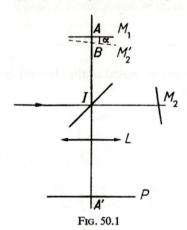

Fig. 50.1

1. The interferometer is illuminated with light of wavelength λ. Explain why one sees linear equidistant fringes on the plane P. Given that one finds 250 fringes in a distance of 5 cm, give the fringe spacing i and the angle α for $\lambda = 3009.14$ Å.

2. The monochromatic source is now replaced by a continuous source. Given that e_0 is of the order of 1 mm, what does one see on the plane P?

3. Plane P contains the entry slit of a spectrograph. The slit is extremely fine and parallel to the fringes observed in monochromatic light. It is situated at the point A', the image of

A given by L. What does one see on a photographic plate when the interferometer is illuminated with a continuous spectral source? For this question and those that follow, make the following simplifying assumptions:

the dispersion of the spectrograph is linear in wave number in the region under study; one will locate the positions of maximum illumination on the plate by the corresponding wave numbers σ;

the magnification, normal to the direction of dispersion, is unity.

Find e_0, knowing that between $\lambda = 3009.14$ Å and $\lambda_2 = 3034.12$ Å, one observes 50.2 times the distance between two consecutive illumination maxima.

4. The spectrograph is inclined so that the entry slit, which will always pass through A', makes an angle θ with the direction of the monochromatic fringes. The position of a point M on the slit is determined by its distance $y = A'M$ from A' (Fig. 50.2). On the figure the

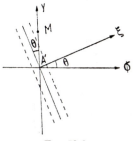

FIG. 50.2

fringes are vertical and the interference order increases from left to right. Find the path difference for rays interfering at M as a function of y, e_0, θ, σ and i, the fringe spacing for radiation σ in the plane P. What does one observe in the plane of the photographic plate when the source is continuous over an interval such that one can neglect the variation of i with σ.

5. In each arm of the interferometer is placed a sealed cylindrical cell of length d with transparent parallel windows and set parallel to the arms. The two cells are identical except that one is evacuated while the other contains heated tin vapour. The positioning of the cells is such that the path difference for the rays which interfere is lessened in absolute value for radiation in which the index of the gas is greater than one.

Recall that the dispersion of the index of a gas is given by Sellmeier's equation, which, in the case of an isolated absorption line and for low pressure, can be written

$$n - 1 = \frac{-B\lambda_0}{\tilde{\nu} - \tilde{\nu}_0}$$

where $\lambda_0 \, (= 1/\tilde{\nu}_0)$ is the wavelength of the absorption line and B is a quantity related to the number of atoms per unit volume and the the charge and mass of the electron. (This expression is not valid for $\tilde{\nu} = \tilde{\nu}_0$ where $\tilde{\nu}$ is very close to $\tilde{\nu}_0$.)

The interferometer is always illuminated by a continuous spectral source.

(a) What happens to the expression derived in the previous question?

15*

(b) What does one find on the photographic plate about $\tilde{v}_0$? Show that each fringe generates two points on a horizontal tangent (parallel to the dispersion direction of the spectrograph). It will be useful to make the change of variables $x = \tilde{v} - \tilde{v}_0$.

(c) Find the values of x and y at these points. Find their height separation Δy and express it as a function of i, B, d, and e_0.

Numerical application:

$$\sin \theta = 0.1$$
$$d \quad = 20 \text{ cm}$$
$$B \quad = 600 \text{ m} \quad \text{for} \quad \lambda_0 = 3009.14 \text{ Å}.$$

SOLUTION

1. One has the localized fringes of a wedge of air (§ 6.5). The fringe spacing in monochromatic light is

$$i = \frac{50}{250} = 0.2 \text{ mm}.$$

Additionally, one has

$$i = \frac{\lambda}{2\alpha}, \quad \text{hence} \quad \alpha = \frac{\lambda}{2i} = \frac{3009.14}{2 \times 0.2 \times 10^7} \simeq \frac{3 \times 10^{-4}}{0.4} \text{ rad} = 2.5'. \tag{1}$$

2. *Observation in white light.* The interference order

$$p \simeq \frac{2e_0}{\lambda} \approx 4000$$

is high. One sees white.

3. *Spectrograph slit parallel to the fringes.* One sees a channelled spectrum having about fifty bright bands between 3009 and 3034 Å.

The path difference on the axis of the interferometer is such that

$$\delta = 2e_0 = k_1\lambda_1 = k_2\lambda_2 = (k_1 - 50.2)\lambda_2, \tag{2}$$

hence

$$k_1 = \frac{50.2\lambda_2}{\Delta\lambda} = 50.2 \times \frac{3 \times 10^3}{24.98}$$

and

$$e_0 = \frac{1}{2} k_1\lambda_1 = \frac{50.2}{2} \times \frac{9 \times 10^6}{24.98} = \frac{45.18}{49.96} \times 10^7 \text{ Å},$$

$$e_0 = 0.94 \text{ mm}.$$

4. *Slit of the spectrograph inclined at an angle θ to the fringes.* Let $A'\xi$ be an axis normal to the direction of the fringes. The path difference of a point on the abscissa ξ is

$$\delta_1 = 2(e_0 + \alpha\xi).$$

Replacing α by the expression obtained in the first question and ξ by $y \sin \theta$, one has

$$\delta_1 = 2\left(e_0 + \frac{\lambda}{2i} y \sin \theta\right) = 2e_0 + \frac{y \sin \theta}{i\tilde{\nu}}.$$

Thus, the path difference at M is

$$\delta_1 = 2e_0 + \frac{y \sin \theta}{i\tilde{\nu}}. \tag{3}$$

If the spectral interval is small, one can take i as a constant. One gets constructive interference for

$$\delta_1 = k\lambda = k/\tilde{\nu} \quad (k \text{ integer}). \tag{4}$$

The dispersion of the spectrograph is linear in wave number (take the coefficient of linearity equal to 1). Combining (3) and (4) allows one to write by taking $i/\sin \theta = \text{const.} = C$

$$y = C(k - 2e_0\tilde{\nu}). \tag{5}$$

This is the equation of a straight line with slope $-2Ce_0$. The y intercept is $y = Ck$ and the intersection with $A'\tilde{\nu}$ is equal to $\tilde{\nu} = k/2e_0$. The bright bands are line segments (Fig. 50.3). The interference order at A' for the wavelength λ_0 is denoted k_0.

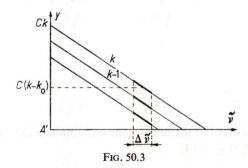

FIG. 50.3

5. *Dispersion in tin vapour.* (a) The cell leads to an additional path difference equal to $2(n-1)d$. Hence, the new path difference on M (taking into account the sign conventions of the text) is

$$\delta_2 = 2\left[e_0 - (n-1)d + \frac{\lambda}{2i} y \sin \theta\right]$$

Sellmeier's equation gives

$$n - 1 = -\frac{B\lambda_0}{\tilde{\nu} - \tilde{\nu}_0}. \tag{6}$$

In the spectral interval where i can be considered constant, one finds constructive interference for

$$\frac{k}{\tilde{\nu}} = 2\left(e_0 + \frac{B\lambda_0 d}{\tilde{\nu} - \tilde{\nu}_0}\right) + \frac{y}{C} \frac{1}{\tilde{\nu}}.$$

Taking $\tilde{v} - \tilde{v}_0 = x$, this becomes

$$y = C\left\{(k - 2\sigma_0\, e_0 - 2B\lambda_0 d) - 2\left(xe_0 + \frac{Bd}{x}\right)\right\}. \tag{7}$$

(b) This equation allows one to describe the appearance of a band of order k in the spectral interval about $\tilde{v}_0$ (but for $\tilde{v} \neq \tilde{v}_0$).

The points P_1 and P_2 with horizontal tangent are given by

$$\frac{dy}{dx} = 0, \quad \text{so that} \quad e_0 - \frac{Bd}{x^2} = 0 \quad \text{or} \quad x = \pm\sqrt{\frac{Bd}{e_0}}.$$

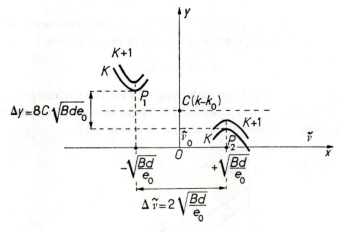

FIG. 50.4

In substituting the values of x in equation (7), one has obtained the slant separation between the two points P_1 and P_2 (Fig. 50.4)

$$\Delta y = 4C\left(e_0\sqrt{\frac{Bd}{e_0}} + \sqrt{\frac{e_0}{Bd}} \times \sqrt{Bd}\right) = 8C\sqrt{Bde_0}.$$

Numerical application:

$$x = \pm\sqrt{\frac{Bd}{e_0}} = \pm\sqrt{\frac{600 \times 0.2}{10^{-3}}} = \pm\sqrt{12} \times 10^2 = \pm 347 \text{ m}^{-1},$$

$$x = \pm 3.47 \text{ cm}^{-1}$$

$$\Delta y = 8 \times \frac{0.2}{0.1}\sqrt{600 \times 0.2 \times 10^{-3}} = 16\sqrt{12} \times 10^{-1} = 5.5 \text{ mm}.$$

In equation (7), when $x \to \pm\sqrt{Bd/e_0}$, the term

$$2B\lambda_0 d = 2 \times 600 \times 0.3 \times 10^{-6} \times 0.2 = 72 \times 10^{-6}$$

is negligible with respect to $2\sqrt{Bde_0} = 2\sqrt{12} \times 10^{-1}$. Near points P_1 and P_2 the portions of the bright band of order k have the equation:

$$y = C\left[(k-k_0)-2\left(xe_0+\frac{Bd}{x}\right)\right]. \tag{8}$$

For other orders the bright fringes are resolved from one another by a translation parallel to Oy (Fig. 50.4). Figure 50.5 shows a picture taken of bands observed for a gas having many absorption lines.

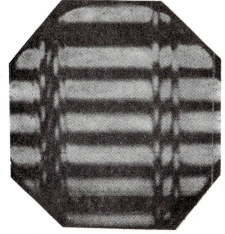

FIG. 50.5

PROBLEM 51

Scattering of Electromagnetic Radiation Using a Classical Atomic Model

Consider an atomic model involving an electron of mass m_e and charge e elastically bound to the atom and capable of undergoing harmonic vibration. Let ω_0 be its natural angular frequency.

I

When this oscillator is put in free vibration it radiates energy. The energy loss per unit time is given by the expression:

$$\frac{dW}{dt} = -\frac{1}{12\pi\varepsilon_0 c^3}\langle \ddot{p}^2\rangle,$$

W being the energy, t the time, c the velocity of light, p the electric dipole moment of the oscillator, and $\langle \ddot{p}^2\rangle$ the mean value of the square of the second derivative of p with respect to time. The energy loss induces a damping of the oscillation. Find the damping coefficient g resulting from the oscillation radiation.

II

1. N identical, independent atoms per unit volume, described by the model above, are placed in a parallel radiation field of angular frequency ω where the electric field oscillates along the Oy direction. The incident radiation propagates in the Ox direction. Under the influence of this field the atomic oscillators undergo forced oscillation. Calculate the atomic electric dipole moment.

The atomic oscillators in forced oscillation themselves radiate. Find the expression giving the total flux scattered by a unit volume of the material as a function of the incident radiant energy.

2. Consider the case where $\omega_0^2 \gg \omega^2$ or $\lambda_0^2 \ll \lambda^2$. Show that the intensity of the scattered light is proportional to λ^{-4} where λ is the wavelength of the incident and scattered radiation. Find the ratio of the intensity of the incident light to that of the scattered light per unit volume. Assume

$$N = 10^{28} \text{ m}^{-3}, \quad \lambda_0/\lambda = 0.1.$$

Also find the ratio of the scattered intensities in the red for $\lambda = 7000$ Å and in the violet for $\lambda = 4000$ Å assuming that the incident radiation intensity is the same in each case. Discuss these results and show that they explain the blue colour of the sky and the red colour of the setting sun.

3. Consider finally the case $\omega_0^2 \ll \omega_2$, useful in the X-ray region. Find the expression for radiation scattering in this case. Calculate the ratio of scattered to incident radiation intensity here taking $N = 10^{22} \text{ cm}^{-1}$.

Compare this ratio for $^{64}_{29}\text{Cu}$ and $^{207}_{82}\text{Pb}$ assuming the number of atoms per cm^3 in the first case is 8×10^{22} and 3×10^{22} for the second. All of the electrons in the atom are assumed to participate in X-ray scattering.

III

1. Consider now conductors with conductivity γ. Assuming the medium to be continuous, write Maxwell's equations for this case. Assuming the electric field to vary sinusoidally with angular frequency ω and to be propagating in the Ox direction, find the real and imaginary parts of the complex permittivity.

2. Establish the dispersion relationship for a metal. In this case one can assume that the electrons giving rise to the optical properties are free so that $\omega_0 = 0$. Assume the damping coefficient g to be different from zero.

To determine the expressions giving the real and imaginary parts of the complex permittivity.

(a) Examine the case $\omega \gg g$. Find the real and imaginary parts of the permittivity. Compare this with the results obtained in III.1 and find the damping coefficient as a function of γ. What is noteworthy about this expression?

(b) Finally, look at the case $\omega \ll g$. Discuss the dispersion expression in this case. Find the wavelength λ_0 for which the complex permittivity vanishes. Discuss the behaviour of the substance for $\lambda > \lambda_0$ and for $\lambda < \lambda_0$.

3. In the case III.2 (b) express the complex permittivity as a function of λ_0 and λ and calculate the reflection factor R of the substance for $(\lambda/\lambda_0)^2 \ll 1$ and for $(\lambda/\lambda_0)^2 \gg 1$.

4. Find the oscillator intensities which arise in the calculations above for Cs, Rb, K, Na, and Li given the respective values of λ_0 are 4400, 3600, 3150, 2100, and 2050 Å, and the number of free electrons/m³ as 0.85×10^{28}, 1.1×10^{28}, 1.3×10^{28}, 2.5×10^{28}, and 4.5×10^{28}, respectively.

SOLUTION

I

The oscillator has an electric dipole moment $p = es$ where s is the amplitude of the electron motion. Since the motion is harmonic, one has $\ddot{p} = -\omega_0^2 p$. Since ω_0 is of the order of 10^{15} sec^{-1}, one can take the mean value $\langle \ddot{p} \rangle^2 = \frac{1}{2} \ddot{p}^2$ in the expression for the energy lost per unit time

$$\frac{\mathrm{d}W}{\mathrm{d}t} = -\frac{\omega_0^4 e^2 s_m^2}{12\pi\varepsilon_0 c^3}. \tag{1}$$

The energy of the oscillator is

$$W = \tfrac{1}{2} m_e \omega_0^2 s_m^2 \tag{2}$$

and (1) becomes

$$\frac{\mathrm{d}W}{\mathrm{d}t} = -\frac{\omega_0^2 e^2}{6\pi\varepsilon_0 c^3 m_e} W. \tag{3}$$

Thus, the energy decrease is exponential in time

$$W = W_0 \exp\left(-\frac{t}{\tau}\right),$$

$$\tau = \frac{6\pi\varepsilon_0 c^3 m_e}{\omega_0^2 e^2}.$$

τ is the time constant. After time τ, the energy is reduced to the fraction $1/e = 0.368$ of its initial value W_0. For visible radiation ($\omega_0 \approx 10^{15}$), this time is of the order of 10^{-7} sec. It therefore contains a large number of periods. The motion of the oscillator is not sinusoidal but rather sinusoidal with an exponential decay. It can be thought of as the solution of the equation

$$m_e\left(\frac{\mathrm{d}^2 s}{\mathrm{d}t^2} + g\frac{\mathrm{d}s}{\mathrm{d}t} + \frac{\omega_0^2 s}{m_e}\right) = 0 \tag{4}$$

where an artificial frictional force $m_e g(\mathrm{d}s/\mathrm{d}t)$, proportional to the speed, has been introduced. g is the damping coefficient which is required. To find it we equate the loss in energy (3) to the work done by the artificial force over unit time

$$m_e g \frac{\mathrm{d}s}{\mathrm{d}t} \times \frac{\mathrm{d}s}{\mathrm{d}t} = m_e g s_m^2 \omega_0^2 \cos^2 \omega t.$$

ω is the pseudo angular frequency and $\omega^2 = \omega_0^2 - g/2$. Since the motion is only slightly damped, one finds, quite reasonably $\omega = \omega_0$.

The mean value of this work, which acts over a large number of pseudo-periods is

$$\tfrac{1}{2} m_e g s_m^2 \omega_0^2 = gW$$

hence, by equating to (3),

$$g = 1/\tau. \tag{5}$$

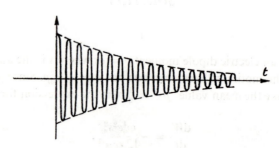

FIG. 51.1

II

1. Under the influence of the field $E_y = E_m \sin \omega t$, the electron is subjected to a force eE_y and undergoes forced movement with its equation, derived from that of the natural motion (4), being

$$m_e\left(\frac{d^2y}{dt^2} + g\,\frac{dy}{dt} + \omega_0^2 y\right) = +eE. \tag{6}$$

Corresponding to the sinusoidal field E_y take the complex field $E_m \exp(j\omega t)$ and to the displacement y the complex function $y = y_m \exp(j\phi) \exp(j\omega t)$ with the same angular frequency as the imposed field. Equation (6) yields

$$\mathbf{y}_m = \frac{-eE_m}{m_e(\omega_0^2 - \omega^2 + jg\omega)}$$

and the moment induced has as its complex amplitude

$$\mathbf{p}_m = -e\mathbf{y}_m = \frac{e^2 E_m}{m_e(\omega_0^2 - \omega^2 + jg\omega)} \tag{7}$$

where the modulus is

$$p_m = \frac{e^2 E_m}{m_e \sqrt{(\omega_0^2 - \omega^2)^2 + g^2\omega^2}}.$$

The radiation from the sinusoidal moment p is equivalent to that of a Hertzian dipole. The mean flux which it radiates into all space is given by (§§ 10.3 and 17.5)

$$\langle \Phi \rangle = \frac{\omega^4 p_m^2}{12\pi\varepsilon_0 c^3} = \frac{\omega^4 e^4 E_m^2}{12\pi\varepsilon_0 c^3 m_e^2[(\omega_0^2 - \omega^2)^2 + g^2\omega^2]}.$$

If the atoms are independent as in the case of an ideal gas, their radiations are incoherent and the mean flux radiated per unit volume is $N\langle\Phi\rangle$. On the other hand, the energy of the incident monochromatic radiation which crosses a unit surface per unit time is given by (§ 2.3)

$$\langle\mathscr{E}\rangle = \tfrac{1}{2}\,\varepsilon_0 c E_m^2.$$

One finds then

$$\frac{N\langle\Phi\rangle}{\langle\mathscr{E}\rangle} = \frac{N\omega^4 e^4}{6\pi\varepsilon_0^2 c^4 m_e^2[(\omega_0^2-\omega^2)^2+g^2\omega^2]}\,. \tag{8}$$

2. If $\omega_0^2 \gg \omega^2$, expression (8) reduces to

$$\frac{N\langle\Phi\rangle}{\langle\mathscr{E}\rangle} = \frac{N\omega^4 e^4}{6\pi_0^2\varepsilon c^4 m_e^2\omega_0^4} = \frac{16\pi^3 e^4 N}{6\varepsilon_0^2 m_e^2 \omega_0^4}\times\frac{1}{\lambda^4}, \tag{9}$$

since $\omega = 2\pi c/\lambda$. One sees that the scattering intensity, all other things being equal, is proportional to λ^{-4}.

For $N = 10^{28}$ m^{-3}, $\omega/\omega_0 = 0.1$, and the ratio (8) has the value

$$\frac{N\langle\Phi\rangle}{\langle\mathscr{E}\rangle} = \frac{10^{28}\times(1.60\times10^{-19})^4\times10^{-4}\times(36\pi\times10^9)^2}{6\pi\times(3\times10^8)^4\times(9.1\times10^{-31})^2} = 6.63\times10^{-5}$$

The ratio of the scattered intensities in the blue and red is equal to $(\lambda_R/\lambda_B)^4$ if $\langle\varepsilon\rangle$ has the same value for both so that

$$\left(\frac{\lambda_R}{\lambda_B}\right)^4 = \left(\frac{7000}{4000}\right)^4 = 9.38.$$

For the atmospheric phenomena involved see § 17.5.

3. If $\omega_0^2 \ll \omega^2$, as is the case for X-rays, since all the atomic electrons, Z in number, participate in the scattering, a unit volume contains NZ oscillators and (7) becomes (§ 10.4)

$$\frac{N\langle\Phi\rangle}{\langle\mathscr{E}\rangle} = \frac{NZe^4}{6\pi\varepsilon_0^2 c^4 m_e^2} = 3.3\times10^{-33}NZ. \tag{10}$$

For $^{64}_{29}$Cu:

$$\frac{N\langle\Phi\rangle}{\langle\mathscr{E}\rangle} = 3.3\times10^{-33}\times8\times10^{28}\times29 = 7.65\times10^{-4}.$$

For $^{207}_{82}$Pb:

$$\frac{N\langle\Phi\rangle}{\langle\mathscr{E}\rangle} = 3.3\times10^{-33}\times3\times10^{28}\times82 = 8.12\times10^{-4}.$$

These two values are nearly equal. If one looks back at (10) for the scattering coefficient per unit mass by dividing by the density ϱ of the element under study, $\varrho = N A/\mathscr{N}$, where A is the atomic mass and $\mathscr{N}$ Avogadro's number, the ratio

$$\frac{N\langle\Phi\rangle}{\varrho\langle\mathscr{E}\rangle} = \frac{\mathscr{U}\langle\Phi\rangle}{A\langle\mathscr{E}\rangle} = 3.3\times10^{-33}\mathscr{U}\,\frac{Z}{A} \tag{11}$$

depends only on the ratio Z/A of the atomic number to the atomic mass of the element considered. This ratio is equal to 0.453 for Cu and 0.395 for Pb. For the light elements it is close to 0.5 so that the mass scattering coefficient for X-rays is constant. This regularity, however, is only true for medium length X-rays (§ 15.14).

III

1. Maxwell's equation for an ohmic conductor are (§ 2.5)

$$\operatorname{curl} H = \gamma E + \varepsilon \frac{\partial E}{\partial t}, \quad \operatorname{curl} E = -\mu_0 \frac{\partial H}{\partial t}.$$

One gets for the equation of propagation of an electric field varying along Oy and propagating in the Ox direction

$$\frac{\partial^2 E}{\partial x^2} = \mu_0 \left(\varepsilon \frac{\partial^2 E}{\partial t^2} + \gamma \frac{\partial E}{\partial t} \right), \tag{12}$$

which has as a solution a sinusoidal function of time. One has then

$$\frac{\partial E}{\partial t} = j\omega E, \quad \frac{\partial^2 E}{\partial t^2} = -\omega^2 E$$

and (12) becomes

$$\frac{\partial^2 E}{\partial x^2} = \mu_0 \left(\varepsilon - j \frac{\gamma}{\omega} \right) \frac{\partial^2 E}{\partial t^2}.$$

This equation has the form of the free-space wave equation

$$\frac{\partial^2 E}{\partial x^2} = \varepsilon_0 \mu_0 \frac{\partial^2 E}{\partial t^2},$$

but the permittivity ε_0 of free space is replaced by the complex permittivity

$$\varepsilon = e - j \frac{\gamma}{\omega} = \varepsilon_0 \left(\varepsilon_r - \frac{j\gamma}{\varepsilon_0 \omega} \right). \tag{13}$$

ε_r is the real part of the complex permittivity and $\gamma/\varepsilon_0\omega$ its imaginary part.

2. Under our assumptions, equation (6) for the motion of an electron in a metal reduces to

$$m_e \left(\frac{d^2 y}{dt^2} + g \frac{dy}{dt} \right) = -eE_m \sin \omega t. \tag{14}$$

Its solution (§ 10.7) is

$$y_m = \frac{-eE}{m_e(-\omega^2 + jg\omega)}.$$

One finds, as in part II.1, the induced moment p, then the electric polarization of the medium $P = Np$, and finally, the relative permittivity by the formula

$$\varepsilon_r = 1 + \frac{P}{\varepsilon_0 E}.$$

One finds

$$\varepsilon_r = 1 + \frac{Ne^2}{\varepsilon_0 m_e \omega} \times \frac{1}{jg - \omega}. \tag{15}$$

This complex permittvity $\varepsilon_r = \varepsilon_r' + j\varepsilon_r''$ has a real part

$$\varepsilon_r' = 1 - \frac{Ne^2}{\varepsilon_0 m_e} \times \frac{1}{g^2 + \omega^2} \tag{16}$$

and an imaginary part

$$\varepsilon_r'' = \frac{Ne^2 g}{\varepsilon_0 m_e \omega} \times \frac{1}{g^2 + \omega^2}. \tag{17}$$

(a) For $g \gg \omega$, expressions (16) and (17) reduce to

$$\varepsilon_r' = 1 - \frac{Ne^2}{\varepsilon_0 m_e g^2},$$

$$\varepsilon_r'' = \frac{Ne^2}{\varepsilon_0 m_e \omega g}.$$

Comparison of this last expression with (13) shows that

$$\frac{\gamma}{e_0 \omega} = \frac{Ne^2}{\varepsilon_0 \omega m_e g}$$

and this allows one to find g as a function of γ

$$g = \frac{Ne^2}{m_e \gamma} \tag{18}$$

a value independent of ω.

(b) For $g \ll \omega$, the dispersion expression (15) becomes

$$\varepsilon_r = 1 - \frac{Ne^2}{\varepsilon_0 m_e \omega^2} = 1 - \frac{Ne^2 \lambda^2}{4\pi^2 c^2 \varepsilon_0 m_e}. \tag{19}$$

This expression vanishes for a wavelength λ_0 such that

$$\lambda_0^2 = \frac{4\pi^2 c^2 \varepsilon_0 m_e}{Ne^2}. \tag{20}$$

For $\lambda > \lambda_0$, the metal has a complex permittivity which represents its absorption properties. For $\lambda < \lambda_0$, the permittivity is real and less than 1. The metal becomes transparent.

3. Expression (19) can be written, taking (20) into account

$$\varepsilon_r = 1 - \frac{\lambda^2}{\lambda_0^2}.$$

The reflection factor at normal incidence is given by

$$R = \left(\frac{\sqrt{\varepsilon_r}-1}{\sqrt{\varepsilon_r}+1} \right)^2 = \left(\frac{\sqrt{\lambda_0^2-\lambda^2}-\lambda_0}{\sqrt{\lambda_0^2-\lambda^2}+\lambda_0} \right)^2.$$

For $\lambda^2 \ll \lambda_0^2$:

$$R \simeq 0$$

For $\lambda^2 \gg \lambda_0^2$:

$$R \simeq 1.$$

This last expression is good when applied to a strongly absorbing body.

4. The oscillator intensity (or oscillator strength) f is the number by which it is necessary to multiply the theoretical terms in the dispersion relationship to make it agree with the experimental values (§ 10.10). Expression (19) becomes

$$\varepsilon_r = 1 - \frac{Ne^2\lambda^2 f}{4\pi^2 c^2 \varepsilon_0 m_e}$$

and equation (20) gives

$$f = \frac{1}{N\lambda_0^2} \times \frac{4\pi^2 c^2 \varepsilon_0 m_e}{e^2}. \qquad (21)$$

Now, one has

$$\frac{4\pi^2 c^2 \varepsilon_0 m_e}{e^2} = \frac{4\pi^2 \times 9 \times 10^{-16} \times 9.1 \times 10^{-31}}{4\pi \times 9 \times 10^9 \times (1.60 \times 10^{-19})^2} = 11.15 \times 10^{14}.$$

The data given in the problem lets one calculate the products $N\lambda_0^2$, and hence the values of using f (21). One finds:

Metal:	Cs	Rb	K	Na	Li
$N\lambda_0^2 \times 10^{-14}$:	16.45	14.25	12.89	11.02	18.90
f:	0.68	0.78	0.86_5	1.02	0.59

PROBLEM 52

Dispersion and Reflection of an Ionic Crystal in the Infrared

Consider here a binary crystal of the NaCl type formed of Na^+ and Cl^- ions situated at the nodes of a cubic lattice which is to be thought of as being infinite (Fig. 52.1). Thermal agitation produces, among other things, the vibration in which the two partial lattices of ions are displaced as a unit with respect to one another. Assume that the interaction of two neighbouring ions (action over a short distance) can be represented near the equilibrium configuration by an elastic restoring force with coefficient k_0. The relative displacements of the ions create an electric dipole moment on each mesh of the lattice. The crystal acquires a homogeneous polarization P and electric field E uniform over the interior of the crystal as a result and this subjects each ion to an additional force (long-distance force).

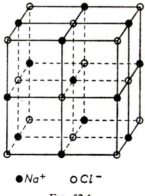

$\bullet Na^+$ $\circ Cl^-$

FIG. 52.1

I

1. Show that if u is the relative displacement of the positive and negative ions whose charge is $\pm e$ and reduced mass μ, the equation of motion is

$$\mu \frac{d^2u}{dt^2} = -k_0 u + eE \tag{1}$$

and that the polarization P is given by

$$P = \frac{1}{V}[(\alpha_+ + \alpha_-)E + eu] \tag{2}$$

α_+ and α_- being the polarizabilities of the two ions and V the volume containing a pair of ions.

Recall to begin with, the definition of the electric displacement D in an isotropic medium of relative permittivity ε_r:

$$D = \varepsilon_0 \varepsilon_r E = \varepsilon_0 E + P. \tag{3}$$

2. The solution of equation (1) is a harmonic motion with angular frequency ω. This angular frequency will have the value ω_r if only actions at a short distance exist.

Find the dispersion expression $\varepsilon_r(\omega)$ from equations (1), (2), and (3). Show that this can be written

$$n^2 = \varepsilon_r = \varepsilon_\infty + \frac{\varepsilon_s - \varepsilon_\infty}{1 - (\omega/\omega_t)^2} \tag{4}$$

where ε_s is the relative permittivity in an electrostatic field ($\omega \to 0$) and ε_∞ the relative permittivity for high frequencies ($\omega \gg \omega_t$).

Draw the curve $\varepsilon_r(\omega)$. Let ω_1 be the frequency for which $\varepsilon_r = 0$. Find the ratio ω_1/ω_t as a function of ε_s and ε_∞.

II

Draw the curve representing the reflection factor R of the crystal under normal incidence as a function of ω.

Let ω_r be the frequency for which $R = 0$. Find a relationship between ω_r, ω_t, ε_s, and ε_∞.

Application: for NaCl the experimental values are $\lambda_r = 31 \mu$, $\lambda_t = 61.1 \mu$, $\varepsilon_s = 5.62$, and $\varepsilon_\infty = 2.25$.

With what precision is the preceding relationship satisfied?

III

The experimental curves $\varepsilon_r(\omega)$ and $R(\omega)$ differ significantly from the theoretical curves found in parts I and II. In particular, ε_r remains finite for $\omega = \omega_t$ and one finds a maximum of R which is less than unity for a certain frequency ω_m. To interpret these results, one adds a damping term $-\gamma \, du/dt$ in equation (1) for the motion. What happens with this equation for periodic complex solutions? What happens with (4)? Assume the ratio γ/ω_t small and examine how this modifies the $R(\omega)$ curve.

Havelock has shown that the expression approaches

$$\frac{\omega_m}{\omega_t} = \left(1 + \frac{\varepsilon_0 - \varepsilon_\infty}{6\varepsilon_\infty - 4}\right)^{1/2}. \tag{5}$$

With the aid of this expression, find λ_m for NaCl and compare it to the experimental value of 52μ for NaCl.

SOLUTION

I

1. In the cubic crystal under consideration, the motion clearly has the same equation for displacements along a direction parallel to any one of the axes of the cube, $x'x$, $y'y$, or $z'z$ (Fig. 52.2). As a result of the cubic symmetry of the ion sites, the coefficient k_0 is the same regardless of the displacement direction. The equation of motion of the two ion species is then

$$m_+ \frac{d^2u_+}{dt^2} = -k_0(u_+ - u_-) + eE \quad \text{and} \quad m_- \frac{d^2u_-}{dt^2} = -k_0(u_- - u_+) - eE.$$

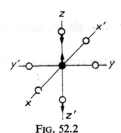

FIG. 52.2

Multiplying these equations by m_- and m_+, then subtracting and dividing by $m_+ + m_-$ one gets equation (1). The dipole moments due to the two species of ions are, respectively,

$$eu_+ + \alpha_+ E \quad \text{and} \quad -eu_- + \alpha_- E.$$

The first term is due to the displacement (§ 10.3.1) of the ion and the second to its deformation. Since one has $1/V$ ion pairs per unit volume, the polarization P is obtained by multiplying the sum of the two dipole moments above by $1/V$ and then one gets equation (2).

2. The harmonic solution of equation (1) is

$$-\omega^2 u = -\frac{k_0}{\mu} u + \frac{e}{\mu} E.$$

If the field E is zero, that is, if one has no long-distance actions, this equation reduces to

$$-\omega^2 u = -\frac{k_0}{\mu} u = -\omega_t^2 E$$

hence

$$\frac{u}{E} = \frac{e}{\mu} \frac{1}{\omega_t^2 - \omega^2}. \tag{6}$$

The permittivity ε_r is obtained from (3), so that, using (2),

$$\varepsilon_r = 1 + \frac{P}{\varepsilon_0 E} = 1 + \frac{\alpha_+ + \alpha_-}{\varepsilon_0 V} + \frac{e}{\varepsilon_0 V} \frac{u}{E}. \tag{7}$$

At optical frequencies, the ions are too heavy to follow the field (§ 17.6) and only react to the first two terms of (7)

$$\varepsilon_\infty = 1 + \frac{\alpha_+ + \alpha_-}{\varepsilon_0 V}. \tag{8}$$

In the electrostatic field $E_s(\omega = 0)$, equation (7) gives

$$\frac{u}{E_0} = \frac{e}{\mu \omega_t^2}$$

hence

$$\varepsilon_s = 1 + \frac{\alpha_+ + \alpha_-}{\varepsilon_0 V} + \frac{e^2}{\varepsilon_0 V} = \varepsilon_\infty + \frac{e^2}{\varepsilon_0 V}. \tag{9}$$

One gets from (7), (8), and (9) the dispersion relation

$$\varepsilon_r = \varepsilon_\infty + \frac{\varepsilon_s - \varepsilon_\infty}{1 - (\omega/\omega_t)^2}. \tag{4}$$

One has $\varepsilon_r = \infty$ for $\omega = \omega_t$. ε_r is positive for $0 < \omega < \omega_t$ (the second term on the right-hand side of (4) is then positive) and for $\omega_l < \omega$, ω_l being the value of ω above which the second term takes a negative value less than ε_∞ in absolute value. ω_l then corresponds to $\varepsilon_r = 0$, so that

$$\omega_l = \omega_t \sqrt{\frac{\varepsilon_s}{\varepsilon_\infty}}. \tag{10}$$

The index $\boldsymbol{n} = \sqrt{\varepsilon_r}$ is therefore complex: $\boldsymbol{n} = n - jk$. n is the index of refraction and k the absorption index. It has real values n below ω_t and above ω_l. Between these limits the values k are purely imaginary (Fig. 52.3).

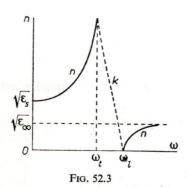

FIG. 52.3

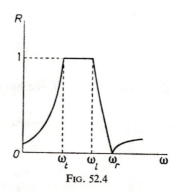

FIG. 52.4

II

The reflection factor under normal incidence is found by using Fresnel's generalized expression (§ 3.5):

$$R = \left(\frac{\boldsymbol{n}-1}{\boldsymbol{n}+1}\right)^2 = \frac{(n-1)^2+k^2}{(n+1)^2+k^2}. \tag{11}$$

Between ω_t and ω_l, where n is purely imaginary, $R = k^2/k^2 = 1$ (Fig. 52.4). It has total reflection (§ 3.3). R vanishes for $n = 1$ or $\varepsilon_r = 1$, so that for a given frequency ω_r, using (4),

$$\left(\frac{\omega_r}{\omega_t}\right)^2 = \frac{\varepsilon_s-1}{\varepsilon_\infty-1}. \tag{12}$$

Application. One has $\lambda = 2\pi c/\omega$, thus

$$\left(\frac{\lambda_t}{\lambda_r}\right)^2 = \left(\frac{\omega_r}{\omega_t}\right)^2 = \left(\frac{61.1}{31}\right)^2 = 3.88.$$

One the other hand, one has

$$\frac{\varepsilon_s-1}{\varepsilon_\infty-1} = \frac{4.62}{1.25} = 3.69.$$

Expression (12) is thus satisfied to within 5%.

III

For sinusoidal solution written in complex form, the equation of motion

$$\mu\, \frac{d^2 u}{dt^2} + \gamma\, \frac{du}{dt} - k_0 u = eE$$

becomes

$$-\omega^2 u = \left(-\omega_t^2 + j\, \frac{\gamma}{\mu}\, \omega\right) u + \frac{e}{\mu}\, E.$$

It is then sufficient to replace ω_t^2 by $\omega_t^2 + j\gamma/\mu\,\omega$ in the calculations of part II of this problem to get in place of (4) the dispersion expression

$$n^2 = (n - jk)^2 = \varepsilon_r = \varepsilon_\infty + \frac{\varepsilon_s - \varepsilon_\infty}{1 - \left(\dfrac{\omega}{\omega_t}\right)^2 + \dfrac{j\gamma}{\mu\omega_t}\left(\dfrac{\omega}{\omega_t}\right)} = \varepsilon_\infty + \frac{A}{\omega_t^2 - \omega^2 + j\, \dfrac{\gamma}{\mu}\, \omega} \tag{13}$$

where $A = (\varepsilon_s - \varepsilon_\infty)\omega_t^2$ represents a constant. By separating the the real and imaginary parts of the complex index in (13), one finds

$$n^2 - k^2 = \varepsilon_\infty + \frac{A(\omega_t^2 - \omega^2)}{(\omega_t^2 - \omega^2)^2 + \dfrac{\gamma^2}{\mu^2}\, \omega^2},$$

$$2nk = \frac{A\gamma\omega}{(\omega_t^2 - \omega^2)^2 + \dfrac{\gamma^2}{\mu^2}\, \omega^2}.$$

One sees that the presence of the term in γ assures that n^2 will always have a finite value and never be purely imaginary. In addition, R is always less than unity. Reflection is never total. R can only attain high values (0.8–0.95) between ω_t and ω_l (Fig. 52.5).

Application: with the given numerical values, equation (5) is written

$$\left(\frac{\omega_m}{\omega_t}\right)^2 = \left(\frac{61.1}{\lambda_m^2}\right)^2 = 1 + \frac{5.62 - 2.25}{6 \times 2.25 - 4} = 1.355$$

hence

$$\lambda_m = \frac{61.1}{\sqrt{1.355}} = 52.5\ \mu.$$

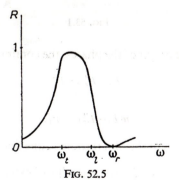

FIG. 52.5

PROBLEM 53

Transmission of an Absorbing Thin Film

Calculate the transmission factor T of an absorbing film with parallel faces, thickness d, and complex index n, placed in air and normally traversed by a plane wave of frequency ω. What happens to the expression for T when the thickness d is much smaller than the wavelength of the incident radiation?

If the dispersion of the index n can be written

$$n^2 = n_0^2 + \frac{A}{\omega^2 - \omega_0^2 + jg\omega}$$

(cf. Problem 49) show that the transmission of a very thin film has a minimum for the frequency ω_0.

SOLUTION

Let E_i be the complex amplitude of the electric field of the incident wave (Fig. 53.1), E_r the reflected field off the first face Σ, E_t the transmitted field, E_r' the reflected field off the second face Σ', and E_t' the field transmitted by the second face.

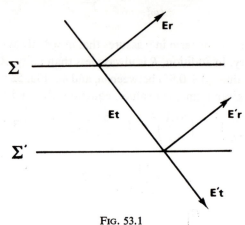

FIG. 53.1

At Σ', where one places the origin of the phases, the continuity conditions for E give

$$E_r + E_r' = E_t' \tag{1}$$

and for $H = (n/c\mu_0)E$

$$n(E_t - E_r') = E_t'. \tag{2}$$

Likewise at Σ, taking into account the film thickness, one finds

$$E_i + E_r = E_t \exp(-j\sigma nd) + E_r' \exp(j\sigma nd) \tag{3}$$

and

$$E_i - E_r = nE_t \exp(-j\sigma nd) - nE'_t \exp(j\sigma nd).$$ (4)

Adding (1) and (2) term by term one finds

$$2E'_t = (1+n)E_t + (1-n)E'_r.$$

By subtracting (2) from (1)

$$(1-n)E_t = -(1+n)E'_r.$$

Adding (3) and (4)

$$2E_i = (1+n)E_t \exp(-j\sigma nd) + (1-n)E'_r \exp(j\sigma nd).$$

Hence

$$t = \frac{E'_t}{E_i} = \frac{(1+n)E_t + (1-n)E'_r}{(1+n)E_t \exp(-j\sigma nd) + (1-n)E'_r \exp(j\sigma nd)}.$$

$$t = \frac{(1+n)^2 - (1-n)^2}{(1+n)^2 \exp(-j\sigma nd) - (1-n)^2 \exp(j\sigma nd)}.$$

If $d \ll \lambda$, $\sigma d = 2\pi d/\lambda \ll 1$. Taking the first two terms of the exponential expansion, this yields

$$t = \frac{4n}{4n - 2j\sigma nd(1+n^2)} = \frac{1}{1 - \dfrac{j\sigma d}{2}(1+n^2)}.$$

The transmission factor is given by

$$T = tt^* = \frac{1}{1 - \dfrac{j\sigma d}{2}(1+n^2)} \times \frac{1}{1 + \dfrac{j\sigma d}{2}(1+n^{*2})} = \frac{1}{1 + \dfrac{j\sigma d}{2}(n^{*2} - n^2)}$$

[ignoring terms in $(\sigma d/2)^2$]. T is minimal when $j\sigma d(n^{*2} - n^2)$ is maximal. Now if

$$n^2 = n_0^2 + \frac{A}{\omega^2 - \omega_0^2 + jg\omega},$$

one has, since $\sigma = \omega/c$,

$$j\sigma(n^{*2} - n^2) = \frac{2gA}{c} \times \frac{\omega^2}{(\omega^2 - \omega_0^2)^2 + g^2\omega^2}.$$

The derivative

$$\frac{d}{d\omega}\left(\frac{\omega^2}{(\omega^2 - \omega_0^2)^2 + g^2\omega^2}\right) = \frac{2\omega(\omega^2 + \omega_0^2)(\omega_0^2 - \omega^2)}{[(\omega^2 - \omega_0^2)^2 + g^2\omega^2]^2}$$

actually vanishes for $\omega = \omega_0$.

PROBLEM 54

Electromagnetic Waves in a Plasma

A gaseous atmosphere is made up of positive and negative ions and contains (per unit volume) N ions with charge $+e$ and mass M and N ions with charge $-e$ and mass m_0.

I

Neglecting the effect of the neutral molecules, one wants:

1. The expression for the relative permittivity of such a medium for electromagnetic waves of frequency v. Since the mass M is many times greater than m_0, show that the role of the positive ions is negligible.

2. The phase velocity v_ϕ of electromagnetic waves of frequency v when the medium has the magnetic permeability of free space. The minimal value of the frequency v_c for which v_ϕ has a real value.

3. The relationship between the velocity v_ϕ, the group velocity v_g of the waves in such a medium and the velocity c of waves propagating in free space.

4. What is the radius of curvature of the trajectories of the electromagnetic waves propagating in a direction perpendicular to the vertical, if one assumes that the gaseous atmosphere undergoes a 6% decrease in the relative value of N for an increase in altitude of 100 metres, the frequency of the waves involved being $v = 2v_c$. Determine the sense of the curvature of the trajectory.

II

1. An infinite plane separates two regions of space where one finds in the first (1) the ionized atmosphere described initially and in the second (2) the same atmosphere but free of ions. Give the value of the energy reflection factor (normal incidence) for the waves of frequency $v = 2v_c$.

2. For the same waves falling on the plane of separation, what is the value of incidence for which one would obtain reflected waves with their electric field vibrating perpendicular to the plane of incidence when the field of the incident wave has some given orientation. Treat both cases where the waves propagate in the sense (1) to (2) and (2) to (1).

Numerical values: $N = 1.226 \times 10^{12}$ m^{-3} (one is also given the values of e, m_e, and c).

SOLUTION

I

1. The equation of motion of the ions is (§ 10.4)

$$m_0 \frac{d^2s}{dt^2} = -eE_m \sin \omega t, \qquad M \frac{d^2S}{dt^2} = eE_m \sin \omega t,$$

$$s = \frac{eE_m}{m_0 \omega^2} \sin \omega t, \qquad S = \frac{eE_m}{M \omega^2} \sin \omega t,$$

s and S have opposite directions, however, $S/s = m_0/M$, and therefore, S is negligible with respect to s if m_0 is negligible with respect to M.

The polarization and the permittivity are then given by (§ 10.6)

$$\varepsilon_r = 1 - \frac{Ne^2}{\varepsilon_0 m_0 \omega^2} = 1 - \frac{Ne^2}{4\pi^2 \varepsilon_0 m_0 v^2},$$

$$\varepsilon_r = 1 - \frac{1.226 \times 10^{12} \times 2.56 \times 10^{-38} \times 4\pi \times 9 \times 10^9}{4\pi^2 \times 9 \times 10^{-31} \, v^2} = 1 - \frac{10^{14}}{v^2}.$$

2. The phase velocity v_ϕ is related to the velocity of light in free-space by

$$v_\phi = \frac{c}{\sqrt{\varepsilon_r}} = \frac{c}{\sqrt{1 - \dfrac{Ne^2}{4\pi^2 \varepsilon_0 m_0 v^2}}} = \frac{c}{\sqrt{1 - \dfrac{10^{14}}{v^2}}}.$$

So that v_ϕ has a real value, the quantity under the radical must be positive, that is

$$\frac{Ne^2}{4\pi^2 \varepsilon_0 m_0 v^2} < 1.$$

The frequency v must then have a value greater than the cut-off frequency v_c defined by

$$v_c = \sqrt{\frac{Ne^2}{4\pi^2 \varepsilon_0 m_0}} = 10^7.$$

One can then write

$$\varepsilon_r = 1 - \frac{v_c^2}{v^2}.$$

3. The relationship between the group velocity v_g and the phase velocity v_ϕ (§ B.3) can be written

$$\frac{1}{v_g} = \frac{\mathrm{d}}{\mathrm{d}\omega}\left(\frac{\omega}{v_\phi}\right) = \frac{\mathrm{d}}{\mathrm{d}v}\left(\frac{v}{v_\phi}\right) = \frac{1}{c}\frac{\mathrm{d}(nv)}{\mathrm{d}v} = \frac{1}{c}\frac{\mathrm{d}(v\sqrt{\varepsilon_r})}{\mathrm{d}v}$$

so that

$$\frac{c}{v_g} = \frac{\mathrm{d}}{\mathrm{d}v}\left(v\sqrt{1 - \frac{v_c^2}{v^2}}\right) = \frac{\mathrm{d}(\sqrt{v^2 - v_c^2})}{\mathrm{d}v} = \frac{v}{\sqrt{v^2 - v_c^2}} = \frac{1}{\sqrt{1 - \dfrac{v_c^2}{v^2}}} = \sqrt{\varepsilon_r} = \frac{v_\phi}{c}.$$

Thus

$$v_g \cdot v_\phi = c^2.$$

4. Formula (5.5) of § 5.1 gives the curvature of the wave normal with $\cos \theta = 1$

$$\frac{1}{\varrho} = -\frac{1}{n}\frac{\mathrm{d}n}{\mathrm{d}z} = -\frac{1}{2n^2}\frac{\mathrm{d}(n^2)}{\mathrm{d}z} = \frac{1}{2\varepsilon_r}\frac{\mathrm{d}\varepsilon_r}{\mathrm{d}z},$$

$$\frac{1}{\varrho} = \frac{1}{2\varepsilon_r}\frac{e^2}{4\pi^2 \varepsilon_0 m_0 v^2} \cdot \frac{\mathrm{d}N}{\mathrm{d}z} = \frac{1}{2\varepsilon_r} \cdot \frac{v_c^2}{Nv^2} \cdot \frac{\mathrm{d}N}{\mathrm{d}z}.$$

For $v = 2v_c$:

$$\frac{1}{\varrho} = \frac{1}{8\varepsilon_r} \cdot \frac{1}{N} \cdot \frac{dN}{dz} .$$

and with $\varepsilon_r = 1 - \frac{1}{4} = \frac{3}{4}$,

$$\frac{1}{\varrho} = \frac{1}{6} \cdot \frac{1}{N} \cdot \frac{dN}{dz} = \frac{1}{6} \cdot \frac{6}{10^4} = 10^{-4} \text{ m}^{-1}$$

$$\varrho = 10^4 \text{ m}.$$

Since N decreases as the altitude increases, ε_r increases and $d\varepsilon_r/dz$ is positive. The curvature is upward.

II

1. The reflection factor for energy at normal incidence is

$$R = \left(\frac{n-1}{n+1}\right)^2 = \left(\frac{\sqrt{\varepsilon_r}-1}{\sqrt{\varepsilon_r}+1}\right)^2$$

for any sense of the propagation.

$$R = \left(\frac{\sqrt{0.75}-1}{\sqrt{0.75}+1}\right)^2 = \left(\frac{0.866-1}{0.866+1}\right)^2 = \left(\frac{-0.134}{1.866}\right)^2 = (-0.073)^2 = 53 \times 10^{-4}.$$

2. The desired angle is the Brewster angle i_B defined by

$$\tan i_B = n = \sqrt{\varepsilon_r} = 0.866$$

for waves moving into the ionized medium

$$i_B = 40°54'.$$

For waves propagating in the opposite sense

$$\tan i_B' = \frac{1}{n} = \cot i_B,$$

$$i_B' = 49°6'.$$

PROBLEM 55

Plasma Oscillations

I

Show that in an isotropic medium, Maxwell's equations have as solutions longitudinal plane waves, that is, plane waves in which the electric vector E is parallel to the wave vector σ. What conditions must be satisfied by the index of refraction of the medium so that these waves can propagate?

II

Establish the equation giving the index of refraction of a plasma for a monochromatic wave. Neglect the collisions of the ions and electrons and, after indicating why, the effect of the ions relative to the electrons. Under what conditions can one establish longitudinal oscillations of the electric field in the plasma? Determine the phase and group velocities of the corresponding waves.

SOLUTION

I

Maxwell's equations for a plane monochromatic wave with angular frequency ω are written (§ 2.3)

$$E \times \sigma = -\mu_0 \omega H \tag{1}$$

$$H \times \sigma = \omega D. \tag{2}$$

D is the displacement vector related to E by

$$D = \varepsilon E = \varepsilon_r \varepsilon_0 E = \varepsilon_0 E + P, \tag{3}$$

so that the relative permittivity ε_r and the polarization P of the medium are joined by the relationship

$$P = (\varepsilon_r - 1)\varepsilon_0 E. \tag{4}$$

The condition which defines a longitudinal wave, $E \| \sigma$ involves from (1), $H = 0$, hence from (2) $D = 0$ and from (3)

$$\varepsilon_0 E = -P. \tag{5}$$

The vectors σ, E, and P are directed as shown in Fig. 55.1. Equation (5) introduced into (4) leads to

$$\varepsilon_r = 0. \tag{6}$$

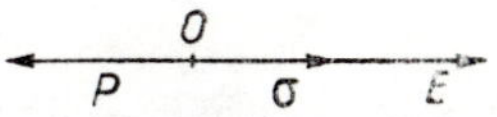

FIG. 55.1

Following Maxwell's relationships, the index $n = \sqrt{\varepsilon_r}$ of the medium should vanish so that longitudinal electromagnetic waves can exist.

II

Neglecting the effects of collisions amounts to treating the ions and electrons as free. Under these conditions, the displacement s of one of them with mass m_i and (algebraic) charge e_i is given by the equation (§ 10.4)

$$m_i \frac{d^2 s}{dt^2} = e_i E_m \sin \omega t.$$

To this corresponds an electric dipole moment

$$d = es = \frac{e^2}{m\omega^2} E_m \sin \omega t$$

and the polarization of the medium due to these charged particles, numbering N per unit volume and without mutual interaction, is $P = Nd$. Equation (4) gives

$$\varepsilon_r = 1 - \frac{e^2 N_e}{\varepsilon_0 m_e \omega^2} - \sum_i \frac{e_i^2 N_i}{\varepsilon_0 m_i \omega^2} . \tag{7}$$

N_e and m_e refer to the electrons and N_i, e_i, and m_i to the various positive and negative ions. The sum is taken over all of the ions. Since the plasma is neutral, the concentration of the positive ions is equal to the sum of the concentrations of the negative ions and electrons. The term relative to the ions in (7) is negligible since their mass is much greater than that of an electron. Taking this simplification into account, equation (6) is satisfied if

$$\omega = \sqrt{\frac{e^2 N_e}{\varepsilon_0 m_e}} . \tag{8}$$

This expression shows that ω is independent of σ. In addition, the phase velocity $v_\phi = \omega/\sigma$ is not subject to the relationships one finds in the study of waves. The group velocity $v_g = d\omega/d\sigma$ is zero. One sees that here one is dealing with oscillations of the electric field and of the electrons, rather than with the so-called waves.

QUANTUM MECHANICS

PROBLEM 56

Electromagnetic and de Broglie Waves

I

Consider a beam of particles with rest mass m_0 moving in a straight line in a vacuum with uniform velocity u.

1. Assume that all the particles have the same velocity. Calling the energy W and the momentum of the particles p, write an expression for the de Broglie wave associated with them. Write an expression for the phase velocity, v_ϕ, in terms of W and p. Also express v_ϕ as a function of u and c, the free-space velocity of light.

2. Assume that there is a distribution of velocities about u, the form of which is not known. Can one write an expression for a de Broglie wave? Show that the group velocity, v_g, can be expressed generally as

$$v_g = \mathrm{d}W/\mathrm{d}p.$$

Give v_g as a function of the mean velocity, u, of the particles.

What relationship exists between the phase velocity and the group velocity?

3. The particles are electrons whose velocity u is such that their wavelength λ is equal to half the Compton wavelength λ_c. Calculate (a) their velocity, u; (b) the phase and group velocities of the associated wave; (c) their mass, m; and (d) the potential difference V necessary to produce this velocity if it is assumed that the electrons are emitted from a hot cathode with zero initial velocity.

Compton wavelength: $\lambda_c = 2h/m_0 c$.

II

Replace the beam of particles by a beam of photons having the same frequency ν as the de Broglie wave associated with the particles. The photons are travelling through a non-absorbing medium of index n. Call λ' and $\tilde{\nu}'$ the wavelength and the wave number respectively in the medium.

1. Assuming the beam to be monochromatic, give the phase velocity v_ϕ as a function of ν and $\tilde{\nu}'$.

2. A photon beam cannot be rigorously monochromatic. Show that by using v and $\tilde{v}'$, the group velocity v_g can be put in a form analogous to that in question I.2 for the group velocity of particles.

3. What must be the dispersion law for the medium, $n = f(\lambda)$, so that between v_g and v_ϕ the same relationship is obtained as is obtained in question I.2 for particle waves?

4. In what region of the electromagnetic spectrum would you expect to find such a dispersion law in a material medium? Show that this law is a limiting case of the general dispersion law valid in regions remote from absorption bands:

$$n^2 = 1 + \sum_i \frac{A_i}{\tilde{v}_{0i}^2 - \tilde{v}_0^2}$$

where $\tilde{v}_0$ is the wave number in vacuum, $\tilde{v}_{0i}$ is the wave number of the centre of the absorption band, and A_i is a constant. Absorption bands are found in the visible, the infrared, and the near ultraviolet.

SOLUTION

I

1. For a particle in uniform rectilinear motion, the momentum $p = mu$ is a constant. The energy W is also constant. It is purely kinetic aside from the rest energy which requires the theory of relativity for its evaluation. The fundamental relationships

$$W = hv \tag{1}$$

$$p = h\tilde{v} \tag{2}$$

allow one to associate with the motion of the particle a sinusoidal plane wave (§ 13.5)

$$\Psi(x, t) = A \exp\left\{\frac{j}{\hbar}(Wt - px)\right\}, \tag{3}$$

A being a constant and x the direction of the wave. Hence the wavelength

$$\lambda = \frac{h}{p}.$$

The phase velocity is given by

$$v_\phi = \lambda v. \tag{4}$$

To express v_ϕ as a function of u and c, one must choose for v the value which corresponds to the total energy defined by the theory of relativity, namely $W = mc^2$, where m is the inertial mass. The momentum p is here equal to mu and (4) can be written

$$v_\phi = \frac{h}{p} \cdot \frac{W}{h} = \frac{c^2}{u}. \tag{5}$$

2. A distribution of the speeds of the particles corresponds to a distribution of momenta. The monochromatic wave (3) is replaced by a group of waves in the form

$$\Psi(x, t) = \int_{-\infty}^{+\infty} f(p) \exp\left\{\frac{j}{\hbar}(Wt - px)\right\} dp. \tag{6}$$

This expression indicates that Ψ has a maximum within the domain Δp surrounding p. The largest value of Ψ is found when the phase of the wave

$$\phi = Wt - px$$

remains almost constant in this domain, that is, when

$$\frac{d\phi}{dp} = t\frac{dW}{dp} - x = 0.$$

The centre of the group of waves moves with a uniform motion, whose velocity is the group velocity

$$v_g = \frac{x}{t} = \frac{dW}{dp}. \tag{7}$$

From relativity: $W^2 = m^2c^4 + p^2c^2$, hence $W\, dW = c^2p\, dp$, and from (5)

$$v_g = \frac{pc^2}{W} = \frac{c^2}{v_\phi} = u. \tag{8}$$

The group velocity is the velocity of the particle.

3. The wavelength of the electrons under consideration is

$$\lambda = \frac{\lambda_c}{2} = \frac{h}{m_0 c}.$$

A non-relativistic calculation then gives

$$u = \frac{p}{m_0} = \frac{h}{\lambda m_0} = c,$$

which is an unacceptable result. It is necessary to take into account the relativistic variation of the mass

$$u = \frac{p}{m} = \frac{h}{\lambda m} = \frac{h\sqrt{1 - \dfrac{u^2}{c^2}}}{\lambda m_0} = c\sqrt{1 - \frac{u^2}{c^2}},$$

hence,

$$u = c/\sqrt{2}.$$

From (8) one finds

$$v_g = c/\sqrt{2},$$

and from (5)

$$v_\phi = c\sqrt{2}.$$

The mass of electrons with velocity u is

$$m = \frac{m_0}{\sqrt{1 - \dfrac{u^2}{c^2}}} = m_0\sqrt{2}.$$

Their kinetic energy is given to them entirely by the action of electrostatic forces

$$eV = W_k = (m-m_0)c^2 = 0.414 \, m_0 c^2$$

hence

$$V = 0.414 \frac{m_0 c^2}{e} = 0.414 \times \frac{0.9 \times 10^{-30}}{1.6 \times 10^{-19}} \times 9 \times 10^{16} = 2.22 \times 10^4 \text{ V.}$$

II

1. By definition

$$v_\phi = \lambda' v = \frac{v}{\tilde{v}'} \,.$$

2. With the fundamental expressions (1) and (2), the wave group (6) can be written in the form

$$\Psi(x, t) = \int_{-\infty}^{+\infty} f(\tilde{v}') \exp \{2\pi j(vt - \tilde{v}'x)\} \, d\tilde{v}' \,, \tag{9}$$

which suits electromagnetic waves. The same reasoning which led from (6) to (7) in this case gives

$$v_g = \frac{dv}{d\tilde{v}'} \,. \tag{10}$$

3. Equations (5) and (8) show that

$$v_g \cdot v_\phi = c^2$$

which when combined with (10) and $\tilde{v}' = v/v_\phi$ give

$$\frac{1}{v_g} = \frac{v_\phi}{c^2} = \frac{d}{dv}\left(\frac{v}{v_\phi}\right)$$

or

$$\frac{v}{c^2} \, dv = \frac{v}{v_\phi} \, d\left(\frac{v}{v_\phi}\right) = \frac{1}{2} \, d\left(\frac{v}{v_\phi}\right)^2 \,,$$

hence

$$\left(\frac{v}{v_\phi}\right)^2 = \frac{v^2}{c^2} + K,$$

K is a positive or negative constant, thus finally

$$\frac{c^2}{v_\phi^2} = n^2 = 1 + \frac{Kc^2}{v^2} = 1 + K\lambda_0^2,$$

λ_0 being the wavelength in vacuum.

4. This equation agrees with the dispersion law given in the text for $\tilde{v}_{0i}^2 \gg \tilde{v}_i^2$; one has then in effect

$$n^2 = 1 - \frac{\sum A_i}{\tilde{v}_{0i}^2}$$

or

$$n^2 = 1 - \lambda_0^2 \sum A_i .$$

This expression is valid for X-rays or whenever the wavelengths are very short compared to the wavelengths of the atomic or molecular absorption bands.

PROBLEM 57

Five Exercises on Uncertainty Relationships

I

In a Michelson interferometry experiment it would seem possible at first inspection to determine if the photon associated with a wave train is reflected from one or the other of the two mirrors by measuring the recoil of the mirror. Show, using the uncertainty relationships, that this measurement is incompatible with the preservation of the coherence of the wave trains which interfere.

SOLUTION

In order for there to be coherence, it is necessary that the uncertainty in the position of the mirrors is much less than the wavelength of the light. If $\Delta x \gg \lambda$, the corresponding recoil should be:

$$\Delta p \simeq \frac{h}{\Delta x} \gg \frac{h}{\lambda} .$$

where h/λ is the momentum of the photon. The experiment is therefore impossible.

II

Starting from the gedanken experiment on the measurement of position using the "Heisenberg microscope" (§ 13.7), determine the short wavelength limit on length measurement imposed by the relationship $W = h\nu$.

SOLUTION

Use of radiation with wavelength λ_0 allows one to measure a minimal length of the order of $\Delta l \simeq \lambda_0/2 \sin u$, with a microscope with numerical aperture $\sin u$. The lower limit λ_l of λ useful for this corresponds to the annihilation of the reference particle, that is, to a proper energy m_0c^2 of the particle. Thus

$$\frac{hc}{\lambda_l} = m_0c^2.$$

λ_l is then the Compton wavelength for the case where the particle is an electron (§ 11.4).

$$\lambda_l = \frac{h}{m_e c} = 2.42 \times 10^{-12} \text{ m}.$$

For heavier particles such as atomic nuclei, the limit decreases but only to about 10^{-15} m. Measurement of lengths much smaller than this is without physical meaning (see L. Brillouin, *Science and Information Theory*, chap. 16).

III

In quantum mechanics a harmonic oscillator of mass m and frequency v has in its ground state a residual, zero-point energy $W_0 = \frac{1}{2}hv$ with corresponding normalized eigenfunction:

$$\psi_0 = \sqrt[4]{\frac{1}{\pi a^2}} \exp\left(-\frac{1}{2}\frac{x^2}{a^2}\right),$$

where x is the oscillator extension and $a = \sqrt{h/4\pi^2 vm}$ the amplitude. Calculate the mean value $\langle x \rangle$ of x and $\langle (\Delta x)^2 \rangle = \langle (x - \langle x \rangle)^2 \rangle$ and show that if the energy is well known and has value W_0 the uncertainty relationship $\langle (\Delta x)^2 \rangle \langle (\Delta p_x)^2 \rangle \geqslant \hbar^2/2$ results.

SOLUTION

The mean value of x is

$$\langle x \rangle = \int_{-\infty}^{+\infty} x\psi_0^2 \, dx = \sqrt{\frac{1}{\pi a^2}} \int_{-\infty}^{+\infty} x \exp\left(-\frac{x^2}{a^2}\right) dx = 0$$

since it is the integral of an odd function. The mean quadratic value of the variation of x is

$$\langle (\Delta x)^2 \rangle = \langle (x - \langle x \rangle)^2 \rangle = \langle x^2 \rangle,$$

since $\langle x \rangle = 0$, so that

$$\langle (\Delta x)^2 \rangle = \int_{-\infty}^{+\infty} x^2 \psi_0^2 \, dx = \sqrt{\frac{1}{\pi a^2}} \int_{-\infty}^{+\infty} x^2 \exp\left(-\frac{x^2}{a^2}\right) dx.$$

Integrating by parts this becomes

$$\langle (\Delta x)^2 \rangle = \sqrt{\frac{1}{\pi a^2}} \left[-\frac{a^2}{2} x \exp\left(-\frac{x^2}{a^2}\right) \right]_{-\infty}^{+\infty} + \sqrt{\frac{1}{\pi a^2}} \frac{a^2}{2} \int_{-\infty}^{+\infty} \exp\left(-\frac{x^2}{a^2}\right) dx.$$

The first term on the right-hand side is zero like (1). The value of the integral is $a\sqrt{\pi}$ and

$$\langle(\Delta x)^2\rangle = \frac{a^2}{2} = \frac{h}{8\pi^2 vm}.$$

The energy expression is

$$W = \frac{p_x^2}{2m} + 2\pi^2 v^2 m x^2.$$

If this is fully determined, one can write

$$W = \frac{\langle p_x^2\rangle}{2m} + 2\pi^2 v^2 m \langle x^2\rangle,$$

and since $\langle p\rangle = 0$

$$W = \frac{\langle(\Delta p_x)^2\rangle}{2m} + 2\pi^2 v^2 m \langle(\Delta x)^2\rangle.$$

The uncertainty relationship

$$\langle(\Delta p_x)^2\rangle \times \langle(\Delta x)^2\rangle \geqslant \hbar^2/4$$

gives

$$W \geqslant \frac{\hbar^2}{8m\langle(\Delta x)^2\rangle} + 2\pi^2 v^2 m \langle(\Delta x)^2\rangle.$$

This expression has a minimum for

$$\langle(\Delta x)^2\rangle = \frac{\hbar}{4\pi vm}.$$

The minimum value of W_0 is

$$W_0 = \hbar\pi v = hv/2.$$

IV

Starting from the uncertainty relationship between the momentum of a particle and the corresponding coordinate, evaluate the ground state energy of the hydrogen atom.

SOLUTION

The energy of an electron at a distance r from the nucleus is given by

$$W = \frac{p^2}{2m_e} - \frac{e^2}{4\pi\varepsilon_0 r}.$$

The minimal energy is obtained by taking the smallest possible values for p and r. However, according to the required uncertainty relationship

$$\Delta p \cdot \Delta r \simeq \hbar.$$

(Note, in writing this expression, that the velocity is radial; one says in effect that the angular momentum of the electron is zero in the ground state.) The mean values $\langle r \rangle$ of r and $\langle p \rangle$ of p cannot be less than Δr and Δp respectively. Thus one has for the minimal mean values

$$\langle r \rangle \times \langle p \rangle \simeq \hbar$$

so that:

$$\langle W \rangle = \frac{\hbar^2}{2m_e \langle r^2 \rangle} - \frac{e^2}{4\pi \varepsilon_0 \langle r \rangle} \cdot$$

This expression has a minimum for

$$\langle r \rangle = \frac{\varepsilon_0 h^2}{\pi m_e e^2} \cdot$$

This is the value of the first Bohr orbit or the most probable distance of the electron from the nucleus. One finds for the energy minimum the value

$$W_0 = -\frac{me^4}{8\varepsilon_0^2 h^2} \cdot$$

V

One proposes to measure the magnetic moment M due to the spin of the electron by measuring the magnetic field H which it produces at a distance r. In order for this experiment to have meaning, one must be able to localize the electron in a domain $\Delta r \ll r$. It is also necessary that the magnetic field H' due to the motion of the electron (velocity v) is negligible compared to the field H. Show that these conditions are not consistent with the uncertainty relations.

SOLUTION

The maximal field H is given by (in ampere-metres)

$$H = \frac{1}{4\pi} \frac{M}{r^3} \cdot$$

The maximal field H' due to the motion of the electron is

$$H' = \frac{1}{4\pi} \frac{ev}{r^2} \cdot$$

One knows in addition that M is equal to a Bohr magneton (§ 15.6), so that

$$M = \frac{e\hbar}{2m_e} \cdot$$

The condition $H \gg H'$ thus leads to

$$\hbar \gg 2pr,$$

with $p = m_e v$. The uncertainty relationship leads to

$$\Delta r \times \Delta p \simeq \hbar,$$

which, with $\Delta r \ll r$, gives

$$\hbar \ll r \, \Delta p,$$

an inequality incompatible with (1).

PROBLEM 58

Potential Barrier

A flux of single velocity particles of mass m and total constant energy W moving from x' to x encounter a potential barrier of width d.

I

The cross-section of the barrier is shown in Fig. 58.1.

$$W_p = W_1 = 0 \qquad \text{for} \quad x < 0$$
$$W_p = W_2 \qquad\qquad \text{for} \quad 0 < x < d$$
$$W_p = W_3 \qquad\qquad \text{for} \quad x > d.$$

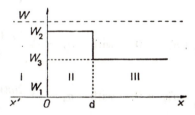

FIG. 58.1

Write the continuity conditions for the wave function associated with these particles and for its first derivative. Derive the transmission coefficient T for the potential barrier (ratio of the transmitted to incident flux) as a function of d and of the wave vectors σ_1, σ_2, and σ_3 corresponding to the regions I, II, and III. Assume that W is greater than W_p. (Initially the energy of the particles is purely kinetic.)

II

Consider the special case where $W_3 = W_1$ and where $W > W_p$ (Fig. 58.2). Give the transmission coefficient T for this barrier as a function of the reflection coefficient R_1 at the potential discontinuity O. Show the analogy between this expression and that for the transmission of electromagnetic waves falling on a glass plate with plane parallel faces. The glass plate has index n_2 and is situated in a homogeneous medium of index n_1.

17*

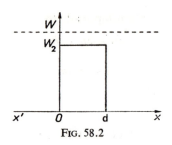

FIG. 58.2

III

Use the potential barrier above but consider the case where $W < W_p$. Calculate T for $W = 1$ eV, $W_2 = 2$ eV and $d = 1$ Å for both electrons and protons.

SOLUTION

I

Asymmetric barrier

Schrödinger's equation is written
in region I:

$$\frac{d^2\psi}{dx^2} + \sigma_1^2\psi = 0, \quad \text{with} \quad \sigma_1 = \frac{\sqrt{2mW}}{\hbar} \tag{1}$$

in region II:

$$\frac{d^2\psi}{dx^2} + \sigma_2^2\psi = 0, \quad \text{with} \quad \sigma_2 = \frac{\sqrt{2m(W-W_2)}}{\hbar} \tag{2}$$

in region III:

$$\frac{d^2\psi}{dx^2} + \sigma_3^2\psi = 0, \quad \text{with} \quad \sigma_3 = \frac{\sqrt{2m(W-W_3)}}{\hbar} \tag{3}$$

In regions I and II one has both a direct wave and a reflected wave. In region III, which is assumed to go to infinity, there is only a direct wave.

The solutions of the Schrödinger equation corresponding to these three regions are

$$\left.\begin{array}{l} \psi_{\mathrm{I}} = e^{-j\sigma_1 x} + r\,e^{+j\sigma_1 x} \\ \psi_{\mathrm{II}} = A\,e^{-j\sigma_2 x} + B\,e^{+j\sigma_2 x} \\ \psi_{\mathrm{III}} = t\,e^{-j\sigma_3 x}. \end{array}\right\} \tag{4}$$

Recall the origin of the continuity conditions. The function ψ, whose square measures the probability density for the particles at a point along $x'x$, can only have a single value at a given point. In addition, since the energies W_2 and W_3 are finite, equations (1), (2), and (3) show that the second derivative of ψ is also finite. Thus the first derivative is continuous. Writing the continuity equations on the plane $x = 0$

$$\left.\begin{array}{l} 1 + r = A + B \\ \sigma_1(1-r) = \sigma_2(A-B) \end{array}\right\} \tag{5}$$

and on the plane $x = d*$:

$$A \exp(-j\sigma_2 d) + B \exp(j\sigma_2 d) = t \exp(-j\sigma_3 d)$$

$$\sigma_2[A \exp(-j\sigma_2 d) - B \exp(j\sigma_2 d)] = t\sigma_3 \exp(-j\sigma_3 d). \tag{6}$$

We have already solved an identical system of equations in Problem 14. It is sufficient to replace t by $t \exp(-j\sigma_3 d)$ in equation (25) and on the right-hand side q_0, q, and q_s by σ_1 σ_2, and σ_3 as well as k_0 by σ_2.

Taking the modulus of this expression, one gets the transmission coefficient for the barrier

$$T = \frac{4\sigma_1 \sigma_2^2 \sigma_3}{\sigma_2^2(\sigma_1 + \sigma_3)^2 + (\sigma_2^2 - \sigma_1^2)(\sigma_2^2 - \sigma_3^2) \sin^2 \sigma_2 d}. \tag{7}$$

Note. It is useful to show in parallel some results from physical optics and quantum optics obtained in Problems 14 and 58.

Dielectric films	*Potential barriers*
One writes continuity expressions for the tangential components of the electric and magnetic fields.	One writes continuity conditions for the wave function associated with the particle and for its first derivative.

The same set of equations result:

$$\text{equations (18) and (22)} \leftrightarrow \text{equation (5)}$$
$$\text{equations (19) and (23)} \leftrightarrow \text{equation (6)}$$

The different media are characterized by their index $(n_0, n_1, \ldots)$ hence the different wavelengths of the electromagnetic waves $(\lambda_0 = c/n_0, \lambda_1 = c/n_1, \ldots)$.	The different regions are characterized by their potential energy $W_p(x)$ hence by the different wavelengths associated with the particles and their corresponding wave numbers,

$$\sigma_1 = \frac{2\pi}{\lambda_1} = \frac{\sqrt{2m(W - W_1)}}{\hbar}, \sigma_2 \ldots$$

II

$W > W_2$, $\sigma_1 = \sigma_3$.
The equation for T simplifies to

$$T = \frac{1}{1 + \dfrac{[\sigma_2^2 - \sigma_1^2]^2}{4\sigma_1^2 \sigma_2^2} \sin^2 \sigma_2 d}. \tag{8}$$

---

* It is necessary to write the exponential on the right-hand side and not just t as is the case with thin films (eqn. (2), Problem 14). It is only through this condition that the factor σ_3 appears in the second equation (6). Later on $\exp(-j\sigma_3 d)$ cancels in the calculation of T.

On the other hand, the reflection coefficient at O is obtained by starting with equation (13.45) *(Optics)* which can be written

$$r = \frac{1 - \sqrt{\dfrac{W - W_2}{W}}}{1 + \sqrt{\dfrac{W - W_2}{W}}} = \frac{\sigma_1 - \sigma_2}{\sigma_1 + \sigma_2}, \tag{9}$$

from which:

$$R_1 = r^2 = \left(\frac{\sigma_1 - \sigma_2}{\sigma_1 + \sigma_2}\right)^2. \tag{10}$$

The transmission coefficient at the potential barrier is

$$T = \frac{1}{1 + \dfrac{4R_1}{(1 - R_1)^2} \sin^2 \sigma_2 d}. \tag{11}$$

Notes

1. The transmission coefficient of a thin plate has the same form. In effect, if in equation (31), Problem 14, one makes $q_0 = q_s$ or $n_0 = n_s = n_1$, one has

$$T = \frac{1}{1 + \dfrac{[n_2^2 - n_1^2]^2}{4n_1^2 n_2^2} \sin^2 \sigma_2 d}. \tag{12}$$

Fresnel's formula allows us to write

$$R_1 = \left(\frac{n_1 - n_2}{n_1 + n_2}\right)^2, \tag{13}$$

and equations (11) and (12) are identical.

2. The potential barrier, like the dielectric plate, is perfectly transparent when $\sin \sigma_2 d = 0$ or $d = \lambda_2/2$. An application of this quantum effect is the Ramsauer effect.

A beam of approximately 0.1 eV electrons passes through an inert gas (neon or argon) as if there were no atoms in the path. The atoms appear practically transparent to electrons at this energy. When the electron energy is greater than or less than this value they are scattered away from their path.

III

$W_1 < W_2$, $\sigma_1 = \sigma_3$ (Fig. 58.2).

The solutions of the Schrödinger equation are

$$\left. \begin{aligned} \psi_\mathrm{I} &= e^{-j\sigma_1 x} + r\,e^{+j\sigma_1 x}, \\ \psi_\mathrm{II} &= A\,e^{-\sigma_2 x} + B\,e^{+\sigma_2 x}, \\ \psi_\mathrm{III} &= r\,e^{-j\sigma_1 x}. \end{aligned} \right\} \tag{14}$$

Equations (5) and (6) become

$$(1+r) = A+B, \quad \left.\right\}$$
$$j\sigma_1(1-r) = \sigma_2(A-B); \quad \left.\right\}$$

(15)

$$A\,e^{-\sigma_2 d}+B\,e^{+\sigma_2 d} = t\,e^{-j\sigma_1 d}, \quad \left.\right\}$$
$$\sigma_2[A\,e^{-\sigma_2 d}-B\,e^{+\sigma_2 d}] = j\sigma_1 t\,e^{-j\sigma_1 d}. \quad \left.\right\}$$

(16)

Solution of these equations gives (it is sufficient to replace $j\sigma_2$ by σ_2 in (6))

$$t = \frac{4j\sigma_1\sigma_2\,e^{-j\sigma_2 d}}{(\sigma_2+j\sigma_1)^2\,e^{-\sigma_2 d}-(\sigma-j\sigma_1)^2\,e^{\sigma_2 d}}.$$

(17)

Since the external media are identical, one has (taking into account the identities $2\sinh x = e^x - e^{-x}$ and $j\sin x = \sinh jx$)

$$T = |t|^2 = \frac{4\sigma_1^2\sigma_2^2}{4\sigma_1^2\sigma_2^2+(\sigma_1^2+\sigma_2^2)^2\sinh^2\sigma_2 d}.$$

(18)

Numerical application:

$$\sigma_1^2 = \sigma_2^2 \rightarrow T = \frac{1}{1+\sinh^2\sigma_2 d},$$

$$\sigma_2 d = \frac{\sqrt{2m(W_0-W)}}{\hbar}\,d.$$

For the electron,

$$\sigma_2 d = \frac{2\times3.14\,\sqrt{2\times0.9\times10^{-30}\times1.6\times10^{-19}}}{6.62\times10^{-34}}\,10^{-10},$$

$$\sigma_2 d = 0.51, \qquad \sinh\sigma_2 d = 0.53,$$

$$T = \frac{1}{1+(0.53)^2} = \frac{1}{1+0.28},$$

$$T = 0.77,$$
$$R = 0.23.$$

For the proton, $m = 1840\,m_e$, $\sigma_2 d = 22$, and the term $\exp(\sigma_2 d)$ which arises in:

$$\sinh\sigma_2 d = \tfrac{1}{2}[\exp(\sigma_2 d)-\exp(-\sigma_2 d)]$$

is of the order of 10^{10}. Therefore, $T \simeq 0$.

PROBLEM 59

The Deuteron

The deuterium nucleus (heavy hydrogen), called the deuteron, is made up of a proton and a neutron bound together by an attractive force derived from a central potential $W_p(r)$. Assume the proton and neutron masses are equal (this is valid to within 0.007 parts) and are 1.672×10^{-27} kg.

1. Write the time independent Schrödinger equation for the deuteron in the system of coordinates relative to its centre of mass.

2. When the deuteron is in a spherically symmetric state, write the radial wave equation.

3. Experiment shows that when the deuteron is in its ground state, the absolute value of the binding energy is $|W| = 2.23$ MeV. Give the sign of the energy and tell its significance. One can assume that the interaction energy $W_p(r)$ in a first approximation can be represented by a square well such that $W_p(r) = -W_0$ for $r < r_0$ and $W_p(r) = 0$ for $r > r_0$ (Fig. 59.1). Assuming the ground state to be spherically symmetric, determine the corresponding wave function (which along with its first derivative should be uniform, continuous, and bounded). One takes in the wave equation $r\psi_r(r) = u(r)$.

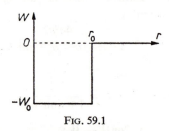

FIG. 59.1

4. Calculate the radius r_0 of the potential well—that is, the nuclear interaction length—for $W_0 = 21$ MeV (one takes for r_0 the smallest possible value).

5. Calculate the probability that r is larger than or less than r_0.

SOLUTION

1. The system is equivalent to a particle of mass $\mu = m/2$. The time independent Schrödinger equation is

$$\Delta\psi + \frac{m}{\hbar^2} [W - W_p(r)]\psi = 0. \tag{1}$$

2. In the spherically symmetric states (S states), $\psi = \text{const.} \times \psi_r(r)$. The laplacian is given by

$$\frac{1}{r^2} \frac{d}{dr} \left(r^2 \frac{d\psi_r}{dr} \right)$$

and the wave equation

$$\frac{d^2\psi_r}{dr^2} + \frac{2}{r} \frac{d\psi_r}{dr} + \frac{m}{\hbar^2} [W - W_p(r)]\psi_r = 0. \tag{2}$$

3. The binding energy is the difference between the energy of the two-component system and the energy of the components at rest at infinite separation. It is, therefore, negative since the proton and neutron attract one another. The value $W = -2.23$ MeV of the ground state is the eigenvalue of the Schrödinger equation for this state.

Taking $r\psi_r = u(r)$, one has

$$\frac{d\psi_r}{dr} = \frac{1}{r}\frac{du}{dr} - \frac{u}{r^2}, \qquad \frac{d^2\psi_r}{dr^2} = \frac{1}{r}\frac{d^2u}{dr^2} - \frac{2}{r^2}\frac{du}{dr} + \frac{2u}{r^3}$$

and equation (2) becomes

$$\frac{1}{r}\frac{d^2u}{dr^2} + \frac{m}{\hbar^2}[W - W_p]\frac{u}{r} = 0. \tag{3}$$

For $r > r_0$, $W_p = 0$, the solution of (3) is an exponential (since $W < 0$), so that

$$u(r) = A\exp(-Kr) + B\exp(Kr), \quad \text{with} \quad K = \frac{\sqrt{m|W|}}{\hbar}.$$

B is zero, since $r^{-1}\exp(Kr)$ is not bounded at infinity.

For $r < r_0$, $W_p < W$. The energy W is the sum of the kinetic and the (non-zero) potential energy. The solution of (3) is sinusoidal,

$$u(r) = C\cos K'r + D\sin K'r, \quad \text{with} \quad K' = \frac{\sqrt{m(W_0 - |W|)}}{\hbar}.$$

C is zero, since $r^{-1}\exp(K'r)$ is unbounded at the origin. The sine solution is acceptable, since as $r \to 0$

$$\frac{1}{r}\sin K'r \to K' \quad \text{and} \quad \frac{d}{dr}\left(\frac{\sin K'r}{r}\right) \to 0.$$

Using the continuity conditions for u and for du/dr at $r = r_0$

$$D\sin K'r_0 = A\exp(-Kr_0),$$

$$DK'\cos K'r_0 = -AK\exp(-Kr_0), \tag{4}$$

from which

$$K'\cot K'r_0 = -K \tag{5}$$

or

$$\sqrt{W_0 - |W|}\cot\frac{\sqrt{m(W_0 - |W|)}}{\hbar}r_0 = -\sqrt{|W|}. \tag{6}$$

4. For $|W| = 2.23$ MeV and $W_0 = 21$ MeV, one has

$$K = \frac{2\times 3.14\sqrt{1.672\times 10^{-27}\times 2.23\times 10^6\times 1.60\times 10^{-19}}}{6.62\times 10^{-34}} = 2.32\times 10^{14}\ \text{m}^{-1},$$

$$K' = K\sqrt{\frac{W_0 - |W|}{|W|}} = 2.32\times 10^{14}\sqrt{\frac{18.77}{2.23}} = 6.72\times 10^{14}\ \text{m}^{-1}.$$

Using (5) one finds

$$\cot K'r_0 = -0.345$$

$$K'r = \frac{\pi}{2} + \text{arc}\tan 0.345 = \frac{\pi}{2} + 0.332 + n\pi \qquad (n\text{ integer}).$$

The smallest value of r_0 is

$$r_0 = \frac{1.571+0.332}{6.72\times10^{14}} = 2.83\times10^{-15} \text{ m}.$$

5. The probability that the proton–neutron separation exceed r_0 is given by

$$P = \int_{r_0}^{\infty} |\psi|^2 4\pi r^2 \, dr = 4\pi A^2 \int_{r_0}^{\infty} \exp\left(-2Kr\right) dr = \frac{2\pi A^2}{K} \exp\left(-2Kr_0\right).$$

The probability that this distance is less than r_0 is

$$P' = \int_0^{r_0} |\psi|^2 4\pi r^2 \, dr = 4\pi D^2 \int_0^{r_0} \sin^2 K'r \, dr = \frac{2\pi D^2}{K'} \left(K'r_0 - \frac{1}{2}\sin 2K'r_0\right).$$

The ratio of these two quantities is, using (4)

$$\frac{P}{P'} = \frac{K'}{K}\frac{A^2}{D^2}\frac{\exp\left(-2Kr_0\right)}{K'r_0-\frac{1}{2}\sin 2K'r_0} = \frac{K'}{K}\frac{\sin^2 K'r_0}{K'r_0-\frac{1}{2}\sin 2K'r_0}.$$

One has: $K'r_0 = 1.903$, $\sin K'r_0 = 0.945$, $\sin 2K'r_0 = -0.615$, and $P/P' = 1.17$.
Since $P+P' = 1$, $P = 0.54$ and $P' = 0.46$.

In the ground state of the deuteron, the proton and the neutron spend the majority of the time outside the range of the nuclear force. This accounts for the small value of the binding force W.

PROBLEM 60

Double Potential Well

I

A particle of mass m can move along the Ox direction in regions where the potential energy has the following values:

$$x < 0, \quad W = \infty;$$
$$0 < x < a, \quad W = 0 \text{ (region 1)};$$
$$a < x < a+b, \quad W = W_0 \text{ (region 2)};$$
$$a+b < x < 2a+b, \quad W = 0 \text{ (region 3)}; \quad \text{and}$$
$$2a+b < x, \quad W = \infty.$$

1. Show that the wave function of the particle can be represented by:

$$\psi_1 = A \sin \sigma x,$$
$$\psi_2 = Be^{\sigma'(x-a-b)}+Ce^{-\sigma'(x-a)},$$
$$\psi_3 = D \sin \sigma[x-(2a+b)].$$

in regions 1, 2, and 3, respectively.
What are the values of σ and σ' when the energy W of the particles is less than W_0?

2. Write the continuity conditions for the wave function at the various interfaces. What relationships can one derive between A and D on one hand, and between B and C on the other?

3. By using these relationships, write the equation defining the possible values of the energy, W, of the particle in the form

$$\tan \sigma a = f(\sigma, \sigma'). \tag{1}$$

For b sufficiently large (how large?) this equation can be put in a more simple form,

$$\tan \sigma a = f_0(\sigma, \sigma'). \tag{2}$$

Resolve (2) graphically using the following numerical values:

$a = 0.4$ Å	$h = 6.6 \times 10^{-34}$ J-sec
$W_0 = 0.20$ eV	$e = 1.6 \times 10^{-19}$ coul
$m = 5 \times 10^{-27}$ kg	b is measured in Å

Derive the possible values of W.

II

Assume now that b is smaller. Show that (1) can be put in the approximate form

$$\tan \sigma a = f_0(\sigma, \sigma')(1 \pm 2\eta) \tag{3}$$

where η is a small quantity depending on b. Show the existence of a doubling of the levels determined in question I.3.

For what value of b does this doubling have a separation of 0.8 cm^{-1} for the first level? What then is the separation of the second level in cm^{-1}?

III

The preceding problem refers to the inversion of NH_3. Indicate quantitatively what happens for ND_3 (it is necessary to take twice the value of m) and for PH_3 (it is necessary to take 3 times the value of W_0)

SOLUTION

I

1. Since the energy W of the particle is less than the height W_0 of the potential barrier, in classical theory the particle can only reside in region I or in region III. In quantum theory it cannot exist to the left of region I or to the right of region III, but it can pass from I to III or from III to I (Fig. 60.1). The wave equation for the stationary states in I or III where $W_p = 0$ is

$$\frac{\hbar^2}{2m} \frac{d^2\psi}{dx^2} + W\psi = 0.$$

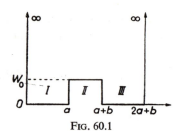

FIG. 60.1

Its solution is sinusoidal and ψ must go to zero for $x = 0$ and for $x = 2a+b$. Hence

$$\psi_1 = A \sin \sigma x,$$
$$\psi_3 = D \sin \sigma[x-(2a+b)],$$

with $\sigma = \sqrt{2mW}/\hbar$. In region II the wave equation is

$$\frac{\hbar^2}{2m} \frac{d^2\psi}{dx^2} + (W-W_p)\psi = 0.$$

Since $W < W_p$, the solution of this equation is sinusoidal between $x = a$ and $x = a+b$, so that

$$\psi_2 = B \exp(\sigma' x) + C \exp(-\sigma'x),$$

with

$$\sigma' = \frac{\sqrt{2m(W_0-W)}}{\hbar}.$$

2. The continuity conditions for ψ and for $d\psi/dx$ are given by

for $x = a$:

$$A \sin \sigma a = B \exp(\sigma'a) + C \exp(-\sigma'a),$$
$$\sigma A \cos \sigma a = \sigma'[B \exp(\sigma'a) - C \exp(-\sigma'a)];$$

for $x = a+b$:

$$B \exp[\sigma'(a+b)] + C \exp[-\sigma'(a+b)] = D \sin \sigma a,$$
$$\sigma'\{B \exp[\sigma'(a+b)] - C \exp[-\sigma'(a+b)]\} = -\sigma D \cos \sigma a.$$

3. By elimination of B and C from the four equations above, one finds

$$\left(\frac{\sigma'}{\sigma} \tan \sigma a + 1\right) A \exp(\sigma'b) = \left(\frac{\sigma'}{\sigma} \tan \sigma a - 1\right) D,$$

$$\left(\frac{\sigma'}{\sigma} \tan \sigma a - 1\right) A \exp(-\sigma'b) = \left(\frac{\sigma'}{\sigma} \tan \sigma a + 1\right) D.$$

The compatibility condition for these two equations is written

$$\left(\frac{\sigma'}{\sigma} \tan \sigma a + 1\right) \exp(\sigma'b) = \pm\left(\frac{\sigma'}{\sigma} \tan \sigma a - 1\right),$$

from which

$$\tan \sigma a = -\frac{\sigma}{\sigma'} \left\{ \frac{1 \pm \exp(-\sigma'b)}{1 \mp \exp(-\sigma'b)} \right\}. \tag{1}$$

This is the required general relationship. If b is sufficiently large that $\sigma'b$ is much greater than one, equation (1) simplifies to

$$\tan \sigma a = -\frac{\sigma}{\sigma'} [1 \pm 2 \exp(-\sigma'b)]. \tag{2}$$

In a first approximation, one sets aside the exponential, it being small with respect to unity, and then gets

$$\tan \sigma a = -\frac{\sigma}{\sigma'}, \tag{3}$$

independent of b.

Equation (3) can be written

$$\sigma a = \arctan \left(-\frac{\sigma}{\sigma'} \right) + k\pi \qquad (k \text{ an integer}).$$

Since

$$\sin \sigma a = \frac{1}{\sqrt{1+\cot^2 \sigma a}} = \frac{\sigma}{\sqrt{\sigma^2+\sigma'^2}} = \frac{\hbar\sigma}{\sqrt{2mW_0}}$$

one has finally

$$\sigma a = k\pi - \arcsin \frac{\hbar\sigma}{\sqrt{2mW_0}}. \tag{4}$$

This transcendental equation, which determines σ and thereby W, can be solved graphically by finding the intersections of the line $y = \sigma a$ with the curve

$$y = k\pi - \arcsin \frac{\hbar\sigma}{\sqrt{2mW_0}} \qquad \text{(Fig. 60.2)}.$$

The energy can be seen to be quantized.

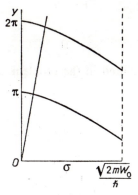

FIG. 60.2

When $W \ll W_0$, that is when $\sigma \ll \sigma'$, the equations (3) and (4) give as a solution

$$\sigma a = k\pi$$

and the energy values are

$$W^{(0)} = \frac{k^2 \pi^2 \hbar^2}{2ma^2}. \tag{5}$$

One finds here again the energy levels for a particle in a box which is natural, since neglecting W with respect to W_0 is equivalent to enclosing the a intervals by infinite potential walls.

With the given numerical values, the energy levels (5) have values

$$W^{(0)} = \frac{k^2 \hbar^2}{8ma^2} = k^2 \frac{43.56 \times 10^{-68}}{8 \times 5 \times 10^{-27} \times 0.16 \times 10^{-20}} = k^2 \times 6.82 \times 10^{-22} \text{ J},$$

while

$$W_0 = 0.20 \times 1.6 \times 10^{-19} = 3.2 \times 10^{-20} \text{ J}.$$

The approximation made by taking $W \ll W_0$ is therefore a good one. A better approximation is given by (3). Since the right-hand side is small, one can take

$$\sigma^{(1)} = \sigma^{(0)} - \frac{\sigma^{(0)}}{\sigma'^{(0)}a} \tag{6}$$

from which

$$W^{(1)} = \frac{\hbar^2 \sigma^{(1)2}}{2m} = \frac{\hbar^2 \sigma^{(0)2}}{2m} - \frac{2\hbar^2 \sigma^{(0)2}}{2m\sigma'^{(0)}a} = W^{(0)} - \frac{2W^{(0)}}{\sigma'^{(0)}a}.$$

This expression relates to the potential curve of Fig. 60.3: $W_p = \infty$ for $x = 0$ and infinite width of the barrier of height W_0.

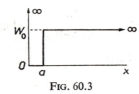

FIG. 60.3

II

Equation (2) has the form required in the problem. By considering σa small, one can make the approximation as in (6)

$$\sigma^{(2)} = \sigma^{(1)} \mp 2 \frac{\sigma^{(0)}}{\sigma'^{(0)}a} \exp(-\sigma'^{(0)}b), \tag{7}$$

from which

$$W^{(2)} = W^{(1)} \mp \frac{4W^{(0)}}{\sigma'^{(0)}a} \exp(-\sigma'^{(0)}b). \tag{8}$$

The last term leads to a doubling of the energy levels. The separation resulting from the splitting of the first $(K = 1)$ level is

$$\Delta W_1^{(2)} = \frac{8W_1^{(0)}}{\sigma'^{(0)}a} \exp\left(-\sigma'^{(0)}b\right).$$

(9)

For

$$\Delta W_1^{(2)} = hc\,\Delta\tilde{\nu}, \quad \text{with} \quad \Delta\tilde{\nu} = 0.8\times10^2\ \text{m}^{-1},$$

it is necessary that

$$\exp\left(-\sigma'^{(0)}b\right) = \frac{\sigma'^{(0)}ahc\,\Delta\tilde{\nu}}{8W_1^{(0)}} = \frac{\sqrt{2m(W_0-W_1^{(0)})}2\pi ahc\,\Delta\tilde{\nu}}{8hW_1^{(0)}}$$

$$\exp\left(-\sigma'^{(0)}b\right) = \frac{\sqrt{3.2\times10^{-46}}\times3.14\times0.4\times10^{-10}\times3\times10^8\times0.8\times10^2}{4\times6.82\times10^{-22}} = 1.98\times10^{-2},$$

from which

$$\sigma'^{(0)}b = 3.92.$$

Since

$$\sigma'^{(0)} = \frac{\sqrt{3.2\times10^{-46}}\times6.28}{6.6\times10^{-34}} = 1.7\times10^{11},$$

one has

$$b = \frac{3.92}{1.7}\times10^{-11} = 2.3\times10^{-11}\ \text{m} \quad \text{or} \quad 0.23\ \text{Å}.$$

The splitting of the second level is

$$\Delta W_2 = \frac{8W_2^{(0)}}{\sigma'^{(0)}a} \exp\left(-\sigma'^{(0)}b\right).$$

$W_2^{(0)}$ is therefore negligible compared to W_0.
   Since

$$W_2^{(0)} = K^2 W_1^{(0)} = 4W_1^{(0)},$$

the splitting of level 2 is 4 times greater than that of level 1, namely 3.2 cm^{-1}.

III

The problem of ammonia, NH_3, inversion is as follows. This molecule has the symmetry of a pyramid with a triangular base. If the atoms are numbered H_1, H_2, and H_3 one can see (Fig. 60.4) that as a result of inversion about the centre of mass the inverted molecule cannot be superimposed on the original molecule by any process of rotation or translation. This leads to two distinct molecular species and to two identical potential minima for the two positions of the N atom with respect to the plane of the hydrogens. One then has the general situation treated in this problem with the more correct potential curve given in

Fig. 60.5. This, however, is difficult to solve. The value 0.8 cm^{-1} adopted for the splitting of the first energy level above is that which is found for NH_3. The transition which one sees between these levels is a dipole transition which gives rise to an absorption at $\lambda = 1.25$ cm.

For ND_3 the form of the potential curve is the same as for NH_3. If one doubles the value of m, $\sigma^{(0)}$ is multiplied by $\sqrt{2}$ and $W_1^{(0)}$ is divided by 2. The value of $-\sigma'^{(0)}b$ changes from 3.92 to $3.92 \times \sqrt{2} = 5.04$ and $\exp(-\sigma'^{(0)}b) = 0.64 \times 10^{-2}$. Equation (9) shows that the value of $\Delta\sigma$ is divided by about 9.

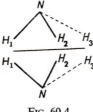

FIG. 60.4

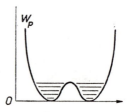

FIG. 60.5

For PH_3 assuming that the width of the barrier b will remain unchanged but that the height W_0 is multiplied by 3, $\sigma'^{(0)}$ is multiplied by $\sqrt{3}$, $\exp(-\sigma'^{(0)}b) = 0.11 \times 10^{-2}$ and equation (9) where only $\sigma'^{(0)}$ varies, shows that the value of $\Delta\sigma$ is reduced to about 0.03 of the value for NH_3. The full calculation is difficult.

PROBLEM 61

Angular Momentum Operators

The one-electron wave function for the state $l = 2$, $m = 2$ is given by:

$$\psi_{n22} = cf(r) \sin^2 \theta \exp(2j\phi),$$

where c is a constant. The function $f(r)$, the radial part of the wave function depends on the principal quantum number n but is not important here. Show that ψ_{n22} is an eigenfunction of the operators $\hat{G}_z$ and $\hat{G}^2$ but not of $\hat{G}_x$ or $\hat{G}_y$ (G being the angular momentum).

SOLUTION

1. The operator $\hat{G}_z$ is given by (§ 14.3) $-j\hbar(\partial/\partial\phi)$, from which

$$\hat{G}_z\psi_{n22} = -j\hbar \frac{\partial\psi_{n22}}{\partial\phi} = -j\hbar 2j\psi_{n22} = 2\hbar\psi_{n22}.$$

ψ_{n22} is thus an eigenfunction of $\hat{G}_z$ with eigenvalue $2\hbar$.

2. The operator $\hat{G}^2$ is given by

$$\hat{G}^2 = -\hbar^2 \left[\frac{1}{\sin\theta} \frac{\partial}{\partial\theta} \left(\sin\theta \frac{\partial}{\partial\theta} \right) + \frac{1}{\sin^2\theta} \frac{\partial^2}{\partial\phi^2} \right].$$

One has:

$$\frac{\partial^2 \psi_{n22}}{\partial \phi^2} = -4\psi_{n22},$$

$$\frac{\partial \psi_{n22}}{\partial \theta} = 2cf(r) \sin \theta \cos \theta \exp (2j\phi),$$

$$\frac{\partial}{\partial \theta} \left(\sin \theta \frac{\partial \psi_{n22}}{\partial \theta} \right) = cf(r) \exp (2j\phi) [4 \sin \theta \cos^2 \theta - 2 \sin^3 \theta],$$

$$\hat{G}^2 \psi_{n22} = -\hbar^2 cf(r) \exp (2j\phi) [4 \cos^2 \theta - 2 \sin^2 \theta - 4]$$
$$= \hbar^2 cf(r) \exp (2j\phi) \cdot 6 \sin^2 \theta = 6\hbar^2 \psi_{n22}.$$

ψ_{n22} is therefore an eigenfunction of $\hat{G}^2$ with eigenvalue $6\hbar^2$ which should be equal to $l(l+1)\hbar^2$ and therefore does have the required value for $l = 2$.

3. The operator $\hat{G}_x$ is given by

$$\hat{G}_x = -j\hbar \left(\sin \phi \frac{\partial}{\partial \theta} + \cot \theta \cos \phi \frac{\partial}{\partial \phi} \right)$$

from which

$$\hat{G}_x \psi_{n22} = -j\hbar(-2 \sin \phi \sin \theta \cos \theta - 2j \cot \theta \sin^2 \theta \cos \phi)\psi_{n22}.$$

The right-hand side of this equation is not equal to the product of a constant with ψ_{n22} and ψ_{n22} is therefore not an eigenfunction of $\hat{G}_x$. The same holds true for $\hat{G}_y$.

PROBLEM 62

The Compton Effect

A monochromatic γ-ray falls on a very thin metallic foil placed in a vacuum and, by the action of a uniform magnetic field B, electrons are extracted. Determine the energy, the frequency, and the wavelength of the incident radiation under the following conditions: R is the radius of curvature of the ejected electrons in a plane perpendicular to the field B; λ_K is the wavelength corresponding to the work function of the metal; and the constants h, c, m_e, and e are given. Neglect relativistic corrections.

Numerically, $B = 15 \times 10^{-4}$ tesla, $R = 0.10$ m, and $\lambda_K = 0.15$ Å.

II

The γ-ray above passes through hydrogen. Derive the theory of the Compton effect using relativistic mechanics. Calculate the wavelength of the photon scattered through an angle of $\theta = 90°$ with respect to the direction Ox of the incident photon beam.

Calculate the kinetic energy W_K of the recoil electrons in a direction making an angle ϕ with respect to the incident direction Ox, as a function of the ratio $\alpha = \lambda_c/\lambda$ (the Compton wavelength over the wavelength of the incident photon). Find a relationship between the angles ϕ and θ. Show on a polar plot, as a function of the angles ϕ and θ, the energy of the scattered photon and the recoil electron.

SOLUTION

I

The energy balance for the action of a γ-photon with frequency v and wavelength λ is written:

$$hv = h\frac{c}{\lambda} = \text{work required to extract the electron} + \text{the electron kinetic energy} \qquad (1)$$

The γ quantum is energetic enough to ionize metal atoms by removing K electrons from them and once this is accomplished the energy necessary to remove the electrons from the mean potential of the metal is negligible. Since one is applying Newtonian mechanics to this electron, equation (1) becomes

$$\frac{hc}{\lambda} = \frac{hc}{\lambda_K} + \frac{1}{2}mv^2. \qquad (2)$$

The velocity v of the electron having charge e and mass m_e is derived from the radius of curvature R which is a result of its trajectory normal to the magnetic field B

$$v = \frac{eBR}{m_e} = \frac{1.6\times10^{-19}}{9\times10^{-31}}\times15\times10^{-4}\times10^{-1} = 27\times10^6 \text{ m/s}.$$

Hence the photon energy

$$\frac{hc}{\lambda} = \frac{6.62\times10^{-34}\times3\times10^8}{0.15\times10^{-10}} + \frac{9}{2}\times10^{-31}\times729\times10^{12} = 165\times10^{-16} \text{ J};$$

its frequency

$$v = \frac{W}{h} = \frac{165\times10^{-16}}{6.62\times10^{-34}} = 25\times10^{18} \text{ Hz},$$

its wavelength

$$\lambda = \frac{c}{v} = \frac{3\times10^8}{25\times10^{18}} = 0.12\times10^{-10} \text{ m} = 0.12 \text{ Å}.$$

II

In relativistic mechanics energy conservation in Compton scattering is written

$$\frac{hc}{\lambda} = \frac{hc}{\lambda'} + m_ec^2\left(\frac{1}{\sqrt{1-\beta^2}} - 1\right), \qquad (3)$$

with $\beta = v/c$, the ratio of the speed of the electron to the speed of light. Conservation of momentum along the axes Ox and Oy (Fig. 62.1) gives

$$\frac{h}{\lambda} - \frac{h}{\lambda'} \cos \theta = \frac{m_e \beta c}{\sqrt{1-\beta^2}} \cos \phi \tag{4}$$

$$\frac{h}{\lambda'} \sin \theta = \frac{m_e \beta c}{\sqrt{1-\beta^2}} \sin \phi. \tag{5}$$

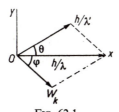

FIG. 62.1

By eliminating ϕ and the velocity βc of the electron from equations (3), (4), and (5), one gets the shift of the Compton ray

$$\lambda' - \lambda = \frac{h}{m_e c} (1 - \cos \theta) = \frac{2h}{m_e c} \sin^2 \frac{\theta}{2} = 0.0485 \sin^2 \frac{\theta}{2}. \tag{6}$$

For $\theta = 90°$

$$\lambda' = 0.12 + \frac{0.0485}{2} = 0.1443 \text{ Å}.$$

Equation (6) can be written

$$\lambda' = \lambda[1 + \alpha(1 - \cos \theta)],$$

with $\alpha = hv/m_e c^2$ one has

$$v' = \frac{v}{1 + \alpha(1 - \cos \theta)}. \tag{7}$$

From (3) and (7) one can get the kinetic energy of the electron

$$W_k = h(v - v') = hv \frac{\alpha(1 - \cos \theta)}{1 + \alpha(1 - \cos \theta)}. \tag{8}$$

By eliminating β from equations (3), (4), (5), and (7) one gets for a relationship between θ and ϕ

$$1 - \cos \theta = \frac{2}{1 + (1 + \alpha)^2 \tan^2 \phi} \tag{9}$$

which when put into (7) yields

$$W_k = hv \frac{2\alpha}{1 + 2\alpha + (1 + \alpha)^2 \tan^2 \phi}. \tag{10}$$

This is the required expression. Also, equation (9) can be written

$$(1+\alpha)^2 \tan^2 \phi = \frac{1+\cos\theta}{1-\cos\theta} = \frac{1}{\tan^2\theta/2},$$

so that by noting that θ and ϕ always have opposite signs

$$\cot\phi = -(1+\alpha)\tan\frac{\theta}{2}. \tag{11}$$

Table 62.1 gives the values of ϕ, $h\nu'/h\nu$, and $W_k/h\nu$ as a function of θ.

TABLE 62.1

θ	0	$\pi/6$	$\pi/4$	$\pi/3$	$\pi/2$	$2\pi/3$	$3\pi/4$	π
$\cos\theta$	1	0.866	0.707	0.500	0	-0.500	-0.707	-1
$\tan\theta/2$	0	0.268	0.414	0.577	1.00	1.732	2.414	∞
$\cot\phi$	0	0.282	0.435	0.605	1.05	1.82	2.53	∞
ϕ^0	90	74.25	66.5	58.8	43.6	28.8	21.55	0
$h\nu'/h\nu$	1	0.875	0.764	0.656	0.488	0.388	0.358	0.322
$W_k/h\nu$	0	0.125	0.236	0.344	0.512	0.612	0.642	0.678

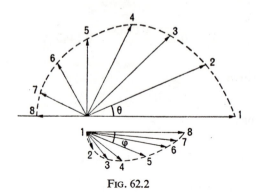

FIG. 62.2

Figure 62.2 represents the various cases found in the table.

PROBLEM 63

Planck Radiation Formula

1. *Derivation.* Consider an isothermal enclosure containing linear harmonic oscillators with eigenfrequency ν_0 and dipole moment d with density N per unit volume. At thermal equilibrium the power radiated by an oscillator in the form of electromagnetic waves is equal to that which it absorbs from the isothermal radiation which is characterized by a density $w(v) \, dv$ in the spectral interval dv. The power is radiated like a Hertzian dipole. To evaluate the absorbed power, one starts from the results of Problem 51. Assuming initially that the total mean energy of one of the oscillators obeys the equipartition principle, show that the expression obtained for $w(v)$—that is, the Rayleigh–Jeans formula—is inacceptable since it does not satisfy Stefan's law. Then making Planck's assumption that the oscillator can only take on values for its total energy which are integral multiples of an energy W, calculate the mean energy of an oscillator and the radiant energy density $w(v)$, assuming $W = hv$ where h is Planck's constant.

2. Show that Plank's formula reduces to the Rayleigh–Jeans expression at low frequencies. What form does it take at high frequencies?

3. By writing $w(v) \, dv = -w(\lambda) \, d\lambda$ express Plank's formula in terms of the wavelength, λ, as a variable. Show that it takes the form $w(\lambda) = \lambda^{-5} f(1/\lambda T)$ and that it satisfies Stefan's law:

$$w = \int w(\lambda) \, d\lambda = C_1 T^4$$

and Wein's law: $\lambda_m T = C_2$.

4. Experiments give for Wein's constant: $C_2 = 2.897 \times 10^{-3}$ m-deg and for Stefan's constant: $C_1 = 7.562 \times 10^{-16}$ W-m^{-3}-deg^{-4}.

Given:

$$\int_0^\infty \frac{x^3 \, dx}{e^x - 1} = \frac{\pi^4}{15}$$

and the solution of $5(e^x - 1) = xe^x$ as $x = 4.965$, find Planck's constant h and Boltzmann's constant k as a function of the velocity of light in free space.

SOLUTION

1. The mean power radiated by a linear Hertzian oscillator is given by (see Problem 51, also § 10.3)

$$\langle \Phi \rangle = \frac{\omega_0^4 d_m^2}{12\pi\varepsilon_0 c^3}.$$

If one introduces the oscillator energy

$$W = \frac{1}{2} m\omega_0^2 s_m^2 = \frac{1}{2} m\omega_0^2 \frac{d_m^2}{q^2}$$

and the frequency $v = \omega/2\pi$, one finds

$$\langle \Phi \rangle = \frac{2\pi v_0^2 q^2}{3\varepsilon_0 c^3 m} W. \tag{1}$$

On the other hand, the expression for the absorbed power is derived from the flux absorbed by a medium made up of classical oscillators (Problem 49). The flux Φ_0 transported by a plane wave of unit area through a distance dx encounters $N\,dx$ oscillators. The power absorbed $d\Phi_0$ is given by equation (9), Problem 49. Each oscillator absorbs power given by

$$\frac{d\Phi_0}{N\,dx} = \frac{q^2}{4\varepsilon_0 m}\,w(v). \tag{2}$$

At thermal equilibrium (1) and (2) are equal, hence

$$w(v) = \frac{8\pi v^2}{3c^3}\,W.$$

But the thermal radiation in an isothermal enclosure being isotropic has a density 3 times greater than that corresponding to a single direction of propagation. Thus

$$w(v) = \frac{8\pi v^2}{c^3}\,W. \tag{3}$$

If one has equipartition of energy (cf. Problem 79) one knows that a linear harmonic oscillator has energy kT. Equation (3) gives

$$w(v) = \frac{8\pi v^2}{c^3}\,kT. \tag{4}$$

The integral $\displaystyle\int_0^\infty w(v)\,dv$ has an infinite value if $w(v)$ is given by (4) instead of being proportional to T^4 as required by the Stefan law. The Rayleigh–Jeans law (4) thus does not represent the spectral distribution of the black body. This expression results from the equipartition of energy obtained in turn by assuming that the energy can vary continuously (see Problem 79). Assume now the converse, that the energy of an oscillator with frequency v can only take on values, 0, hv, $2hv$, ..., vhv (v being a positive integer) and that N_0 is the number of oscillators which are not excited. The number of oscillators of energy nhv in the enclosure at $T°$ is, using the Boltzmann factor,

$$N_v = N_0 \exp\left(-\frac{vhv}{kT}\right).$$

The total number of oscillators is

$$N = N_0\left[1 + \exp\left(-\frac{hv}{kT}\right) + \exp\left(-\frac{2hv}{kT}\right) + \cdots\right] = \frac{N_0}{1 - \exp\left(-hv/kT\right)},$$

and the energy of the assembly is given by

$$U = hvN_0 \exp\left(-\frac{hv}{kT}\right) + 2hvN_0 \exp\left(-\frac{2hv}{kT}\right) + \cdots + vhvN_0 \exp\left(-\frac{vhv}{kT}\right) + \cdots .$$

U is equal to the derivative of N with respect to $1/kT$, hence

$$U = \frac{N_0\,hv \exp\left(-\dfrac{hv}{kT}\right)}{\left[1 - \exp\left(-\dfrac{hv}{kT}\right)\right]^2} = \frac{Nhv}{\exp\left(\dfrac{hv}{kT}\right) - 1}.$$

The mean energy $\langle W \rangle$ of the oscillators is equal to U/N and equation (2) becomes

$$w(v, T) = \frac{8\pi h v^3}{c^3} \frac{1}{\exp(hv/kT) - 1} . \tag{5}$$

This is Planck's radiation law which gives the distribution of energy radiating from a black body with good precision.

2. For low frequencies, if $hv \ll kT$, one can expand the exponential in (5) in a series and limit it to its first two terms, giving

$$w(v, T) = \frac{8\pi h v^3}{c^3} \cdot \frac{kT}{hv} .$$

The constant h vanishes and one again has the classical expression (4) which was unsuitable.

At high frequencies, if $hv \gg kT$, the unity is negligible with respect to the exponential in (5) which leads to

$$w(v, T) = \frac{8\pi h v^3}{c^3} \exp\left(-\frac{hv}{kT}\right). \tag{6}$$

This expression defines a function of v and T analogous to that which had been proposed by Wein before Planck by placing particular conditions on the emission and absorption of radiation. It is a good representation of the isotherms of a black body above 2×10^{14} Hz (see Fig. 63.1).

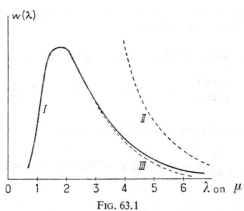

FIG. 63.1

3. Let $w(v)$ be the energy density of radiation on a frequency interval dv and $d\lambda$ the wavelength interval corresponding to dv. One has

$$\lambda v = c, \quad \text{hence} \quad v \, d\lambda + \lambda \, dv = 0$$

and

$$w(v) \, dv = -w(\lambda) \, d\lambda$$

hence

$$w(\lambda, T) = \frac{v}{\lambda} w(v) = \frac{8\pi hc}{\lambda^5} \frac{1}{\exp(hc/k\lambda T - 1)} . \tag{7}$$

This expression has the form $w_\lambda = \lambda^{-5} f(1/\lambda T)$.

Figure 63.1 represents an isotherm of a black body ($T = 1600°K$). Curve I represents formula (7), curve II the expression derived from (4), and curve III represents the formula analogous to (6).

Taking $x = hc/kT\lambda$, one has from (7)

$$w(T) = \int_0^\infty w(\lambda, T)\, d\lambda = \frac{8\pi k^4}{h^3 c^3} T^4 \int_0^\infty \frac{x^3\, dx}{e^x - 1}. \tag{8}$$

The integral is a number independent of T. Thus the total radiation from a black body is proportional to T^4. This is Stefan's law.

The maximum value of $w(\lambda, T)$ at a given temperature is found for the value, λ_m, of λ which minimizes

$$\lambda^5 \left[\exp\left(\frac{hc}{k\lambda T}\right) - 1 \right]$$

or

$$\frac{A(e^x - 1)}{x^5}, \quad \text{with} \quad A = \left(\frac{hc}{kT}\right)^5.$$

The minimum condition can be written

$$5(e^x - 1) = xe^x. \tag{9}$$

The solution of this equation in x gives a value of x as a function of constants. Thus $\lambda_m T$ is equal to a constant. This is Wein's law.

4. Equation (8) using the value of the integral given in the problem yields

$$\frac{8\pi^5 k^4}{15 h^3 c^3} = C_1 = 7.562 \times 10^{-16}\ \text{W m}^{-3}\,\text{deg}^{-4}.$$

Equation (9) has as its solution $x = 4.965$ and one has

$$\frac{hc}{4.965k} = \lambda_m T = C_2 = 2897 \times 10^{-6}\ \text{m deg}.$$

One finds

$$h = \frac{7.652 \times 10^{-16} \times (4.965 \times 2.897)^4 \times 10^{-12} \times 15}{8 \times (3.14)^5 \times 3 \times 10^8} = 6.627 \times 10^{-34}\ \text{J sec}.$$

Also

$$k = \frac{6.627 \times 10^{-34} \times 3 \times 10^8}{4.965 \times 2897 \times 10^{-6}} = 1.382 \times 10^{-23}\ \text{J/deg}.$$

PROBLEM 64

Ground State of a Two-electron Atom

One wants here to estimate the ground state energy of a two-electron atom or ion using the uncertainty relations. To apply these semi-quantitative concepts, one reasons via a one-dimensional system. The nucleus with charge Ze is assumed to be an infinitely heavy point nucleus situated at the origin.

1. Write the expression for the mechanical energy of the system in terms of the coordinates r_1 and r_2 and the momenta p_1 and p_2 of the electrons.

2. R_1 and R_2 are the dimensions of the regions where the position probabilities of the two electrons are appreciable. What is the order of magnitude of the uncertainties Δp_1 and Δp_2 of the momenta?

3. Estimate the order of magnitude of the ground state energy by finding W as a function of R_1 and R_2 then taking the minimum of this expression which is symmetric with respect to the two variables.

4. Compare the values thus obtained for the systems, H^-, He, and Li^+ with the experimental values which are -14.2, -78.4, and -196.6 eV respectively.

SOLUTION

1.

$$W = \frac{1}{2m}(p_1^2 + p_2^2) - \frac{1}{4\pi\varepsilon_0}\left(\frac{Ze^2}{r_1} + \frac{Ze^2}{r_2} - \frac{e^2}{r_{12}}\right), \qquad (1)$$

with $r_{12} = |r_1 - r_2|$.

2. $\Delta r_1 = R_1$, $\Delta r_2 = R_2$, hence

$$\Delta p_1 \approx \frac{h}{R_1} \quad \text{and} \quad \Delta p_2 \approx \frac{h}{R_2}.$$

3. The energy minimum is found classically when the two electrons are at rest ($W_k = 0$) at the origin (W_p minimally negative). This configuration is incompatible with the uncertainty relation. The states of the electrons are described by wave functions on the origin with extensions R_1 and R_2:

$$W \approx \frac{h^2}{2m}\left(\frac{1}{R_1^2} + \frac{1}{R_2^2}\right) - \frac{1}{4\pi\varepsilon_0}\left\{Ze^2\left(\frac{1}{R_1} + \frac{1}{R_2}\right) + \frac{e^2}{R_1 + R_2}\right\}. \qquad (2)$$

The minimal value corresponds to $\partial W/\partial R_1 = \partial W/\partial R_2 = 0$, so that, since W is symmetric in R_1 and R_2 and $R_1 = R_2 = R$

$$W \approx \frac{h^2}{mR^2} - \frac{1}{4\pi\varepsilon_0}\left(\frac{2Ze^2}{R} + \frac{e^2}{2R}\right). \qquad (3)$$

The condition $\partial W / \partial R = 0$ gives

$$R = \frac{4\pi\varepsilon_0 h^2}{m^2} \times \frac{1}{Z - \frac{1}{4}}. \tag{4}$$

By substituting this value in (3)

$$W_{\min} = -\left(Z - \frac{1}{4}\right)^2 \frac{me^4}{16\pi^2\varepsilon_0 h^2}. \tag{5}$$

4. One notes that if one writes the uncertainty relationship in the form $\Delta p \cdot \Delta r \approx \hbar$, equation (4) contains the factor $\varepsilon_0 h^2 / \pi m e^2$, which is the Bohr radius and (5) can be written:

$$W_{\min} = (Z - \tfrac{1}{4})^2 \times 2W_0$$

where W_0 is the ground state energy of the H atom (§ 14.5), namely 13.6 eV, and one finds then in eV

	H$^-$	He	Li$^+$
Z	1	2	3
$W_{\min}$	-15.3	-83	-205

in close agreement with the measured values.

PROBLEM 65

First-order Perturbation. Ground State of Helium

The helium atom is made up of a nucleus with $Z = 2$ and two electrons.

1. Write the Schrödinger equation for the stationary states taking the potential as being coulombic.

To solve this equation, one initially neglects the mutual interaction of the electrons (hydrogenic approximation). The equation is then in the form

$$\hat{H}_0\psi_0 = W_0\psi_0 \tag{1}$$

where $\hat{H}_0$ is the hamiltonian operator. What are the eigenvalues of the energy W_{0n} and the eigenfunctions ψ_{0n}? Show that in the minimum energy state allowed by the Pauli principle the one-electron wave function is

$$\psi_{100} = A \exp\left(-\frac{\varrho}{2}\right) \tag{2}$$

where $\varrho = 4r/r_0$, r being the distance from the nucleus to the electron and r_0 the mean radius of H and A the normalization constant whose value one should calculate.

2. Consider the coulomb repulsion between the electrons as a perturbation, that is, replace equation (1) by

$$(\hat{H}_0 + \varepsilon\hat{H}')\psi = W\psi. \tag{3}$$

$\varepsilon\hat{H}'$ is the operator corresponding to the hamiltonian perturbation which is written this way (ε being a small real number) to indicate that this term is small compared to $\hat{H}_0$. Write the expression for $\hat{H}'$. To find the eigenvalues W_n and the eigenfunctions ψ_n, one makes a series expansion in powers of ε about W_{0n} and ψ_{0n}, respectively,

$$W_n = W_{0n} + \varepsilon W'_n + \varepsilon^2 W''_n + \dots \tag{4}$$

$$\psi_n = \psi_{0n} + \varepsilon\psi'_n + \varepsilon^2\psi''_n + \dots \tag{5}$$

Introducing these expansions in equation (3) one writes the equation giving the effect of the first-order perturbation. Replace ψ'_n by an expansion on the eigenfunctions of the non-perturbed atomic system and show that the variation W'_n which the energy level n undergoes as a result of the first order perturbation is given by

$$W'_n = \int \psi^*_{0n} H' \psi_{0n}\, d\tau. \tag{6}$$

3. Show that in the ground state, the calculation of W'_1 is similar to the energy calculation for an electric charge distribution with spherical symmetry subject to an analogous potential distribution. Calculate W'_1 numerically given the ground state energy of hydrogen as 13.56 eV and thereby find the ground state energy of the helium atom. Compare this with the experimental value of -78.4 eV.

SOLUTION

1. The mass of the electrons are much smaller than the nuclear mass so that the nucleus can be considered as fixed. The hamiltonian is then given by

$$H = \frac{p_1^2}{2m_e} + \frac{p_2^2}{2m_e} - \frac{Ze^2}{4\pi\varepsilon_0 r_1} - \frac{Ze^2}{4\pi\varepsilon_0 r_2} + \frac{e^2}{4\pi\varepsilon_0 r_{12}}.$$

The indices 1 and 2 refer respectively to each of the electrons, r_1 and r_2 are their respective distances from the nucleus, and r_{12} is their mutual separation. If one regards the mutual coulomb repulsion as a perturbation, that is, the term

$$H = \frac{e^2}{4\pi\varepsilon_0 r_{12}}, \tag{7}$$

the hamiltonian for the unperturbed system is given by

$$\hat{H}_0 = \left(-\frac{\hbar^2}{2m_e}\Delta_1 - \frac{2e^2}{4\pi\varepsilon_0 r_1}\right) + \left(-\frac{\hbar^2}{2m_e}\Delta_2 - \frac{2e^2}{4\pi\varepsilon_0 r_2}\right).$$

The variables are separable and the Schrödinger equation can be broken into two parts both of which are for the hydrogen problem with nuclear charge Ze. On can easily see that if, in the solution of the hydrogen atom problem, one multiplies the potential energy by Z the energy W is multiplied by Z^2 and the eigenfunction ψ_{100} becomes

$$\psi_{100} = A \exp\left(-\frac{Zr}{r_0}\right)$$

so that with $Z = 2$ this satisfies equation (2). The constant A can be calculated by writing the normalization condition

$$\int_0^\infty \psi_{100}^2 \, d\tau = 1 = 4\pi A^2 \int_0^\infty \exp(-\varrho)\varrho^2 \, d\varrho.$$

Integrating by parts

$$\int \exp(-\varrho)\varrho^2 \, d\varrho = -\varrho^2 \exp(-\varrho) - 2\varrho \exp(-\varrho) - 2\exp(-\varrho). \tag{8}$$

The value of the integral is 2 and

$$A = \frac{1}{\sqrt{8\pi}}. \tag{9}$$

The Pauli principle allows two electrons to have the same eigenfunction ψ_{100} in the ground state if their spins are opposed. The eigenfunction of the ground state of the unperturbed system is the product of the eigenfunctions of the two electrons.

$$\psi_{100}(r_1, r_2) = \psi_{100}(r_1) \times \psi_{100}(r_2).$$

The energy of this state is the sum of the energies of each of the electrons and this is equal to $Z^2 = 4$ times the ground state energy of the hydrogen atom.

2. When one substitutes the expansions (4) and (5) in equation (3) and equates the zero and first powers of ε in the identity obtained in ε, one gets

$$\hat{H}_0 \psi_{0n} = W_{0n} \psi_{0n}$$
$$\hat{H}_0 \psi_n' + \hat{H}' \psi_{0n} = W_{0n} \psi_n' + W_n' \psi_{0n} \tag{10}$$

The first of these equations is only the unperturbed equation for the system (1). If in the second ψ_n' is replaced by the expansion

$$\psi_n' = \sum_{n'} c_{nn'} \psi_{0n'}$$

one gets, using (1),

$$\sum_{n'} c_{nn'}(W_{0n'} - W_{0n})\psi_{0n'} = (W_n' - \hat{H}')\psi_{0n}. \tag{11}$$

Multiplying both sides of this equation by ψ_{0n}^* and integrating over all space, one finds

$$\sum_{n'} c_{nn'}(W_{0n'} - W_{0n}) \int \psi_{0n}^* \psi_{0n'} \, d\tau = W_n' \int \psi_{0n}^* \psi_{0n} \, d\tau - \int \psi_{0n}^* \hat{H}' \psi_{0n} \, d\tau$$

or by taking into account the orthogonality and normalization

$$W_n' = \int \psi_{0n}^* \hat{H}' \psi_{0n} \, d\tau. \tag{12}$$

This is the required expression. One sees that the shift of the energy level n is equal to the mean value of the perturbation for the unperturbed system in state n.

3. When one applies expression (12) to the problem of the ground state of helium where (7) gives the perturbation term, one finds

$$W_1' = \int \int \psi_{100}^*(r_1) \, \psi_{100}^*(r_2) \, \frac{e^2}{4\pi\varepsilon_0 r_{12}} \, \psi_{100}(r_1) \, \psi_{100}(r_2) \, dr_1 \, dr_2. \tag{13}$$

The product $e\psi_{100}^*$ represents the electric charge density at some point due to the presence of an electron at that point. Equation (13) therefore represents the electrostatic energy of two spherically symmetric distributions each related to one electron. The energy can then be written, using (2) and (9), as

$$W_1' = \frac{4e^2}{r_0(8\pi)^2 \, 4\pi\varepsilon_0} \int \int \frac{\exp(-\varrho_1) \times \exp(-\varrho_2)}{\varrho_{12}} \, d\tau_1 \, d\tau_2 \tag{14}$$

with $\varrho_{12} = (4/r_0)r_{12}$ and the volume elements $d\tau_1$ and $d\tau_2$ being expressed as functions of ϱ_1 and ϱ_2 respectively. To evaluate this integral, one forms at each point the expression for the coulomb potential due to the volume density $\exp(-\varrho_1)$ and then the energy at this point for the volume density $\exp(-\varrho_2)$ due to the other charge. The charge

$$dQ_1 = \exp(-\varrho_1) 4\pi\varrho_1^2 \, d\varrho_1$$

is contained in a spherical shell of thickness $d\varrho_1$. At an interior point this gives a potential which is constant and equal to $dV = dQ_1/\varrho_1$. At an exterior point situated at a distance ϱ_2 from its centre it gives the same potential as if the charge contained in the shell were massed at the centre, namely $dV' = dQ_1/\varrho_2$. The potential due to the density distribution $e\psi(\varrho_1)$ is then

$$V(\varrho_2) = 4\pi \int_0^{\varrho_2} \exp(-\varrho_1) \frac{\varrho_1^2}{\varrho_2} \, d\varrho_1 + 4\pi \int_{\varrho_2}^\infty \exp(-\varrho_1)\varrho_1 \, d\varrho_1.$$

using (8) and the integral

$$\int \exp(-\varrho)\varrho \, d\varrho = -\varrho \exp(-\varrho) - \exp(-\varrho), \tag{15}$$

one finds

$$V(\varrho_2) = \frac{4\pi}{\varrho_2} [2 - 2\exp(-\varrho_2) - \varrho_2 \exp(-\varrho_2)].$$

The integral (14) then is given by

$$\mathfrak{J} = \int V(\varrho_2) \, dQ_2$$

and, by taking for dQ_2 the charge of a spherical shell

$$dQ_2 = 4\pi\varrho_2^2 \exp(-\varrho_2) \, d\varrho_2,$$

$$\mathfrak{J} = (4\pi)^2 \int_0^\infty [2\varrho_2 \exp(-\varrho_2) - 2\varrho_2 \exp(-2\varrho_2) - \varrho_2^2 \exp(-2\varrho_2)] \, d\varrho_2,$$

or, using (8) and (15),

$$\mathfrak{J} = (4\pi)^2 \times \tfrac{5}{4}.$$

The energy (14) then has the solution

$$W_1' = \frac{5}{4} \times \frac{4e^2(4\pi)^2}{(8\pi)^2 (4\pi\varepsilon_0) r_0} = \frac{10}{4} \frac{e^2}{(4\pi\varepsilon_0) 2r_0}.$$

Now the ground state energy of the hydrogen atom is given by

$$W_1 = -\frac{e^2}{(4\pi\varepsilon_0) 2r_0},$$

from which

$$W_1' = -\tfrac{10}{4}W_1 = +\tfrac{10}{4} \times 13.56 = +33.9 \text{ eV}.$$

The energy of the helium ground state in the approximation where one neglects the electronic interactions is, as was seen in no. 1,

$$W_0 = 2 \times 4W_1 = -108.5 \text{ eV},$$

Therefore, the corrected ground state energy becomes

$$W = W_0 + W_0' = -108.5 + 33.9 = -74.6 \text{ eV}.$$

One sees that W_1' is not very small compared to W_0 as should be the case in applying this method. None the less, the result obtained is still correct to within 5%.

PROBLEM 66

Second-order Perturbation. Stark Effect for a Rotor

1. By using the general perturbation method described in the second part of Problem 65, write the equation for the effect of a second-order perturbation. To do this replace the second-order term ψ_n'' of the eigenfunction of state n by a series expansion of the eigenfunctions of the unperturbed system. Use an expansion analogous to that used in Problem 65 and find the coefficients. Show that the second-order correction which must be applied to the energy level n is given by

$$W_n'' = \sum_{n' \neq n} \frac{\left(\int \psi_{0n'}^* \hat{H}' \psi_{0n} \, d\tau \right)^2}{W_{0n} - W_{0n'}}. \tag{1}$$

2. A diatomic molecule, which has a moment of inertia I about an axis passing normally through the line joining the nuclei and through the center of mass of the molecule and a dipole moment d, will be treated as a planar rotor. It is placed in a constant uniform electric field E normal to the axis of rotation. By treating the action of this field as a perturbation, give the first non-zero term correcting the energy levels of this rotor.

SOLUTION

1. By introducing the expansions (4) and (5) from Problem 65 into equation (3) of that problem, the terms in ε^2 are found to be

$$\hat{H}_0 \psi_n'' + \hat{H}' \psi_n' = W_{0n} \psi_n'' + W_n \psi_n' + W_n'' \psi_{0n}$$

or

$$(\hat{H}_0 - W_{0n}) \psi_n'' = (W_n' - \hat{H}') \psi_n' + W_n'' \psi_{0n}. \tag{2}$$

Assuming, as indicated in the statement of the problem

$$\psi_n'' = \sum_{n'} \gamma_{nn'} \psi_{0n'}$$

and introducing this expansion in (2) then multiplying each term by ψ_{0n}^* and integrating over all space, one finds, using equation (10) (Problem 65) and the expansion of ψ_n',

$$\sum_{n'} \gamma_{nn'} (W_{0n'} - W_{0n}) \int \psi_{0n}^* \psi_{0n'} \, d\tau =$$

$$= \sum_{n' \neq n} c_{nn'} W_n' \int \psi_{0n}^* \psi_{0n'} \, d\tau - \sum_{n' \neq n} c_{nn'} \int \psi_{0n}^* \hat{H}' \psi_{0n'} \, d\tau + W_n'' \int \psi_{0n}^* \psi_{0n} \, d\tau.$$

As a result of the orthonormality of the unperturbed eigenfunctions, the integral on the left and the first integral on the right are zero and the last integral on the right is unity. Thus

$$W_n'' = \sum_{n' \neq n} c_{nn'} \int \psi_{0n}^* \hat{H}' \psi_{0n'} \, d\tau.$$

To calculate $c_{nn'}$ one refers again to equation (11) (Problem 65) (changing the present indices n' and n'') and multiplies both sides of the equation by $\psi_{0n'}^*$. One gets

$$\sum_{n''} c_{nn'} (W_{0n'} - W_{0n}) \int \psi_{0n'}^* \psi_{0n''} \, d\tau = \int \psi_{0n'}^* (W_n' - \hat{H}') \psi_{0n} \, d\tau$$

from which, for $n' \neq n$ and using equation (12) (Problem 65),

$$c_{nn'} = \frac{\int \psi_{0n'}^* \hat{H}' \psi_{0n} \, d\tau}{W_{0n} - W_{0n'}}$$

and

$$W_n'' = \sum_{n' \neq n} \frac{\left(\int \psi_{0n'}^* \hat{H}' \psi_{0n} \, d\tau \right)^2}{W_{0n} - W_{0n'}},$$

which is equation (1).

2. The unperturbed rotor has only kinetic energy. The hamiltonian operator is

$$\hat{H}_0 = \frac{\hat{G}_z^2}{2I}$$

G_z being the angular momentum. By replacing the operator $\hat{G}_z$ by its value (§ 14.3)

$$\hat{G}_z = -\frac{\hbar}{j}\frac{\partial}{\partial\phi},$$

where $j = \sqrt{-1}$ and ϕ is the angle variable, Schrödinger's equation

$$H_0\psi_0 = W_0\psi_0$$

becomes

$$-\frac{\hbar^2}{2I}\frac{\partial^2\psi_0}{\partial\phi^2} = W_0\psi_0 \tag{3}$$

which has eigenvalues

$$W_{0J} = \frac{\hbar^2 J^2}{2I} \tag{4}$$

and eigenfunctions

$$\psi_{0J} = \frac{1}{\sqrt{2\pi}}\exp(\pm jJ\phi). \tag{5}$$

J being zero or an integer.

In a uniform electric field E, the potential energy of a rotor with dipole moment d is

$$W_p = -E \cdot d = -Ed\cos\phi$$

since E, normal to the axis of rotation and making the angle ϕ with d, is a suitable coordinate origin reference.

The perturbed hamiltonian is $\hat{H}' = W_p$ and equation (3) is replaced by

$$\frac{d^2\psi}{d\phi^2} + \frac{2I}{\hbar^2}(W_0 + Ed\cos\phi)\psi = 0. \tag{6}$$

Calculating the first order correction using equation (12) (Problem 65), one finds

$$W'_J = \int_0^{2\pi}\psi_{0J}^*\hat{H}'\psi_{0J}\,d\phi = -\frac{Ed}{2\pi}\int_0^{2\pi}\exp[j(J-J)\phi]\cos\phi\,d\phi. \tag{7}$$

The integral is zero and as a result the energy shift of all the levels is zero in the first order.

The second-order correction is given by equation (1) which is written here as

$$W''_J = \sum_{J'\neq J}\frac{\left(\int_0^{2\pi}\psi_{0J'}^*\hat{H}'\psi_{0J}\,d\phi\right)^2}{W_{0J}-W_{0J'}}. \tag{8}$$

The W_{0J} are given by (4) and the ψ_{0J} by (5). One then has

$$\int_0^{2\pi}\psi_{0J'}^*\hat{H}'\psi_{0J}\,d\phi = -\frac{Ed}{2\pi}\int_0^{2\pi}\exp[j(J'-J)\phi]\cos\phi\,d\phi.$$

One knows that this integral is zero for all values of $J'-J$ except for the value ± 1 in which case it is π. The sum (8) then reduces to two terms, in agreement with the selection rule

$\Delta J = \pm 1$ for the rotor (§ 16.2)

$$W''_J = \frac{E^2 d^2}{4\pi^2} \left\{ \frac{\pi^2}{W_{0J} - W_{0, J-1}} + \frac{\pi^2}{W_{0J} - W_{0, J+1}} \right\}$$

$$W''_J = \frac{E^2 d^2}{4\pi^2} \left\{ \frac{2I\pi^2}{\hbar^2[J^2 - (J-1)^2]} + \frac{2I\pi^2}{\hbar^2[J^2 - (J+1)^2]} \right\} = \frac{E^2 d^2 I}{\hbar^2 (4J^2 - 1)} .$$

Thus the energy levels of the perturbed rotor are given by

$$W_J = W_{0J} + W''_J = \frac{\hbar^2 J^2}{2I} + \frac{E^2 d^2 I}{\hbar^2 (4J^2 - 1)} .$$

PROBLEM 67

Intramolecular Potential of Ethane

One assumes that the function $W_p = -W_0 \cos 3\theta$ suitably represents the variation in potential energy of the ethane molecule when the two methyl groups rotate with respect to one another about the carbon–carbon bond. The angle θ is the angle between $C'CH_1$ and $CC'H'_1$ (Fig. 67.1) and W_0 is a constant characteristic of the molecule.

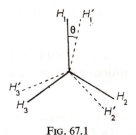

FIG. 67.1

Take the bond lengths $CH_1 = CH_2 = CH_3 = a$ and recall that the direction of the bonds from C to H_1, H_2, H_3, and C' have the symmetry of a regular tetrahedron.

Two separate cases will be considered in this problem:

Case 1: the ethane molecule H_3C—CH_3, and Case 2: the hexachloroethane molecule Cl_3C—CCl_3.

Numerical values: mass of $Cl = 35.5$ times the mass of H; $a = 1.08$ Å (Case 1) and 1.8 Å (Case 2); $W_0 = 0.06$ eV (both cases). Also given are the values of N, e, and h.

1. Show that the Schrödinger equation can be written:

$$\frac{h^2}{8\pi^2 I} \cdot \frac{d^2\psi}{d\theta^2} + W_0 \cos 3\theta \cdot \psi = -W\psi, \tag{1}$$

I being the reduced moment of inertia with respect to the CC' axis. Find I for both Case 1 and Case 2.

2. Assume initially that W_0 is negligible with respect to W.

(a) Solve the Schrödinger equation in this case giving the expression for the eigenvalues and the eigenfunctions of the system by taking into account the periodicity conditions on the function ψ.

(b) What is the probability of finding the various states of the system as a function of the angle θ?

(c) Compare the energy levels found above with those of a rotor having the same moment of inertia about a fixed axis.

(d) Is the condition W_0 negligible with respect to W satisfied? Calculate W in electron-volts for Cases 1 and 2.

3. Assume now that W_0 plays an important role.

(a) Show from purely physical considerations, that ψ takes on significant values for $3\theta = 2k\pi$ (k integer). This leads to the study of the Schrödinger equation with the retention of only the first two terms of the expansion of $\cos 3\theta$.

(b) Is the form of the transformed equation recognizable? With what problem is it associated?

4. One wants to determine the energy levels of the system defined by the preceding equation. Look for solutions of the form

$$\psi(\theta) = \exp\left(-\frac{b\theta^2}{2}\right) \cdot P(\theta)$$

where b is a constant to be determined and $P(\theta)$ is a polynomial. By writing $P(\theta)$ with a finite number of terms, give the conditions which define the possible energy values W of the system.

(a) Apply these conditions and write the general expression for W.

(b) Compare W to W_0. Calculate the first three levels for W in electron-volts in Case 1 and Case 2. Discuss the results.

SOLUTION

1. The rotation of the CH_3 groups with respect to one another takes place about the C—C' axis fixed in the molecule. This axis is perpendicular to the plane of Fig. 67.1 with the group $CH_1H_2H_3$ being in front of $C'H_1'H_2'H_3'$. The relative position of these two groups depends only on the angle θ. The problem is actually analogous to the rotor about a fixed axis (§ 14.2) but differs on two points. First, since the methyl groups are both mobile, the moment of inertia which applies is a reduced moment of inertia analogous to the reduced mass of a linear oscillator, it being in general

$$I = \frac{I_1 I_2}{I_1 + I_2}.$$

Here $I_1 = I_2 = I_0$ and $I = I_0/2$, I_0 being the moment of inertia of a methyl group with respect to the CC' axis. One has (Fig. 67.2)

$$I_0 = \Sigma mr^2 = 3m(a \sin \alpha)^2$$

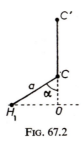

FIG. 67.2

where m is the mass of the hydrogen (or chlorine) atom and α the supplement to the angle

$$\widehat{H_1CC'} = 109°28'.$$

For C_2H_6:

$$I = \frac{3 \times 10^{-26}}{2 \times 6.025} (1.08 \times 0.9428)^2 = 2.66 \times 10^{-47} \text{ kg/m}^2.$$

For C_2Cl_6:

$$I = \frac{3 \times 10^{-26} \times 35.5}{2 \times 6.025} (1.8 \times 0.9428)^2 = 254 \times 10^{-47} \text{ kg/m}^2.$$

Secondly, one has here a potential energy W_p which does not exist for the rotor. Starting from the Schrödinger equation relative to the stationary states for a mass m

$$\frac{\hbar^2}{2m} \Delta\psi + (W - W_p)\psi = 0,$$

a discussion analogous to that of the rotor shows that the first term can be put in the form

$$\frac{\hbar^2}{2I} \frac{d^2\psi}{d\theta^2}.$$

Additionally, taking into account the expression $W_p = -W_0 \cos 3\theta$ given in the problem, one finds the Schrödinger equation to be equation (1) as required.

2. If the potential energy is negligible with respect to the energy of the molecule, equation (1) takes the form of a rotor about a fixed axis. It has a free rotation for the two halves of the molecule one with respect to the other. The eigenfunctions have the form of the rotor functions (§ 14.2):

$$\psi = C \exp \left(\pm j \sqrt{\frac{2IW}{\hbar^2}} \phi + \alpha \right). \tag{2}$$

But here, examination of Fig. 67.1 shows that the function is periodic for period $2\pi/3$

$$\psi(0) = \psi\left(\frac{2\pi}{3}\right).$$

This condition, when applied to (2), gives

$$\exp\left(\pm j\sqrt{\frac{2IW}{\hbar^2}} \cdot \frac{2\pi}{3}\right) = 1$$

which determines the energy eigenvalues

$$W_J = J^2 \frac{9\hbar^2}{2I} \qquad (J \text{ integer}). \tag{3}$$

One out of three of the energy levels of a rotor about a free axis is permitted.
For C_2H_6:

$$W_J = J^2 \frac{9\times(1.054\times10^{-34})^2}{2\times2.66\times10^{-47}} = J^2 \frac{9.999\times10^{-68}}{5.32\times10^{-47}} = 1.90\times10^{-21} \, J^2 \text{ joules}$$

$$W_J = \frac{1.90\times10^{-21}}{1.60\times10^{-19}} J^2 = 1.19\times10^{-2} \, J^2 \text{ eV}.$$

The energy W_J exceeds the value W_0 of the second excited level $(J = 2)$.
For C_2Cl_6:

$$W_J = J^2 \frac{9\times(1.054\times10^{-34})^2}{2\times254\times10^{-47}\times1.60\times10^{-19}} = 1.2\times10^{-4} \, J^2 \text{ eV}.$$

This time, J must be equal to 23 for W_J to reach W_0. Thus it is easy to cause an internal rotation by excitation of the C_2H_6 molecule while in contrast this is difficult in the case of the C_2Cl_6 molecule.

3. Equilibrium exists in the molecule when the potential energy is minimal, that is for $\theta = 2K\pi/3$. The probability of finding such a configuration is maximal. By virtue of the probabilistic interpretation of ψ, this function will have significant values for those values of θ above. For small values of θ, equation (1) becomes

$$\frac{\hbar^2}{2I}\frac{d^2\psi}{d\theta^2} + \left[W + W_0\left(1 - \frac{9}{2}\theta^2\right)\right]\psi = 0. \tag{4}$$

In this form, where the potential energy is a quadratic form of the dependent variable of ψ, it appears similar to the equation for the harmonic oscillator. One has here rotational oscillations which produce a torsion in the molecule about the axis CC′.

4. To solve equation (5) one proceeds in the same way as for the harmonic oscillator (§ 14.4). Taking

$$a = \frac{2I}{\hbar^2}(W + W_0), \qquad b^2 = \frac{9IW_0}{\hbar^2}, \qquad q = \theta\sqrt{b},$$

equation (4) becomes

$$\frac{d^2\psi}{dq^2} + \left(\frac{a}{b} - q^2\right)\psi = 0. \tag{5}$$

For values of q much greater than a/b, this equation has an asymptotic solution

$$\psi = \exp\left(-\frac{q^2}{2}\right) = \exp\left(-\frac{b\theta^2}{2}\right).$$

One then looks for a general solution of the form

$$\psi = \exp\left(-\frac{q^2}{2}\right) \times P(q),$$

for which

$$\frac{d^2\psi}{dq^2} = \exp\left(-\frac{q^2}{2}\right)\left[\frac{d^2P}{dq^2} - 2q\frac{dP}{dq} + (q^2-1)P\right].$$

By substitution in equation (5) one gets

$$\exp\left(\frac{-q^2}{2}\right)\left[\frac{d^2P}{dq^2} - 2q\frac{dP}{dq} + \left(\frac{a}{b} - 1\right)qP\right] = 0. \tag{6}$$

The exponential term only vanishes for $q = \infty$, therefore the expression within the brackets must be zero. Taking:

$$P(q) = a_0 + a_1 q + a_2 q^2 + \ldots \tag{7}$$

substituting in (6) and letting the coefficients of the successive powers of q be zero, one finds

$$2a_2 + \left(\frac{a}{b} - 1\right)a_0 = 0, \quad 6a_3 - 2a_1 + \left(\frac{a}{b} - 1\right)a_1 = 0 \ldots$$

whose general form is

$$(n+1)(n+2)a_{n+2} - 2na_n + \left(\frac{a}{b} - 1\right)a_n = 0.$$

Hence the recurrence relationship

$$\frac{a_n + 2}{a_n} = \frac{a/b - 1 - 2n}{(n+1)(n+2)} \qquad (n \text{ integer or zero}). \tag{8}$$

If $a/b = 2n+1$, the series (7) ends with the term q^n since the coefficient of q^{n+2} and all terms of higher order vanish. The same is true for a^{n+1}. One has then

$$2n+1 = \frac{a}{b} = \frac{2\sqrt{I}}{3\hbar}\frac{W+W_0}{\sqrt{W_0}},$$

hence

$$W = 3\left(n+\frac{1}{2}\right)\hbar\sqrt{\frac{W_0}{I}} - W_0. \tag{9}$$

For C_2H_6 the values of the first three energy levels are

$$n = 0, \quad W = \frac{3}{2} \times 1.054 \times 10^{-34} \sqrt{\frac{0.06 \times 1.60 \times 10^{-19}}{2.66 \times 10^{-47}}} - 0.06 \times 1.60 \times 10^{-19}$$

$$W + W_0 = 0.30 \times 10^{-20} \text{ J} = 1.87 \times 10^{-2} \text{ eV}.$$

The quantity $W + W_0$ is the energy measured from the minimum of the potential energy

$$n = 1, \quad W + W_0 = 0.90 \times 10^{-20} \text{ J} = 5.61 \times 10^{-2} \text{ eV}$$

$$n = 2, \quad W + W_0 = 1.50 \times 10^{-20} \text{ J} = 9.35 \times 10^{-2} \text{ eV}$$

For C_2Cl_6:

$$n = 0, \quad W + W_0 = 0.308 \times 10^{-21} \text{ J} = 0.19 \times 10^{-2} \text{ eV}$$

$$n = 1, \quad W + W_0 = 0.57 \times 10^{-2} \text{ eV}$$

$$n = 2, \quad W + W_0 = 0.95 \times 10^{-2} \text{ eV}$$

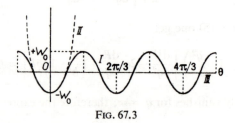

FIG. 67.3

Figure 67.3 shows the value of the assumed potential energy (sinusoidal curve) and its value in the parabolic (II) approximation of equation (4). For large values of W, W_p is negligible and is treated like the line (III).

PROBLEM 68

Vibrational–rotational Energy of a Diatomic Molecule

One will consider in this problem a heteronuclear diatomic molecule in the Born–Oppenheimer approximation. Assume that the potential energy of the two-atom system (due both to the actions of the electrons and the nuclei) can be represented by the expression

$$W_p(r) = -2D\left(\frac{a}{r} - \frac{a^2}{2r^2}\right),$$

where r is the internuclear separation and a and D are constants (D being the dissociation energy of the molecule).

1. Set up the time-independent Schrödinger equation characterizing the motion of the nuclei.

2. Write in polar coordinates—the advantage of which should be shown—the general wave function solutions of the Schrödinger equation.

3. Write the equation which determines the radial part of the wave function $\psi_r(\varrho)$ by using two dimensionless quantities, $\varrho = r/a$ and the parameter $A^2 = (2\mu a^2/\hbar^2)D$. Show the analogy between this problem and the hydrogen atom problem. One will rely on this analogy later on in the problem.

4. Show that the equation giving $\psi_r(\varrho)$ solutions of the form

$$\psi_r(\varrho) = \varrho^p \exp{(-b\varrho)} \times f(\varrho).$$

Determine the values of the constants p and b and write the equation which the function $f(\varrho)$ must satisfy.

5. Find the possible values of the rotation-vibration energy of the molecule as functions of the vibrational quantum numbers n and the rotational quantum numbers J.

6. The parameter A is large compared to unity. For small values of n and J find a suitable expression for the energy inclusive of the second-order terms. Interpret this result.

SOLUTION

1. The Schrödinger equation relative to the molecule whose nuclei are numbers 1 and 2 is

$$\frac{1}{M_1} \Delta_1\psi + \frac{1}{M_2} \Delta_2\psi + \frac{2}{\hbar^2} [W - W_p(r)]\psi = 0.$$

Δ_1 and Δ_2 are the laplacians expressed as functions of the coordinates of the two nuclei. By introducing the reduced mass

$$\mu = \frac{M_1 M_2}{M_1 + M_2}$$

and the internuclear distance $r = r_1 - r_2$, one gets

$$\Delta\psi + \frac{2\mu}{\hbar^2} \left\{ W + 2D\left(\frac{a}{r} - \frac{a^2}{2r^2}\right) \right\}\psi = 0. \tag{1}$$

2. Since the potential is central, polar coordinates allow the separation of the variables (§ 14.7.3). One can assume (§ 14.5.1)

$$\psi(r, \theta, \phi) = \psi_r(r) \cdot \Phi(\theta, \phi),$$

and one finds an equation analogous to the H atom, namely

$$\frac{d}{dr}\left(r^2 \frac{d\psi_r}{dr}\right) + \left\{\frac{2\mu r^2}{\hbar^2}\left[W + 2D\left(\frac{a}{r} - \frac{a^2}{2r^2}\right) - J(J+1)\right]\right\}\psi_r = 0. \tag{2}$$

3. By taking $\varrho = r/a$ and $d^2/dr^2 = 1/a^2 (d^2/d\varrho^2)$, equation (2) becomes

$$\varrho^2 \frac{d^2\psi_\varrho}{d\varrho^2} + 2\varrho \frac{d\psi_\varrho}{d\varrho} + \left\{ \frac{2\mu a^2 W}{\hbar^2} \varrho^2 + \frac{4D\mu a^2}{\hbar^2} \varrho - \frac{2D\mu a^2}{\hbar^2} - J(J+1) \right\} \psi_\varrho = 0$$

or, setting $A^2 = 2D\mu a^2/\hbar^2$,

$$\frac{d^2\psi_\varrho}{d\varrho^2} + \frac{2}{\varrho} \frac{d\psi_\varrho}{d\varrho} + \left\{ \frac{A^2}{D} W + \frac{2A^2}{\varrho} - \frac{A^2 + J(J+1)}{\varrho^2} \right\} \psi_\varrho = 0. \tag{3}$$

4. For $\varrho \to \infty$, the asymptotic equation derived from (3) is

$$\frac{d^2\psi_\varrho}{d\varrho^2} + \frac{A^2 W}{D} \psi_\varrho = 0, \tag{4}$$

For which the proper solution is

$$\exp\left(-A \sqrt{\frac{W}{D}} \varrho \right)$$

hence

$$b = A \sqrt{\frac{W}{D}}. \tag{5}$$

On the other hand, for $\varrho \to 0$, equation (3) is asymptotic to

$$\frac{d^2\psi_\varrho}{d\varrho^2} + \frac{2}{\varrho} \frac{d\psi}{d\varrho} - \frac{A^2 + J(J+1)}{\varrho^2} \psi_\varrho = 0, \tag{6}$$

a second-order differential equation with constant coefficients. A discussion analogous to that for the H atom shows that the solution of this equation must behave as ϱ^p, which gives by substitution in (6), the equation

$$p^2 + p - A^2 - J(J+1) = 0.$$

The only acceptable root is

$$p = -\tfrac{1}{2} + \sqrt{A^2 + (J + \tfrac{1}{2})^2}, \tag{7}$$

since $p > -1$.

Thus, the solution of (3) has the form required, namely

$$\psi_r(\varrho) = \varrho^p \exp(-b\varrho) f(\varrho), \tag{8}$$

where p is given by (7) and b by (5). By substituting (8) in (3), one finds the equation which $f(\varrho)$ satisfies

$$f'' + 2f'\left(\frac{p+1}{\varrho} - A \sqrt{\frac{W}{D}} \right) + \frac{2f}{\varrho} \left[A^2 - A \sqrt{\frac{W}{D}} (p+1) \right] = 0.$$

5. By reasoning similar to that applied in the H atom problem (§ 14.5.1) one finds that he must have

$$A \sqrt{\frac{D}{W}} - (p+1) = n,$$

n being a positive integer. The energy is given by

$$W = -\frac{DA^2}{\left[n+\frac{1}{2}+\sqrt{A^2+(J+\frac{1}{2})^2}\right]^2} \cdot$$

(9)

6. If $A^2 \gg n$ and J, one has

$$\sqrt{A^2+(J+\tfrac{1}{2})^2} \approx A\left(1+\frac{1}{2}\left(\frac{J+\frac{1}{2}}{A}\right)^2\right)$$

and the denominator of (9) becomes

$$A^2\left[1+\frac{n+\frac{1}{2}}{A}+\frac{1}{2}\left(\frac{J+\frac{1}{2}}{A}\right)^2\right],$$

from which the energy approximation is

$$W = -D\left[1-\frac{2(n+\frac{1}{2})}{A}-\left(\frac{J+\frac{1}{2}}{A}\right)^2\right]$$

or:

$$W = -D+\hbar\omega\left(n+\frac{1}{2}\right)+\left(J+\frac{1}{2}\right)^2\frac{\hbar^2}{2\mu a^2}$$

(10)

by taking:

$$\omega = \frac{2D}{A} = \hbar\sqrt{\frac{2D}{\mu a^2}} \cdot$$

The first term of (10) represents the dissociation energy (calculated from a zero value of energy, without taking into account the ground state vibrational energy) (§ 16.3.1). The second term represents the harmonic vibrational energy (one should verify that ω has the dimensions of frequency). The third represents the rotational energy (§ 14.2).

ATOMIC AND MOLECULAR SPECTRA

PROBLEM 69

Spectrum of the Hydrogen Atom

I

Quantum theory of spectral line emission. Bohr theory. Application to light hydrogen. Calculate the Rydberg constant, R_H, in the case of a fixed nucleus and in the case where the motion of the nucleus is taken into account.

Numerical application:

$$h = 6.625 \times 10^{-34} \text{ J sec}, \quad c = 2.998 \times 10^8 \text{ m/s}, \quad e = 1.602 \times 10^{-19} \text{ coulomb}.$$

$$\varepsilon_0 = \frac{1}{4\pi \times 9 \times 10^9}, \quad m = 9.108 \times 10^{-31} \text{ kg}.$$

The ratio of the mass of the electron, m, to the mass of the nucleus of light hydrogen M is equal to 1/1838.

II

Find the general relationship giving the number of waves per centimetre in the radiation emitted by a light hydrogen atom when the electron passes from the level n to the level n'.

Numerical application. Calculate the wavelength in dry air at 15°C and at atmospheric pressure of the first four lines of the Balmer series (the lines H_α, H_β, H_γ, H_δ). What is the wavelength of the series limit? What are the resonance potentials and the ionization potential for the light hydrogen atom?

Take as the index of dry air at 15°C and atmospheric pressure 1.000280 and for R_H the experimental value 109,678 cm^{-1}.

III

The spectrum of ionized helium and deuterium. Rydberg and Pickering series

Numerical application. Calculate the Rydberg constant R_D for deuterium (heavy hydrogen) for the case of a fixed nucleus and in the case where the motion of the nucleus is taken into account. The ratio of the mass of the electron to the mass of the deuterium nucleus is equal to 1/3571.

Assuming that $R_D = 109{,}707$ cm^{-1}, determine the wavelengths in dry air at 15°C and atmospheric pressure of the first four lines of the Balmer series of deuterium analogous to the H_α, H_β, H_γ, and H_δ lines of light hydrogen.

IV

One wants to study the two H_α lines assumed to be emitted with the same intensity from a mixture of light hydrogen and deuterium using a grating spectrograph with a plane reflection grating. The grating is 5 cm long with 500 lines/mm and is illuminated by a vanishingly narrow slit placed in the focal plane of an objective. The slit is parallel to the lines on the grating.

The diffracted ray is observed with a telescope with a 40-cm focal length directed normally at the grating.

What should be the angle of incidence so that the image of the slit, diffracted in third order, is formed on the crosshairs of the telescope? What is the value of the angular dispersion? What is the linear separation in the focal plane of the objective of two lines 1 Å apart? What is the linear separation of the H_α lines of the light hydrogen and deuterium? What is the theoretical power of resolution? What should the power of the ocular be so that the resolving power is effectively used? Assume that the eye can separate 1' of arc and the objective of the collimator and of the telescope have an opening sufficiently large so as not to lessen the characteristic resolving power of the grating.

SOLUTION

I

In the primitive Bohr theory (§ 12.1) one assumes that the electron describes about the nucleus in the ground state of hydrogen (or about the centre-of-mass), a circular orbit of radius r_0 and that the Coulomb attraction

$$F = \frac{e^2}{4\pi\varepsilon_0 r_0^2} \tag{1}$$

provides the required centripetal attraction for this orbit, viz. $F' = mv^2/r_0$ and thus

$$\frac{e^2}{4\pi\varepsilon_0 r_0} = mv^2. \tag{2}$$

The electrostatic potential energy is, using (1),

$$W_p = -\frac{e^2}{4\pi\varepsilon_0 r_0}$$

and the total energy

$$W_0 = W_p + W_k = -\frac{e^2}{4\pi\varepsilon_0 r_0} + \frac{mv^2}{2} = -\frac{e^2}{8\pi\varepsilon_0 r_0} \tag{3}$$

(in accordance with the virial theorem, § 12.2.1). Bohr fixed the radius r_0 by the condition that the angular momentum be equal to $h/2\pi$, thus arbitrarily introducing quantization. In (2) one replaces v with $h/2\pi m r_0$:

$$r_0 = \frac{\varepsilon_0 h^2}{\pi m e^2},$$
(4)

and using this latter value in (3),

$$W_1 = -\frac{m e^4}{8 \varepsilon_0 h^2}.$$
(5)

Equations (4) and (5) are found to coincide with the equation for the ground state of hydrogen found from quantum mechanics (§ 14.5). But the reasoning by which this is obtained here is not satisfactory since it is necessary to postulate a circular motion for the electron so that it has angular momentum. However, one knows that the ground state, like all s states, lacks angular momentum. The Bohr theory in addition fails when one tries to extend it to apply to excited states of hydrogen or to atoms with more than one electron.

Using the definition of spectral term, one has (§ 12.2)

$$W_1 = -hcR,$$
(6)

hence

$$R_\infty = \frac{m e^4}{8 \varepsilon_0^2 h^3 c} = \frac{9.108 \times 10^{-31} \times (1.602 \times 10^{-19})^4 \times 16 \times (3.14)^2 \times 81 \times 10^{18}}{8 \times (6.625 \times 10^{-34})^3 \times 2.998 \times 10^8} = 109{,}915.3 \text{ cm}^{-1}.$$

This value is obtained under the assumption that the electron moves about a fixed nucleus. To take into account the simultaneous motion of the electron and the nucleus with mass—about the centre-of-mass it is necessary to replace the mass m by the reduced mass (§ 14.2)

$$\mu = \frac{Mm}{M+m}.$$

One then finds

$$R_H = R_\infty \frac{\mu}{m} = R_\infty \frac{1}{1+m/M} = 109{,}855.5 \text{ cm}^{-1}.$$
(7)

II

Using the fundamental Bohr postulates—whose validity has survived in spite of its elementary nature—the emission and absorption of radiation of frequency ν and energy $h\nu$ is accomplished by transition between two stationary states whose quantum energies are given for hydrogen by the expression

$$W_n = -\frac{m e^4}{8 \varepsilon_0 h^2} \cdot \frac{1}{n^2},$$
(8)

where n is a positive integer. The conservation of energy in passing from level n to level n' by emission ($n' < n$) is written

$$h\nu = W_n - W_{n'} = \frac{m e^4}{8 \varepsilon_0 h^2} \left(\frac{1}{n'^2} - \frac{1}{n^2} \right).$$

Taking equation (6) into account and the expression $\nu = c\tilde{\nu}_0$ between the frequency and the spectroscopic wave number $\tilde{\nu}_0 = 1/\lambda_0$, this becomes

$$\lambda_0 = R\left(\frac{1}{n'^2} - \frac{1}{n^2}\right)$$

$\lambda_0 = 1/\tilde{\nu}_0$ is the wavelength in free space since equation (6) uses the velocity c. In a medium with index N, the wavelength $\lambda = \lambda_0/N$. For the lines of the Balmer series $n' = 2$ thus

$$\tilde{\nu}_0 = R\left(\frac{1}{4} - \frac{1}{n^2}\right) = 109{,}678\left(\frac{1}{4} - \frac{1}{n^2}\right).$$

$H_\alpha \quad \tilde{\nu}_0 = 109{,}678\left(\frac{1}{4} - \frac{1}{9}\right) = 109{,}678 \times \frac{5}{36} = 15{,}233.1 \text{ cm}^{-1}$

$$\lambda_\alpha = 6562.8 \text{ Å}$$

$H_\beta \quad \tilde{\nu}_0 = 109{,}678\left(\frac{1}{4} - \frac{1}{16}\right) = 109{,}678 \times \frac{3}{16} = 20{,}564.6 \text{ cm}^{-1}$

$$\lambda_\beta = 4861.3 \text{ Å};$$

$H_\gamma \quad \tilde{\nu}_0 = 109{,}678\left(\frac{1}{4} - \frac{1}{25}\right) = 109{,}678 \times 0.21 = 23{,}032.4 \text{ cm}^{-1}$

$$\lambda_\gamma = 4340.5 \text{ Å};$$

$H_\delta \quad \tilde{\nu}_0 = 109{,}678\left(\frac{1}{4} - \frac{1}{36}\right) = 109{,}678 \times \frac{8}{36} = 24{,}372.9 \text{ cm}^{-1}$

$$\lambda_\delta = 4101.8 \text{ Å}.$$

At the series limit, $n = \infty$, hence

$$\tilde{\nu}_0 = \frac{R}{4} = \frac{109{,}678}{4} = 27{,}419.5 \text{ cm}^{-1},$$

$$\lambda_\infty = 3646.0 \text{ Å}.$$

The successive resonance potentials, multiplied by the charge e, represent the energy necessary for the electron to pass from the ground state to the successive levels $n = 1, 2, \ldots$, with n having a finite value.

For the first of these potentials, namely V_1, one finds using (8) and (6)

$$eV_1 = W_1 - W_2 = \frac{me^4}{8\varepsilon_0 h^2}\left(\frac{1}{1} - \frac{1}{4}\right) = hcR\left(1 - \frac{1}{4}\right),$$

$$V_1 = \frac{hcR}{e} \times \frac{3}{4} = \frac{6.625 \times 10^{-34} \times 2.998 \times 10^8 \times 1.096\ 78 \times 10^4 \times 3}{1.602 \times 10^{-19} \times 4} = 10.198 \text{ eV}.$$

For the second resonance potential:

$$V_2 = \frac{hcR}{e}\left(\frac{1}{1} - \frac{1}{9}\right) = \frac{hcR}{e} \times \frac{8}{9} = 12.087 \text{ eV}.$$

The ionization potential V_i is that potential through which the electron with charge e must pass from the level $n_0 = 1$ to the state $n = \infty$, where it is separated from the nucleus. One has

$$eV_i = W_1 = hcR$$

from which, using the value calculated for V_1,

$$V_i = \frac{hcR}{e} = 13.598 \text{ eV.}$$

One notes that the ionization energy hcR corresponds to the Lyman series limit, $n' = 1$, $n = \infty$, whose wave number is then $\tilde{v}_0 = R$.

III

The deuterium atom D has only one electron and a nucleus with the same charge and twice the mass of that of light hydrogen. Its spectrum will be identical to that of H if one neglects the motion of the nucleus. If one takes this motion into account, the value of the reduced mass is modified and so is the Rydberg constant which now becomes:

$$R_D = R_\infty \frac{1}{1 + \dfrac{1}{3571}} = 109{,}884.5 \text{ cm}^{-1}$$

The ionized helium atom He^+ has one electron, a mass 4 times that of hydrogen and twice the nuclear charge. The effect of the mass once again is given by (7). For the effect of the charge difference it is necessary to make the force and the Coulomb potential twice as large which leads to multiplication of the energy levels (8) by the factor 4. The helium lines as a result have wave numbers given by

$$\tilde{v}_0 = 4R_{He}\left(\frac{1}{n_0^2} - \frac{1}{n^2}\right).$$

where

$$R_{He} = R_\infty \frac{1}{1 + m/M_{He}} = 109{,}900.3 \text{ cm}^{-1}.$$

For the Rydberg series, $n_0 = 3$ and for the Pickering $n_0 = 4$.

The calculation of the lines in the Balmer series of deuterium is made in the same way as for those of hydrogen

$$\sigma_0 = 109{,}707\left(\frac{1}{4} - \frac{1}{n^2}\right).$$

One finds

D_α	$\tilde{v}_0 = 15{,}237.08 \text{ cm}^{-1}$	$\lambda_0 = 6561.1 \text{ Å,}$
D_β	$\tilde{v}_0 = 20{,}570.06 \text{ cm}^{-1}$	$\lambda_0 = 4860.1 \text{ Å,}$
D_γ	$\tilde{v}_0 = 23{,}038.47 \text{ cm}^{-1}$	$\lambda_0 = 4339.3 \text{ Å,}$
D_δ	$\tilde{v}_0 = 24{,}379.33 \text{ cm}^{-1}$	$\lambda_0 = 4100.7 \text{ Å.}$

This calculation can be done directly by use of the equation (§ 15.4):

$$\frac{\Delta \tilde{\nu}_0}{\tilde{\nu}_0} = \frac{\Delta R}{R},$$

taking for $\tilde{\nu}_0$ the values found for H, for R the value 109,678 cm^{-1} found for hydrogen and for ΔR (109,707 − 109,678) cm^{-1}.

IV

The assembly is shown in Fig. 69.1. The maximum of the third order is diffracted in the direction $i_K = 0$. The grating equation gives as the angle of incidence i'

$$\sin i' = \frac{3\lambda_0}{d}.$$

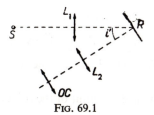

FIG. 69.1

Assuming the desired image is relative to $\lambda_0 = 0.6563 \times 10^{-3}$ mm with $d = 0.002$ mm

$$\sin i' = \frac{3 \times 0.6563 \times 10^{-3}}{2 \times 10^{-3}} = 0.98445, \qquad i' = 79°52'.$$

The angular dispersion (§ 7.9) with $\cos i_K = 1$ is given by

$$\mathcal{D} = \frac{K}{d} = \frac{3}{2 \times 10^{-3}} = 1500 \text{ rad/mm}.$$

The linear separation of two lines separated by 1 Å $= 10^{-7}$ mm is

$$fD \cdot 10^{-4} = 400 \times 1500 \times 10^{-7} = 60 \text{ mm}.$$

The linear separation of the H_α lines is

$$400 \times 1500 \times (6562.8 - 6561.1) \times 10^{-7} = 1.02 \text{ mm}.$$

Theoretical resolving power (§ 7.9)

$$\frac{\lambda}{\delta \lambda} = NK = 5000 \times 5 \times 3 = 75,000.$$

So that two lines will be separated, it is necessary that their images are separated by an angle $\Delta i \geqslant \lambda / Nd$ (§ 7.8). This angle must be transformed by the telescope to an angle at least equal to $1'$ (§ 5.13). The useful magnification of the telescope is

$$G = \frac{3\times10^{-4}\times5}{0.6563\times10^{-4}} \approx 23.$$

Now $G = 40/f'$, hence $1/f' = 23/40 = 0.57$ diopters.

PROBLEM 70

Spectrum of Neutral Lithium

Recall that the lithium atom has one optical electron and that its ground state is an $S(l = 0)$ state.

I

1. Given that its ionization potential is $V_i = 5.390$ V, find:

(a) The ground state term T_0 in cm^{-1}, taking the ionization limit as the origin of the terms.

(b) The free space wavelength λ_0 of the limit of the principal series given $e/hc = 8.0682\times 10^5$ (mks units).

2. By electrical methods the first ionization potential is measured and found to be $V_1 = 1.85$ volts.

(a) Calculate the value of the corresponding term T_1 in cm^{-1}. What can one say about the precision with which T_1 is then fixed? Compare this with the precision which will be found by spectroscopic methods.

(b) Combine this term with the ground state term and find the wavelength of the corresponding emission line.

(c) By what symbol $(S, P, D, F, \ldots)$ should this term be designated?

II

1. One studies the emission treated above using a reflecting grating spectrograph. The grating has 1200 lines/mm. The spectrum is formed in the focal plane of a lens L_1 with a 2-m focal length. The grating is used in a Littrow mounting (that is, the diffracted rays coincide with the incident rays all assumed parallel).

(a) Calculate the sine of the angle i between the rays and the normal to the plane of the grating for which the preceding conditions are satisfied in first order for the wavelength λ_1.

(b) With the grating placed in this position, one sees two lines in the focal plane of L_1 separated by 40 μ. Assuming that they form two lines in the first order, calculate their difference in wavelength and in wave number.

(c) Assuming that the emission lines are vanishingly narrow, what is the minimal width of the ruled grating that will allow resolution of these lines?

(d) How would one explain the doubling of the lines? What is represented by their difference in wave number?

III

Assume that:

the preceding doublet has been obtained by excitation of lithium vapour in thermal equilibrium at temperature T;

the resolving power of the spectroscope is much greater that that considered above;

under these experimental conditions the spectral line width comes only from the Doppler–Fizeau effect.

(a) Find the line profile, that is, the intensity distribution as a function of the wavelength (or wave number).

(b) Give explicitly their half-width.

(c) To what temperature can one heat the lithium vapour without destroying the resolution of these components (that is, that the width of each of them should at most be equal to the distance between the two lines calculated above)?

Note. Recall that the fraction dN/N of atoms whose velocity component in a given direction lies between u and $u+du$ is given by:

$$\frac{dN}{N} = A \exp\left(\frac{-Mu^2}{2RT}\right) du$$

where A is a constant whose value need not be explicitly known. M is the atomic mass (for lithium 7×10^{-3} kg), R is the ideal gas constant (8.32 joules).

Take $\log_e 2 = 0.69$ and $c = 3 \times 10^8$ m/s.

IV

Now work with an atomic beam of lithium (Fig. 70.1) travelling in the $x'Ox$ direction coming from a small opening cut in the side of an oven and collimated by a hole cut in a screen. Assume that the diameter of both openings are vanishingly small. The lithium

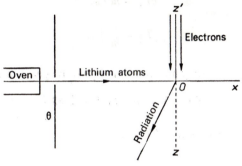

FIG. 70.1

atoms are excited by collision with an electron beam travelling in the $z'Oz$ direction normal to $x'Ox$ and accelerated by a potential difference slightly greater than V_1. Collect in a spectrograph the radiation emitted in a direction Oy normal to both beams.

(a) Explain why here one gets much finer lines.

(b) If one assumes that all other causes of broadening are eliminated (resolving power of the spectrograph, influence of the electric and magnetic fields, etc.) one again finds that the two lines have a width of the order of 0.003 cm^{-1}.

What is the source of this residual width? What characteristic can be deduced about one of the two levels involved in this transition? To what level does it apply?

V

Set aside consideration of the doublet and examine the lines of the "sharp" series ($2P-nS$ series).

(a) Why are all of the lines in this series doublets? What can one say about the wave number separation of the two lines of one of the doublets of this series?

(b) In studying the first line of the sharp series, one discovers that it is a double and that the higher of the two lines has the wavelength

$$\lambda_2 = 8128.75 \text{ Å}.$$

Derive the value in cm^{-1} of the first term T_2 of the corresponding transition and the separation in angstroms of the two components.

SOLUTION

I

1. The ionization energy eV_i corresponds to the difference between the energy origin and the ground-state level. The energy of a term having the value T in cm^{-1} is $W = hcT$ (§ 12.1).

(a) Hence, for the ground state:

$$eV_i = hcT_0,$$

$$T_0 = \frac{e}{hc} V_i.$$

The value of e/hc is given in S.I. units and thus V_i will be given in m^{-1}:

$$T_0 = 8.0682 \times 10^5 \times 5.390 = 4348,700 \text{ m}^{-1} = 43,487 \text{ cm}^{-1}$$

(b) $\lambda_0 = \dfrac{1}{T_0} = 2299 \times 10^{-8}$ cm $= 2299$ Å.

2. (a) The first excitation potential involves the energy necessary for the electron to be excited from the ground state $\bar{T}_0$ to the level T_1. Thus one has

$$T_0 - T_1 = 8.0682 \times 10^5 \times 1.85 = 14{,}920 \text{ cm}^{-1}$$

and

$$T_1 = 43{,}487 - 14{,}920 = 28{,}567 \text{ cm}^{-1}.$$

If one is given $V_1 = 1.85$ volts, this implies, in the absence of other indication, an uncertainty of ± 0.005 volt and thus an uncertainty in T_1 of

$$8.0682 \times 10^5 \times 5 \times 10^{-3} \simeq 4000 \text{ cm}^{-1},$$

thus a relative uncertainty of the order of 0.1 (on the ground-state level). This uncertainty is much greater than that which can be obtained spectroscopically where one can get commonly a precision in wavelength of 0.1 Å and thus a relative uncertainty of the order of 10^{-4} to 10^{-5}.

(b) $$\lambda_1 = \frac{1}{T_0 - T_1} = 6710 \text{ Å}.$$

This line is the first of the principal series and is homologous to the D-line of sodium. It is situated in the red and gives that characteristic coloration to flames charged with lithium salts.

(c) The ground-state term is S $(l = 0)$ and this can only combine with a P $(l = 1)$ term as a result of the selection rule $l = 1$ (§ 15.3). The term T_1 is therefore a P term.

II

1. (a) The grating equation (§ 7.8) gives with $i = i'$ in first order:

$$2 \sin i_1 = \frac{\lambda}{d} \quad \text{or} \quad \sin i_1 = \frac{6710 \times 10^{-7} \times 1200}{2} = 0.40.$$

(b) The distance l which separates the two lines in the focal plane is bound to their angular separation δi by $\delta l = f \cdot \delta i$, where f is the focal length of L_1, hence

$$\delta i_1 = \frac{4 \times 10^{-5}}{2} \text{ rad.}$$

On the other hand, one has (§ 7.9)

$$\frac{\delta i_1}{\delta \lambda} = \frac{1}{d \cos i_1},$$

therefore

$$\delta \lambda = d \times \cos i_1 \times \delta i_1 = \frac{1}{1200} \times 0.9165 \times \frac{4 \times 10^{-5}}{2} = 0.153 \times 10^{-7} \text{ mm} \quad \text{or} \quad 0.153 \text{ Å}.$$

One has $\tilde{\nu} = 1/\lambda$, hence $\delta \tilde{\nu} = -\delta \lambda / \lambda^2$

$$|\delta| = \frac{0.153}{(6710)^2} = \frac{0.153}{0.4502 \times 10^8} = 0.339 \times 10^{-8} \text{ Å}^{-1} \quad \text{or} \quad 0.340 \text{ cm}^{-1}.$$

(c) In the assumption taken, the width of the lines is only due to diffraction and the finite width of the grating (§ 7.8). The theoretical resolution is defined by

$$\frac{\lambda}{\delta\lambda} = NK$$

allowing one to calculate N and therefore the width of the ruled grating. For $K = 1$, one has

$$L = \frac{N}{n} = \frac{\lambda}{n \times \delta\lambda} = \frac{6710 \times 10^{-8}}{1200 \times 0.153 \times 10^{-8}} = 36.6 \text{ mm.}$$

(d) The doublet is due to spin–orbit interaction (§ 15.7). The separation in wave numbers multiplied by hc represents the difference between two spin energy levels of the optical electron in the magnetic field caused by its orbital motion.

III

(a) The transverse Doppler effect is negligible (§ 9.10) and the spreading of the line is due only to the Doppler effect produced by the velocity u along the observational direction. For an atom, the variation in wave numbers which results is given by

$$\frac{\Delta\tilde{\nu}}{\tilde{\nu}} = \frac{\Delta\nu}{\nu} = \pm\frac{u}{c} \tag{1}$$

the $+$ sign relating to the case where the atom is moving toward the observer. In a perfect gas (which is assumed for lithium vapour) the distribution of atomic velocities is statistical and the number of atoms of mass m whose velocity in a given direction and interval lying between u and $u+du$ at the equilibrium temperature T, is given by

$$dN = C \exp\left(-\frac{mu^2}{2kT}\right) du \tag{2}$$

C being a constant and k Boltzmann's constant (§ E.3). This expression is a result of the Maxwell–Boltzmann velocity distribution. With

$$\frac{m}{k} = \frac{\mathcal{N}m}{\mathcal{N}k} = \frac{M}{k},$$

$\mathcal{N}$ being Avogadro's number, equation (2) can be written

$$dN = C \exp\left(-\frac{Mu^2}{2RT}\right) du, \tag{3}$$

which is the expression given in the problem.

To find the intensity distribution in a line as a function of the wave number, one notes that the emitted intensity in an interval $d\tilde{\nu}$ is proportional to the number of atomic emitters whose wave number lies within that interval, since the emission is incoherent. Additionally, the interval $d\tilde{\nu}$ is related to the interval du of atomic velocities. Putting in (3) the value of u

derived in (1), one gets the expression for the relative intensity $I(\tilde{\nu})$

$$I(\tilde{\nu}) = I_0 \exp\left[-\frac{Mc^2}{2RT\tilde{\nu}_0^2}(\tilde{\nu}-\tilde{\nu}_0)^2\right]. \tag{4}$$

$\tilde{\nu}_0$ is the wave number of the centre of the line corresponding to zero velocity. The curve (Fig. 70.2) representing $I(\tilde{\nu})$ is symmetric about $\tilde{\nu}_0$ as a result of the random distribution of velocities. This is a gaussian curve.

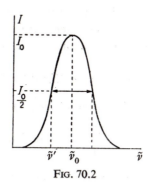

FIG. 70.2

(b) The intensity at half the maximum I_0 has wave number f' such that

$$\frac{1}{2} = \exp\left[-\frac{Mc^2}{2RT\tilde{\nu}_0^2}(\tilde{\nu}'-\tilde{\nu}_0)^2\right].$$

hence

$$(\tilde{\nu}'-\tilde{\nu}_0)^2 = \frac{2RT\tilde{\nu}_0^2}{Mc^2}\ln 2.$$

The half-width is given by the wave-number interval:

$$2\,|(\tilde{\nu}'-\tilde{\nu}_0)| = 2\,\Delta\tilde{\nu}',$$

$$2\,\Delta\tilde{\nu}' = \frac{2\tilde{\nu}_0}{c}\sqrt{\frac{2RT\ln 2}{M}}. \tag{5}$$

(c) The two components considered in part II cease to be resolved when the width of each of them is greater than the wave number interval by which they are separated, so that, using the results of question II.1 (b) of 0.34 cm^{-1}, one has

$$0.34 = 2\times\frac{14920}{3\times 10^8}\sqrt{\frac{2\times 8.32T\times 0.59}{7\times 10^{-3}}}$$

hence, the maximal temperature which allows resolution

$$\sqrt{T} = 83.8, \quad T = 7022° \text{ K}.$$

IV

(a) The atoms in an atomic beam (§ E.4) have a velocity variable about a mean but the direction is well defined. Since this is perpendicular to the observational direction the longitudinal Doppler effect is eliminated, it does not exist for the transverse Doppler effect as a cause of line spreading and the effects of collision and atomic interaction are small in the rarified gas which constitutes the beam. Hence the lines are narrow.

(b) If the preceding causes of spreading, those associated with the apparatus, are eliminated there remains only the residual natural line width, a consequence of the uncertainty relationship $\Delta W \times \Delta t \simeq h$. The energy uncertainty is, using the parameters from the problem statement,

$$\Delta W = h \times c \times \Delta \tilde{\nu} = h \times 3 \times 10^{10} \times 3 \times 10^{-3}.$$

One therefore has a uncertainty in the period of the corresponding transition

$$\Delta t \simeq \frac{\Delta W}{h} \simeq 9 \times 10^{-7} \text{ s}.$$

Since the transition returns the lithium atom to its ground state, its lifetime is infinite in the absence of perturbation. The interval Δt characterizes the upper level and is its mean lifetime.

V

(a) The lines of the sharp series are emitted by transitions from an nS level ($n = 3, 4, \ldots$) to the $2P$ level. Moreover, this latter level is doubled as a result of spin–orbit interaction into the levels $2P'$ and $2P''$ as seen in II (d) (Fig. 70.3); however, the S levels are not since

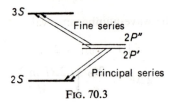

FIG. 70.3

their orbital moment is zero. The wave number separation of the sharp series doublets is

$$\Delta \tilde{\nu} = (2P' - nS) - (2P'' - nS) = 2P' - 2P''.$$

This value is constant. On the other hand, the lines of the principal series arising from transitions between the nP ($n = 2, 3, 4, \ldots$) levels and the ground state $2S$ level of the optical electron lead to

$$\Delta \tilde{\nu} = (nP' - 2S) - (nP'' - 2S) = nP' - nP''$$

and this separation is variable since it decreases as n increases.

(b) The wave number of the radiation studied is

$$\tilde{\nu}_2 = \frac{1}{\lambda_2} = \frac{1}{8128.75} = 12,302 \text{ cm}^{-1}.$$

The separation of the two components of this line is the same as that of the components of the first line of the principal series (Fig. 70.3) that is, to the separation calculated in question II.1 (b), namely $\Delta \tilde{\nu} = 0.34$ cm^{-1}. This is, in wavelength,

$$|\Delta \lambda| = |\Delta \tilde{\nu}| \times \lambda^2 = 0.34 \times (8130)^2 = 0.223 \text{ Å}.$$

PROBLEM 71

Doppler Effect. Spectral Line Width

I

The first spectral lines of the first two series of the hydrogen atom have the wavelengths respectively (in Å):

Lyman: 1215.7...

Balmer: 6563.07; 4861.33...

In addition the limit of the Balmer series is 3645.9.

1. Recall the expression which leads to the calculation of these wavelengths and briefly explain its significance.

2. Calculate the ionization potential V_i of the hydrogen atom.

3. Calculate the wavelength of the first line of the third (Paschen) series of hydrogen using the combination principle.

II

1. Hydrogen atoms, excited by an electric discharge in a low-pressure tube, escape through a channel in the cathode C (Fig. 71.1) and move into an evacuated space along the Cx direction. Examine with a spectroscope placed first at O_1 and then at O_2, the H_β-line. One

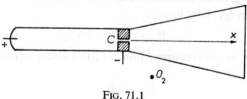

FIG. 71.1

will find two different values for the wavelength, at O_1, $\lambda_1 = 4855.45$ Å and at O_2, $\lambda_2 = 4861.33$ Å. Briefly interpret these results. Calculate the velocity of the atoms along Cx using the Lorentz transformation. Use a coordinate system $S(xyz)$ fixed to the atoms moving with constant velocity u with respect to the laboratory system $S'(x'y'z')$.

The argument of the sinusoidal frequency function, $\sin 2\pi\nu(t - x/c)$ is an invariant in the Lorentz transformation. Neglect u^2/c^2 in finding u.

2. The slit of the spectroscope is placed at O_2 and receives light emitted normally to Cx by the atoms in a segment of the atomic beam lying between x and $x+dx$. By displacing the spectroscope parallel to Cx one sees an exponential decrease in the intensity of the H_β radiation which follows the law, $I = I_0 \exp(-Kx)$ with $K = 138 \text{ m}^{-1}$. I_0 is the intensity at the exit of the cathode. Calculate the mean lifetime τ of the hydrogen atom in the excited state, that is, the time required for the emitted intensity to fall to $1/e$ of its initial value. Derive the natural line width of H_β assuming the lifetime of the lower level is very much longer than that of the excited state.

III

Now observe the emission of H_β by the atoms in an ordinary discharge tube at 27°C. Analyse the light using a Fabry–Pérot interferometer where the plate separation has been increased to the point where the fringe visibility goes to zero. The limiting interference order is 50,000. What is the width of the line if one assumes the line profile to be rectangular? One finds a width much greater than the natural width. Assuming that the atoms all move with the same velocity, their mean thermal velocity $\langle v \rangle$, calculate the line width due to the Doppler effect and compare this to the experimental value.

Recall that in the kinetic theory of gases, $\langle v \rangle = \sqrt{8RT/\pi A}$, where R is the ideal gas constant, T the absolute temperature, and A the atomic mass.

SOLUTION

I

$$\frac{1}{\lambda} = R\left(\frac{1}{n_0^2} - \frac{1}{n^2}\right). \tag{1}$$

R is the Rydberg constant, n_0 is an integer which characterizes the spectral series ($n_0 = 1$ for the Lyman series, $n_0 = 2$ for the Balmer series, etc.), and n is a member of the infinite series of integers which characterize the lines of each series and which begin with the integer immediately greater than n_0.

2. The ionization energy is that energy which must be given to the H atom to remove the electron situated in the ground state from the field of the nucleus to infinity and place it there without kinetic energy. One has

$$W_i = eV_i = h\nu_l$$

e is the electron charge, ν_1 the frequency limit of the Lyman series, so that, using (1) with $n_0 = 1$ and $n = \infty$,

$$V_i = hRc/e.$$

R can be calculated from the series limit of the Balmer series given above, with $n_0 = 2$ and $n = \infty$, equation (1) yields

$$\frac{1}{\lambda} = \frac{R}{4}$$

hence

$$V_i = \frac{6.62 \times 10^{-34} \times 4 \times 3 \times 10^8}{3645.9 \times 10^{-10} \times 1.60 \times 10^{-19}} = 13.6 \text{ V}.$$

3. The first line in the Paschen series ($n_0 = 3$) has wavelength

$$\frac{1}{\lambda} = R\left(\frac{1}{9} - \frac{1}{16}\right)$$

since the first two lines of the Balmer series have wavelengths

$$\frac{1}{\lambda_1} = R\left(\frac{1}{4} - \frac{1}{9}\right) \quad \text{and} \quad \frac{1}{\lambda_2} = R\left(\frac{1}{4} - \frac{1}{16}\right).$$

one sees that

$$\frac{1}{\lambda} = \frac{1}{\lambda_2} - \frac{1}{\lambda_1}$$

hence

$$\lambda = \frac{\lambda_1 \lambda_2}{\lambda_1 - \lambda_2} = \frac{6563.07 \times 4861.33}{1701.74} = 18{,}748.0 \text{ Å.}$$

II

1. The H^+ ions accelerated by the electric field between A and C capture electrons at C and exit into the vacuum with their acquired velocity. In the coordinate system of the emitting atoms the frequency of the H_β radiation is ν, in the laboratory it is ν'. The monochromatic wave propagating along Cx is given by the expression in the S system

$$\mathbf{E} = \mathbf{E}_m \exp\left[2\pi j \nu \left(t - \frac{x}{c}\right)\right].$$

Its phase $\nu(t - x/c)$ is an invariant (§ 9.10) and thus

$$\nu'\left(t' - \frac{x'}{c}\right) = \nu\left(t - \frac{x}{c}\right).$$

the primed letters referring to the system S'. The Lorentz expressions

$$x' = \gamma(x - ut) \quad t' = \gamma\left(t - \frac{ux}{c^2}\right)$$

with

$$\gamma = \frac{1}{\sqrt{1 - \dfrac{u^2}{c^2}}},$$

give

$$t' - \frac{x'}{c} = \gamma\left(t - \frac{ux}{c^2}\right) - \gamma\left(\frac{x}{c} - \frac{ut}{c}\right) = \gamma\left(1 + \frac{u}{c}\right)\left(t - \frac{x}{c}\right),$$

hence

$$\nu = \nu'\gamma\left(1 + \frac{u}{c}\right). \tag{2}$$

The frequency is higher for the observer at O_1 due to the longitudinal Doppler effect. If one neglects u^2/c^2, and thus if $\gamma = 1$, the transverse Doppler effect, which exists in principle for the observer O_2 although always very small (§ 9.10), vanishes. Thus

$$\nu' = c/\lambda_1, \qquad \nu = c/\lambda_2,$$

and

$$\lambda_2 = \lambda_1\left(1+\frac{u}{c}\right)$$

from which

$$u = c\left(\frac{\lambda_2-\lambda_1}{\lambda_1}\right) = \frac{(4861.33-4855.45)3\times10^8}{4855.45} = 3.635\times10^5 \text{ m/s}.$$

2. The luminous intensity observed is proportional to the number of atoms contained in the volume limited by planes dx apart and which de-excite per unit time (since the emission is incoherent). This number is also proportional to the number of atoms still excited in the same volume (§ 15.1). The spontaneous emission decreases with time according to the law

$$I = I_0 \exp\left(-\frac{t}{\tau}\right),$$

and comparison with the expression given in the statement of the problem shows that

$$Kx = t/\tau$$

or

$$\tau = \frac{t}{Kx} = \frac{1}{Ku} = \frac{1}{138\times3.635\times10^5} \approx 5\times10^{-7} \text{ Hz}.$$

In solution of the Schrödinger equation, one assumes that the excited levels of the atom have a precise energy. The uncertainty relationship $\Delta W \cdot \Delta t \approx h$ shows that since $\Delta W = 0$, $\Delta t = \infty$ and these states are stationary. However, only the ground state has an infinite lifetime. The preceding experiment shows that the lifetime τ imposes on the excited energy level an uncertainty $\Delta W \approx h/\tau$. Since $\Delta W = h\,\Delta \nu$, the natural line width of H_β is

$$\Delta\nu \approx \frac{1}{\tau} \approx 5\times10^5 \text{ Hz}$$

or

$$|\Delta\lambda| = \frac{\lambda^2}{c}|\Delta\nu| = \frac{0.236\times5}{3}\times10^{-13} \text{ m} = 0.39\times10^{-3} \text{ Å}.$$

III

Fringe visibility vanishes when for the path difference $\delta = p\lambda$ the interference order p varies by 1 unit for the extremes of the radiation contained in the ray (§ 6.10), so that

$$\Delta\delta = p\,\Delta\lambda = \lambda\,\Delta p,$$

$$\Delta\lambda = \frac{\lambda}{p} = \frac{4861.33}{50,000} = 0.097 \text{ Å}.$$

This width is much greater than the preceding natural line width.

In calculating the broadening produced by the Doppler effect taking as the velocity equation (2) given in the problem for the thermal motion, one finds

$$\langle v \rangle = \sqrt{\frac{8RT}{\pi A}} = \sqrt{\frac{8 \times 8.32 \times 10^3 \times 300}{3.14}} = 2530 \text{ m/s}.$$

The velocity $\langle v \rangle$, unlike the velocity u considered in part II, is related both to the motion of the atom toward the observer and away from him. The line width is therefore

$$\frac{\Delta v}{v} = \frac{2\langle v \rangle}{c} = \frac{\Delta \lambda}{\lambda}$$

hence

$$\Delta \lambda = \lambda \frac{2\langle v \rangle}{c} = 4861.33 \times \frac{2 \times 2530}{3 \times 10^8} = 0.082 \text{ Å},$$

a value much closer to the experimental value than the natural line width.

PROBLEM 72

Polarization of Resonance Radiation

I

The resonance radiation $\lambda = 2537$ Å emitted by a mercury atom is produced when an electron drops from the 6^3P_1 excited state to the 6^1S_0 ground state. This latter level is normally occupied by two electrons.

1. Explain the significance of the symbols representing the two levels. Give the values of their Lande factors.

2. Consider the gas $^{198}_{80}Hg$ (in order to avoid certain complications due to nuclear spin) contained in a transparent tube at a sufficiently low pressure that atomic interactions are negligible and placed in a uniform magnetic field B. Show the splittings of the $^3P_1$ and $^1S_0$ states produced by the Zeeman effect, the statistical weight of each of the sublevels and the transitions permitted by the selection rules.

3. Emission of the resonance line is excited in the gas discussed above and this radiation is observed with an apparatus capable of separating the various components of the radiation spectrum. Assume that the separation of the levels produced by B is small with respect to the thermal energy of the atoms. Determine the frequencies, the polarization states, and the relative intensities of the radiation observed in the following cases: (a) normal to the field lines B; (b) parallel to these lines; and (c) at 30° to them.

To do this, make use of the analogy between a quantum emitter of dipole radiation and a Hertzian oscillator. Recall that a circular oscillator is composed of two linear oscillators of the same amplitude.

Compare the power emitted in the various transitions and derive the relative probabilities for these spontaneous transitions.

II

Mercury vapour, subject to no excitation and held at a temperature sufficiently low that all the atoms can be thought of as being in their ground state, is now illuminated by a parallel beam of $\lambda = 2537$ Å radiation propagating in the Ox direction. One can linearly polarize this beam to give the electric field of the wave a fixed direction in the yOz plane. One observes the re-emitted radiation from the resonance in the Oz direction through an analyser. The gas is placed in a uniform magnetic field B, with adjustable direction and an intensity such that the Zeeman splitting is small compared to the thermal energy and with respect to the resonance line width formed by the excitation source. One observes the radiation along the Oz direction using a non-dispersive analyser. Determine the polarization state for the radiation with the following orientations:

$$E: \quad Oy \quad Oy \quad Oy \quad Oz \quad Oz$$

$$B: \quad Oy \quad Ox \quad Oz \quad Oy \quad Ox$$

When B is parallel to Oy, determine the angle, θ, which E must take with B in the yO plane so that the radiation observed along Oz will be depolarized.

III

Now consider sodium vapour. With the knowledge that the transition of the single optical electron which gives rise to the D_1 transition is produced for the $3^2P_{1/2}$ to $3^2S_{1/2}$ transition, draw the emission diagram for these levels when they are subjected to a Zeeman field. Determine the relative intensity of the emission lines and the relative probabilities of the lines which are produced. Note that when $B = 0$, the emitted radiation contains only a single non-polarized line and, reasoning by continuity, show that for transverse observation there exists a simple relationship between the sum of the intensities of the π components and those of the σ components. Take into account in addition that the Zeeman spectra are symmetric in both intensity and frequency about the frequency ν_0.

Place the sodium vapour in an arrangement analogous to that of the mercury vapour in part II and illuminated by a parallel beam of D_1 radiation propagating along Oy. What is the polarization state of the resonance radiation in the presence of a B field directed along Oy?

If the exciting radiation is circularly polarized, all things being equal, show that the irradiation of the vapour leads, after an unlimited time, to all of the sodium atoms populating the $m_J = +\frac{1}{2}$ sublevel of the lower state.

SOLUTION

I

1. Using the nomenclature of spectral terms (§ 15.8), the number n designates the electron shell of the optical electron. In the ground state, the two valence electrons of the mercury atom occupy the $6s$ subshell. They have an orbital angular momentum of zero, $l = 0$ (S state) and the total orbital angular momentum L is zero. They have opposite spins $s = \pm\frac{1}{2}$ and the total spin angular momentum S is zero. The total angular momentum $J = L+S$ is zero. The general symbol $^{2S+1}L_J$ is $^1S_0$ for the ground-state term. In the excited state, one of the electrons remains in the initial s ($l = 0$) state; the other moves into the state $l = 1$, since the total angular momentum $L = 1$ and the symbol is P; the spins of the two electrons are parallel and the resultant spin, $S = 1$, since $2S+1 = 3$. The total angular momentum, which can take on the values $L+1$, L, and $L-1$ is 1.

The Lande factor is given by (§ 15.11)

$$g = 1 + \frac{J(J+1)+S(S+1)-L(L+1)}{2J(J+1)}$$

In the ground state $J = S = L = 0$, $g = 1$.
In the excited state, $J = S = L = 1$, $g = 2$.

2. The magnetic field does not act on the ground state which lacks angular momentum and thereby a magnetic moment. The projection of the moment $J = 1$ of the excited state on the **B** direction can take the values $m_J = +1$, 0, or -1. The degeneracy of the level corresponding to the various values of m_J for a given J is removed by the action of B and each sublevel has the same statistical weight.

The allowed transitions obey the rule (§ 15.10)

$$\Delta m_J = +1 \quad \text{or} \quad 0 \quad \text{or} \quad -1.$$

Figure 72.1 shows the diagram for these transitions. The Grotrian diagram (Fig. 72.1a) shows the energy levels. The Heisenberg diagram (Fig. 72.1b) displays on the same horizontal the levels with the same m_J, but allows immediate distinction to be made between the vertical π ($\Delta m_J = 0$) transitions and the σ_+ ($\Delta m_J = +1$) and σ_- ($\Delta m_J = -1$).

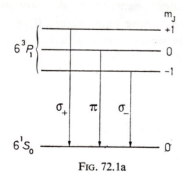

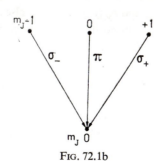

FIG. 72.1a   FIG. 72.1b

3. Figure 72.1 a and b shows that the decay of the $^3P_1$ state to the $^1S_0$ state corresponds to the emission of the normal Zeeman triplet: one line (π) whose frequency v_0 is the same as in the absence of the field and two others (σ) which are symmetrically displaced by an amount

$$\pm\Delta v = \frac{e}{4\pi m_e}B. \tag{1}$$

Normal to the direction of the field one observes three linearly polarized lines: the π-line vibrating parallel to $\boldsymbol{B}$ and the σ_+ and σ_- lines normal to $\boldsymbol{B}$. Parallel to the field the π line disappears and the σ lines are circularly polarized in opposite senses. These results, which follow from quantum theory (§ 15.10.4) can, in simple cases be tied to the classical theory of electric dipole emission. The vibration of a linear sinusoidal dipole $\boldsymbol{d}$ can in effect (Fig-72.2) be decomposed into a vibration π along a general Oz axis and a vibration σ in the xOy plane. This latter vibration can be thought of as the resultant of two circular vibrations in opposite senses σ_+ and σ_- (see the Fresnel theory on Optical rotary power).

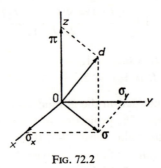

FIG. 72.2

The preceding resolution is shaped to the symmetry of the field $\boldsymbol{B}$, an axis of revolution and a plane normal to that axis. Take the direction of $\boldsymbol{B}$ for the Oz axis. The vibration along Oz is not affected by the field $\boldsymbol{B}$ which is parallel to the displacement of the charge of the dipole. The circular vibrators are subjected to the Lorentz force in opposite senses for σ_+ and σ_-. One therefore concludes that their frequencies must change by an amount symmetric with respect to the unaltered π-radiation. (The classical calculation which gives rise to equation (1) is not of interest here.)

If one now considers a large set of dipole oscillators oriented at random, the three components π, σ_x, and σ_y (Fig. 72.2) or else π, σ_+ and σ_-—have equal values. If one observes the radiation emitted in a given direction when the dipoles are subjected to some directed action, one observes the transverse electric field emitted by the two components normal to the direction of observation while the component directed along this direction is zero (§ 10.3). The observed intensity is thus $2I$, where I is the maximal radiation intensity emitted by each of the three components.

If the oscillators are placed in a uniform magnetic field, by observing in the direction Oz of the field lines, one observes the radiation emitted by the σ_+ and σ_- components but not that of the π component. The intensity is then $2I$. By observing normally to the field lines, for example along Ox, one see the maximal radiation with intensity I from the π component and the linear components parallel to Oy of each of the σ components which have then the intensity $I/2$. The total observed intensity is then $2I$ (Fig. 72.3).

When one observes in a direction OP making an angle θ with Oz and situated, for example, in the zOx plane (Fig. 72.3), one sees the intensity $I \sin^2 \theta$ of the π-component, the intensity $I/2 \cos^2 \theta$ of the components of σ parallel to Oy, and the intensity $I/2 \cos^2 \theta$ of the components of σ parallel to Ox. The π radiation appears linearly polarized in the xOz plane and normal to OP. The σ radiation is elliptically polarized and the ratio of the elliptical axes has the value $\cos \theta$.

For $\theta = 30°$, $\sin^2 \theta = \frac{1}{4}$, $\cos^2 \theta = \frac{3}{4}$. The line with frequency ν_0 has intensity $I/4$ and each of the lines with frequency $\nu_0 + \Delta\nu$ have intensity:

$$\frac{I}{2} + \frac{I}{2} \times \frac{3}{4} = \frac{7}{8} I.$$

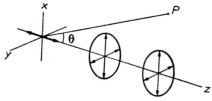

FIG. 72.3

The ratio

$$\frac{I_\pi}{I_\sigma} = \frac{2}{7}.$$

The total intensity is therefore $2I$.

One has seen that the statistical weight of the Zeeman sublevels are equal. If in addition the magnetic energies are small compared to the thermal energy kT, all of the sublevels are equally populated. The orientation of the orbital cannot affect the velocity of the emission of energy and the power radiated in the transition from any Zeeman sublevel is the same, whatever may be the number m_J which characterizes the level. On the other hand, a σ-radiation corresponds, all things being equal, to a power emission twice that of a π-radiation since it has two linear components. Since the intensities of the π-line and each of the σ-lines are equal, the relative transmission probabilities are

$$\frac{A_\pi}{A_\sigma} = 1.$$

II

The exciting radiation has a width which by hypothesis contains all of the Zeeman components. The action of the B field separates all of the Zeeman sublevels of the 6^3P_1 level in the absorbing vapour. If the B and E fields are both parallel to Oy and normal to the direction of observation Oz, only the π-component of the resonance line is absorbed and only the $m_J = 0$ sublevel of the $6\,^3P_1$ level is excited. The light re-emitted by the transition between this sublevel and the ground state is a π-vibration (Fig. 72.1) and the resonance radiation is then linearly polarized parallel to Oy.

If $E = E_y$ and $B = B_x$, this time the two σ-components are absorbed and re-emitted and the resonance radiation is circularly polarized with Ox as the axis. One sees in addition linear polarization parallel to Oy for observations made along Oz.

If $E = E_y$ and $B = B_z$, one no longer observes the π component, but together the components σ_+ and σ_-, the light is not polarized.

If $E = E_z$ and $B = B_y$, only the levels $m_J = \pm 1$ are excited, the σ_+ and σ_- components are re-emitted. One sees a linearly polarized radiation parallel to Ox. Likewise one sees a vibration parallel to Oy if $E = E_z$ and $B = B_x$.

When in the yOz plane E makes an angle θ with B (along Oy), one sees, in comparison to the first and fourth cases which have been studied, that the energies of excitation of the π-component and the σ-components are respectively proportional to $E^2 \cos^2 \theta$ and $E^2 \sin^2 \theta$. Consequently, for $\sin^2 \theta = \cos^2 \theta$, the linear components relative to Ox and Oy are equal when observed along Oz and the light is not polarized. For this $\theta = 54°44'$.

III

The optical electron has a spin $\frac{1}{2}$ with the result that the ^3S and ^3P levels are doublets $(2s+1 = 2)$. For the S level, $l = 0$ and $j = \frac{1}{2}$. For the P level, $l = 1$ and $j = \frac{1}{2}$ or $\frac{3}{2}$. The transition $P_{1/2} \rightarrow S_{1/2}$ produces the D_1 line and the $P_{3/2} \rightarrow S_{1/2}$ transition the D_2 line. Here we are interested with the first of these. Since the Lande factor is equal to 2 for the S level and to $\frac{2}{3}$ for the P level, the four possible transitions allowed by the selection rules give four lines of different frequency numbered 1, 2, 3, and 4 on the diagram in Fig. 72.4a.

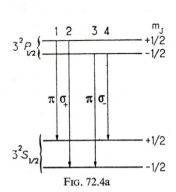

FIG. 72.4a

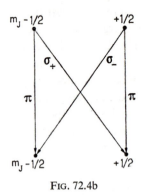

FIG. 72.4b

The π-component and the σ-components each appear in transverse observation as linearly polarized with two perpendicular azimuths. As at the limit for $B = 0$, the radiation is not polarized and it is necessary that the intensity of the set of π-components is equal to that of the σ-components:

$$\Sigma I_\pi = \Sigma I_\sigma. \tag{2}$$

The result obtained in I.3 is consistent with this. For the D_1-line, since Zeeman spectra are symmetric, one has

$$I_1 = I_3 \quad \text{and} \quad I_2 = I_4.$$

On the other hand, equation (2) gives

$$I_1 + I_2 = I_3 + I_4.$$

One gets

$$I_1 = I_2 = I_3 = I_4.$$

The probabilities of the spontaneous transitions are thus in the ratio

$$A_1 : A_2 : A_3 : A_4 = 1 : 2 : 1 : 2$$

since 1 and 3 are π-lines and 2 and 4 are σ-lines.

When the sodium vapour is illuminated by an E_y radiation in a B_y field, the π-transitions are excited. The two upper sublevels are populated equally and the reradiation to the ground state occurs through π- and σ-transitions of the same intensity. Therefore the observed radiation is not polarized.

Excitation by circularly polarized light with a given sign, right for example, excites the atoms of the $m_J = -\frac{1}{2}$ sublevel of the S level to the $m_J = +\frac{1}{2}$ sublevel of the upper state.

Hence, decay to the ground state is made between the two sublevels, but, as indicated by the transition probability, two-thirds of the excited atoms decay to the original sublevel with emission of a σ line and one-third to the $m_J = +\frac{1}{2}$ sublevel with emission of a π-line. These latter atoms can never be re-excited by the incident radiation and in principle their number continually increases.

PROBLEM 73

Spectral Terms of Two-electron Atoms

Consider an atom which has several electrons each of which has an orbital angular momentum G_l and a spin angular momentum G_s. Allow the orbital moments, on one hand, and the spin moments, on the other, to combine to give the total angular momenta G_L and G_S. Significantly different atomic energy states correspond to the different values of L and S. Again the moments G_L and G_S combine to give the total angular momentum G_J. Each atomic energy level—or each spectral term—is designated by a symbol $^{2S+1}L_J$.

I

Determine the possible terms for two "non-equivalent" electrons, that is, electrons whose quantum numbers n and l are different and which belong to different subshells as a result. Consider the various cases where each of the two electrons can be s, p, or d.

II

Determine the possible term symbols for two "equivalent" electrons, that is, electrons having the same values of n and l. Take into account the Pauli principle. Consider the cases where the electrons are s, then p.

III

Determine the symbols of the ground state terms for the elements from $H (Z = 1)$ to Ne $(Z = 10)$. Do this taking into account the following empirical rules due to Hund: (a) the lowest terms correspond to the maximal value of S and the maximal value of L compatible with the preceding value of S taking into account the Pauli principle, and (b) $J = L - S$ for the elements having less than half the electrons of a group of equivalent electrons and $J = L + S$ for those having more than half.

SOLUTION

The angular momentum coupling is Russell–Saunders coupling (§ 15.12.3).

The resultants $\overset{\smile}{G_S} = \sum \overset{\smile}{G_s}$, $\overset{\smile}{G_L} = \sum \overset{\smile}{G_l}$, and $\overset{\smile}{G_J} = \overset{\smile}{G_L} + \overset{\smile}{G_S}$ follow the Russell–Saunders scheme and the results are quantized.

I

In the case where n and l are different for the two electrons, all the combinations are possible. For example, for two d-electrons ($l_1 = l_2 = 2$), the possible values of L are different integers and lie between $l_1 - l_2 = 0$ and $l_1 + l_2 = 4$. The possible terms are S, P, D, F and G.

Figure 73.1 shows the results in the inexact vector representation of $\overset{\smile}{G}$.

One finds without any difficulty that for all of the electron couplings considered the L values are as follows:

$$ss : 0; \qquad sp : 1; \qquad sd : 2; \qquad pp : 0, 1, 2; \qquad pd : 1, 2, 3.$$

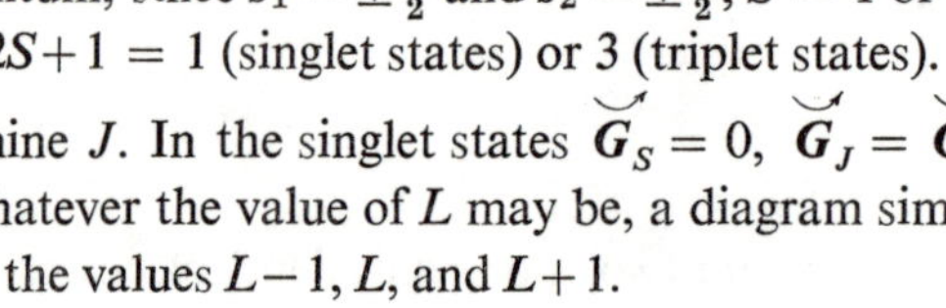

FIG. 73.1

As to the spin momentum, since $s_1 = \pm \frac{1}{2}$ and $s_2 = \pm \frac{1}{2}$, $S = 1$ or 0^* and the multiplicity of each of the terms is $2S + 1 = 1$ (singlet states) or 3 (triplet states).

It remains to determine J. In the singlet states $\overset{\smile}{G_S} = 0$, $\overset{\smile}{G_J} = \overset{\smile}{G_L}$ and $J = L$. For the triplet states, $S = 1$, whatever the value of L may be, a diagram similar to Fig. 73.1 shows that J can only take on the values $L - 1, L$, and $L + 1$.

---

* One must be careful not to confuse S with the symbol for the $L = 0$ state.

The possible terms symbols are collected in Table 73.1.

TABLE 73.1

Electrons	Singlet term	Triplet term
ss	$^1S_0$	$^3S_1$
sp	$^1P_1$	$^3P_0\ ^3P_1\ ^3P_2$
sd	$^1D_2$	$^3D_1\ ^3D_2\ ^3D_3$
pp	$^1S_0\ ^1P_1\ ^1D_2$	$^3S_1\ ^3P_0\ ^3P_1\ ^3P_2\ ^3D_1\ ^3D_2\ ^3D_3$
pd	$^1P_1\ ^1D_2\ ^1F_3$	$^3P_0\ ^3P_1\ ^3P_2\ ^3D_1\ ^3D_2\ ^3D_3\ ^3F_2\ ^3F_3\ ^3F_4$
dd	$^1S_0\ ^1P_1\ ^1D_2\ ^1F_3\ ^1G_4$	$^3S_1\ ^3P_0\ ^3P_1\ ^3P_2\ ^3D_1\ ^3D_2\ ^3D_3\ ^3F_2\ ^3F_3\ ^3F_4\ ^3G_3\ ^3G_4\ ^3G_5$

II

The two equivalent electrons having the same quantum numbers n and l must differ either in their quantum number m or in s.

For s-electrons, $l = 0$ and $m = 0$ thus the quantum numbers s of the two electrons under consideration are $+\frac{1}{2}$ and $-\frac{1}{2}$ and therefore $S = 0$ and since $L = 0$ the only term possible is $^1S_0$. The triplet state is not permitted.

For two p-electrons the situation is more complex. Table 73.2 lists the quantum cases of an np subshell and the possible distributions of two electrons taking into account their spin. The configurations ↑↓ and ↓↑ are equivalent.

TABLE 73.2

Configu-ration	$m = +1$	$m = 0$	$m = -1$	$M = \sum m$	$S = \sum s$	L	Term
1		↑↓		0	0	0	^1S
2	↑↓			+2	0	2	^1D
3			↑↓	-2	0	2	^1D
4	↑	↓		+1	0	2	^1D
5	↑		↓	-1	0	2	^1D
6	↑		↓	0	0	2	^1D
7	↑	↑		+1	1	1	^3P
8	↑		↑	-1	1	1	^3P
9	↑		↑	0	1	1	^3P

One has, as in I, $L = 0$, 1, or 2 thus S, P, or D states. The quantum number $M = \sum m$ is associated with the projection of $\boldsymbol{G_L}$ on the z-axis.

For $L = 0$, $M = 0$ can only be obtained for $m_1 = m_2 = 0$. Therefore the spins are opposed and one has a $^1S_0$ state (configuration number 1).

For $L = 2$, the possible values of M are $+2$, $+1$, 0, -1, and -2. The first is obtained from $m_1 = m_2 = +1$ (configuration 2) and the last by $m_1 = m_2 = -1$ (configuration 3). Thus the spins are opposed and one has a $^1D_2$ state.

21*

This same term arises again with opposing spins for $M = +1$ by the combination $m_1 = +1$ and $m_2 = 0$ (no. 4), for $M = 0$ with $m_1 = +1$ and $m_2 = -1$ (no. 6), and for $M = -1$ by $m_1 = -1$ and $m_2 = 0$ (no. 5).

For $L = 1$, the values of M are $+1, 0$, and -1. The configurations 7, 8, and 9 correspond to this with $S = 1$ this time and thus ^3P terms which subdivide into $^3P_1$, $^3P_2$, and $^3P_3$ as has been seen in part I.

Figure 73.2 shows the geometric construction relative to the number M.

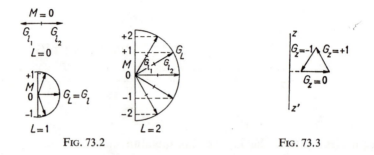

FIG. 73.2   FIG. 73.3

III

For H with one $1s$ electron, the ground state is ^2S.

For He, the two $1s$ electrons form a complete K-shell with ground state ^1S.

For Li the third electron is $2s$ and the ground state ^2S.

For Be the two $2s$ electrons have opposing spins and the term is ^1S.

For B the fifth electron is $2p$ hence a ^2P term.

For C which has two $2p$ electrons, the possible terms have been determined in part II. Hund's rules give a triplet term for the ground state, thus ^3P.

For N the maximum value of S is $\frac{3}{2}$ corresponding to parallel spins for the three $2p$ electrons, which as a result each occupy one of the three $2p$ quantum states. The G_L is then zero (Fig. 73.3). The ground state term is ^4S.

For O the fourth p-electron can only occupy one of the p-levels which already has an electron hence $S = 1$. The total orbital angular momentum is due to the fourth electron since, as has been seen, the other three give a result of zero. The ground state term is ^3P.

To treat F it is best first to examine Ne which has a complete p sublevel and thus both the orbital angular momentum and the spin angular momentum are zero and the ground state

TABLE 73.3

Element	H	He	Li	Be	B	C	N	O	F	Ne
Atomic number	1	2	3	4	5	6	7	8	9	10
S	$\frac{1}{2}$	0	$\frac{1}{2}$	0	$\frac{1}{2}$	1	$\frac{3}{2}$	1	$\frac{1}{2}$	0
L	0	0	0	0	1	1	0	1	1	0
J	$\frac{1}{2}$	0	$\frac{1}{2}$	0	$\frac{1}{2}$	0	$\frac{3}{2}$	2	$\frac{3}{2}$	0
Ground-state term	$^2S_{1/2}$	$^1S_0$	$^2S_{1/2}$	$^1S_0$	$^2P_{1/2}$	$^3P_0$	$^4S_{3/2}$	$^3P_2$	$^2P_{3/2}$	$^1S_0$

is ^1S. If one then removes one electron to return to the F structure, one sees that $L = l = 1$ and $S = s = \frac{1}{2}$ and thus the ground state is ^2P.

The value of J is equal to S for the first four elements since $L = 0$. From B to Ne where the $2p$ shell is being completed, J is given by the second of Hund's rules, for B and C $J = L-S$ and for O and F, $J = L+S$.

Finally, one gets the symbols shown in Table 73.3 for the ground state terms (Fig. 73.4):

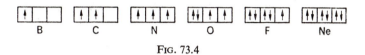

FIG. 73.4

PROBLEM 74

Zeeman Effect in a Two-electron Atom

In the many-electron atoms discussed in the previous question, the angular momentum components follow the Russell–Saunders rules. Use $\overset{\smile}{S}$ and $\overset{\smile}{L}$ for the spin and orbital angular momenta and $\overset{\smile}{J}$ for the total angular momentum. The magnetic moment of the atom is related to its total angular momentum, just as in the case where there is only one electron, through the Lande factor whose existence results in the magnetic spin anomaly.

I

Give the general expression for the magnetic moment of an atom in a fixed stationary state. Find, taking the Bohr magneton as unity, the magnetic moment of the following atoms in their ground state: H, He, Li, Ne, and Na. How can one verify these values experimentally?

II

1. Study the emission spectrum of strontium (the atom homologous to Be with the two optical electrons in the O-shell) when it is placed in a weak magnetic field. Show graphically the Zeeman levels for the following transitions and the permitted transitions:

(a) $\lambda = 4678$ Å $10\ _1F_3 \rightarrow 7\ ^1D_2$,

(b)   4962 $8\ ^3D_3 \rightarrow 6\ ^3P_2$,

(c)   4892 $8\ ^3F_4 \rightarrow 7\ ^3D_3$.

Recall that the selection rules for the quantum number m_J are the same as for a one electron atom but that ΔJ cannot be zero. Among the transitions considered are those which give a normal Zeeman effect?

One finds that the structure of the Zeeman spectra line (c) is very close to that of (a) but not to that of line (b). Give the reason for this.

2. Calculate the separation in wave numbers of the π- and σ-components of line (c) in a magnetic field of 30,000 oersteds. The speed of light is $c = 3\times10^8$ m/s and $e/m_e = 1.67\times10^{11}$ C/kg.

SOLUTION

I

In Russell–Saunders coupling, one has $\vec{S} = \sum \vec{s}$, $\vec{L} = \sum \vec{l}$ and $\vec{J} = \vec{S} + \vec{L}$. The modulus of $\vec{J}$ is $\sqrt{J(J+1)}\hbar$. Thus

$$\vec{M} = g\sqrt{J(J+1)}\,\vec{\mu}_B \tag{1}$$

g is the Lande factor and $\vec{\mu}_B$ the Bohr magneton.

Refer to Problem 73 which gives the ground states of the atoms and calculate g using the expression

$$g = 1 + \frac{J(J+1)+S(S+1)-L(L+1)}{2J(J+1)}. \tag{2}$$

One finds (§ 15.11)

TABLE 74.1

Atom	H	He	Li	Be	Na
Z	1	2	3	4	11
S	$\frac{1}{2}$	0	$\frac{1}{2}$	0	$\frac{1}{2}$
L	0	0	0	0	0
J	$\frac{1}{2}$	0	$\frac{1}{2}$	0	$\frac{1}{2}$
Term	$^2S_{1/2}$	$^1S_0$	$^2S_{1/2}$	$^1S_0$	$^2S_{1/2}$
$\sqrt{J(J+1)}$	$\sqrt{\frac{3}{4}}$	0	$\sqrt{\frac{3}{4}}$	0	$\sqrt{\frac{3}{4}}$
g	2	1	2	1	2
M	$\sqrt{3}$	0	$\sqrt{3}$	0	$\sqrt{3}$

In the calculation for Na, take into account the fact that the L-shell is complete and its moments $\vec{S}$, $\vec{L}$, and $\vec{J}$ are zero as has been shown for the K-shell of He (§ 15.12).

The Stern–Gerlach experiment (and its modern variation due to Rabi) allows one to measure $\vec{M}$ in the ground state.

II

1. J, m_J, and g have the values shown in Table 74.2 for the levels under consideration (the principal quantum number is not important).

The number of Zeeman sublevels is the same as the number of values of m_J and their splitting is proportional to g. The upper parts of Fig. 74.1, 74.2, and 74.3 show these levels graphically for the three lines under consideration.

The allowed transitions obey the selection rules

$$\Delta m_J = 0 \quad \text{or} \quad +1 \quad \text{or} \quad -1.$$

The three lines behave differently.

TABLE 74.2

Level	$^1F_3$	$^1D_2$	$^3P_2$	$^3D_3$	$^3F_4$
J	3	2	2	3	4
m_J	-3 to $+3$	-2 to $+2$	-2 to $+2$	-3 to $+3$	-4 to $+4$
g	1	1	$\frac{3}{2}$	$\frac{4}{3}$	$\frac{5}{4}$

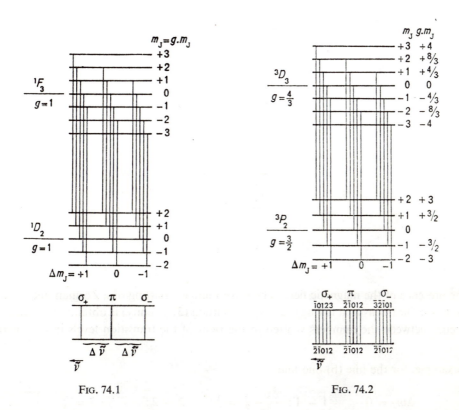

FIG. 74.1   FIG. 74.2

For line (a), the Lande factor is the same for both levels with the result that the splitting of the Zeeman levels is the same. As a result the five transitions permitted by the selection rules have the same wave number and one observes a normal triplet. Each of the $\Delta m_J = \pm 1$ lines are split symmetrically about the $\Delta m_J = 0$ line by a number of cm^{-1} equal to

$$\Delta \tilde{v} = 0.467B, \qquad (3)$$

where B is the magnetic field acting on the atom.

For the lines (b) and (c), the initial and final states have different Lande factors. All of the lines corresponding to permitted transitions are distinct. The $0 \rightarrow 0$ line is not shifted

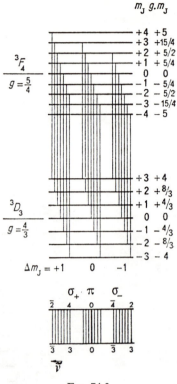

FIG. 74.3

by the presence of the magnetic field. The wave number splitting of a Zeeman line relative to the $0 \to 0$ line (measured taking the normal splitting (3) as unity) is obtained by taking the difference between the numbers written to the right of the transition levels in the figures.

For example, for the line (b) one finds

$$\Delta m_J = 0 \qquad 1 \to 1: -\tfrac{4}{3}+\tfrac{3}{2} = \tfrac{1}{6}; \qquad 2 \to 2: \quad -\tfrac{8}{3}+3 = \tfrac{2}{6}$$

$$\Delta m_J = +1 \qquad \bar{3} \to \bar{2}:4-3 = \tfrac{6}{6}; \qquad \bar{2} \to \bar{1}:\tfrac{8}{3}-\tfrac{3}{2} = \tfrac{7}{6};$$

$$\bar{1} \to 0:\tfrac{4}{3}-0 = \tfrac{8}{6};$$

$$0 \to 1:0+\tfrac{3}{2} = \tfrac{9}{6}; \qquad 1 \to 2:-\tfrac{4}{3}+3 = \tfrac{10}{6}.$$

One can easily verify that the splittings for $\Delta m_J = -1$ are symmetrical with those for $\Delta m_J = +1$. On the whole, to each of the three permitted values of m_J there corresponds a group of five lines separated by one-sixth the normal splitting.

For the line (c), one also finds a symmetric distribution in three groups of seven lines with the splitting of two lines of a group being one-twelfth the normal splitting (3). The structure of this line is very close to that of line (b) of the normal triplet since the Lande factors of the levels involved are close to unity.

2. The splitting of the π and σ components of the (c) line is $\frac{1}{12}$ of the normal splitting. In a field of 0.3 tesla, equation (3) gives

$$\Delta \tilde{\nu} = 0.467 \times 0.3 \times \tfrac{1}{12} = 0.0117 \text{ cm}^{-1}$$

a splitting difficult to detect.

PROBLEM 75

X-ray Spectra

Here one looks at the molybdenum absorption and emission spectra in the X-ray region. Assume for this purpose an X-ray spectrograph.

I

Describe briefly the construction and function of an X-ray spectrograph.

II

1. Explain the mechanism of X-ray absorption in an element with high atomic number Z. Why does one require a screening constant? Find the wave numbers of the first three absorption limits (or discontinuities) for molybdenum. Initially neglect the fine structure of the limits and take the Rydberg constant to be approximately $R = 1.1 \times 10^5$ cm^{-1}. For X-ray absorption one can assume in the calculation the screening constant to be $C_K = 3.5$ for the K limit, $C_L = 14$ for the L limit, and $C_M = 25.4$ for the M limit.

2. Give the theory for line emission in the X-ray region for atoms of high atomic number. Set up the Moseley equation. Find the wave number of the K_α- and K_β-lines in molybdenum. With the Moseley law one assumes for a transition from an M or L-level to the K-level one can take a single mean screening constant $C_K = 1$.

Show that one gets approximately the same values for the wave numbers of the K_α- and K_β-lines when one takes the wave number differences of the K, L, and M absorption edges (neglecting the fine structure).

3. Show how one can perfect the preceding theory explaining the fine structure of the emission lines and the X-ray edge absorption. Show that one can represent the fine-structure energy levels by the expression

$$\frac{W}{Rch} = -\left[\frac{(Z-C)^2}{n^2} + \frac{\alpha^2(Z-C_x')^4}{n^3} \left(\frac{1}{J+\frac{1}{2}} - \frac{3}{4n} \right) \right] \tag{1}$$

where C and C_x' are empirical screening constants which depend on the quantum numbers of the transitions involved. In this expression α is the fine-structure constant and the other notations have their usual significance.

Diagram the K, L, and M levels of molybdenum. Give the selection rules for transitions between these levels and indicate the permitted transitions on the diagram.

4. Find the wave numbers of the L_I, L_{II}, and L_{III} molybdenum absorption limits. Take

$$C = 14, \quad C'_{L_I} = 2, \quad C'_{L_{II}, L_{III}} = 3.5, \quad \alpha^2 = 5.3 \times 10^{-5}.$$

Explain the origin of the screening doublet.

5. Find the splitting of the spin doublet in molybdenum:

$$\Delta \nu_{K\alpha_1 K\alpha_2} = \nu_{K\alpha_1} - \nu_{K\alpha_2}.$$

To do this make use of the fine-structure equation.

III

Explain the origin of the anode continuous emission spectrum of X-rays.

Consider a flux of electrons accelerated by a 100 kV potential difference falling on a molybdenum anode. Find the wave-number limit of the continuous emission spectrum under these conditions.

SOLUTION

I. See § 7.15.

II

1. When the atomic number Z is large the K, L, M, ..., electron shells corresponding to the values 1, 2, 3, ... of the principal quantum number n are filled in accordance with the Pauli principle. The absorption of an X-ray photon with considerable energy $h\nu$ of the order of 10^4 eV does not lead to a transition of an electron from a deep level to an already occupied higher level, but to the removal of this electron, that is, to the ionization of the atom. The work necessary to accomplish this is equal to the coulomb energy of the electron in the nuclear field if there are no other electrons and this is given by the hydrogen atom energy level expression (Problem 69):

$$W_n = \frac{\mu e^4 Z^2}{8 \varepsilon_0 h^2} \cdot \frac{1}{n^2} = \frac{ch Z^2 R}{n^2}.$$

μ is the reduced mass of the electron-nucleus system. This is roughly the same as the electron mass, m_e, when the atomic number—and therefore the mass number—is high. The Rydberg constant takes the value R_∞ as a result.

In practice, the electrons present in deeper and shallower energy levels create a potential which detracts from that of the nucleus and lessens the value of W_n, a quantity which depends essentially on n. This is taken into account by using

$$W'_n = \frac{chR}{n^2}(Z-C_n)^2. \tag{2}$$

C_n is the screening constant which depends on the level n in a first approximation neglecting the fine structure which will be considered in question 4.

Use the Hartree method (§ 14.7) to evaluate the constant C_n. The potential energy of an electron at a distance r_0 from the atomic nucleus is given by

$$4\pi\varepsilon_0 W_p(r_0) = \frac{Ze^2}{r_0} - \frac{4\pi e^2}{r_0}\int_0^{r_0}\psi^2(r)r^2\,\mathrm{d}r - \frac{4\pi e^2}{\langle r\rangle}\int_{r_0}^\infty\psi^2(r)r^2\,\mathrm{d}r.$$

The last two terms give the screening effect assuming that the density of electronic charge $\varrho(r)$ has a symmetric spherical distribution about the nucleus. The second term gives the internal screening effect which follows from the well-known electrostatics theorem which states that the potential does not depend on the details of the internal distribution of charges. The third term, relative to the external screening, has a different dependence. It is always much smaller than the second. For example, for a K-electron, the screening effect of the other electron in the K-shell is less than 1 (in units of Z) which it would be equal to if its charge were concentrated at a distance less than r_0. As to the eight L-electrons, since their mean distance $\langle r\rangle$ to the nucleus is of the order of $4r_0$ their contribution to the screening effect (§ 14.6), which would be zero if their distance remained always beyond r_0 and their symmetry spherical, is small. The same is true for the eighteen M electrons, etc.

The wave numbers are given by the expression

$$\tilde{\nu}_n = \frac{1}{\lambda_n} = \frac{R(Z-C_n)^2}{n^2}$$

Using the given numerical values

$$\tilde{\nu}_K = \frac{1.1\times10^5(42-3.5)^2}{1^2} = 16{,}305\times10^4\ \mathrm{cm^{-1}},$$

$$\tilde{\nu}_L = \frac{1.1\times10^5(42-14)^2}{2^2} = 2156\times10^4\ \mathrm{cm^{-1}},$$

$$\tilde{\nu}_M = \frac{1.1\times10^5(42-25.4)^2}{3^2} = 313\times10^4\ \mathrm{cm^{-1}}.$$

2. The K_α-line arises in the transition $L \to K$ ($n_1 = 2 \to n_2 = 1$) and the K_β in the transition $M \to K$ ($n_1 = 3 \to n_2 = 1$). Thus, with $\bar{C}_K = 1$,

$$\tilde{\nu}_{K_\alpha} = R(Z-\bar{C}_K)^2\left(\frac{1}{n_2}-\frac{1}{n_1}\right) = 1.1\times10^5\times(42-1)^2\times\left(1-\frac{1}{4}\right) = 13{,}868\times10^4\ \mathrm{cm^{-1}}$$

$$\tilde{\nu}_{K_\beta} = 1.1\times10^5\times(42-1)^2\left(1-\tfrac{1}{9}\right) = 16{,}436\times10^4\ \mathrm{cm^{-1}}.$$

Also, one finds

$$\tilde{\nu}_K - \tilde{\nu}_L = 14{,}146 \times 10^4 \text{ cm}^{-1},$$

$$\tilde{\nu}_K - \tilde{\nu}_M = 15{,}989 \times 10^4 \text{ cm}^{-1}.$$

Thus one gets similar values for $\tilde{\nu}_{K_\alpha}$ and $\tilde{\nu}_K - \tilde{\nu}_L$ and for $\tilde{\nu}_{K_\beta}$ and $\tilde{\nu}_K - \tilde{\nu}_M$ (§ 15.13.3).

3. Proving the expression given in the statement of the problem is extremely difficult. Here we limit ourselves to the statement that the first term between the brackets comes from the energy (2) and the second is a correction term. This is introduced when one takes into account simultaneously the relativistic form of the equations of motion and the spin–orbit interaction (§ 15.7). The various values of the internal quantum number J corresponding to a given value of the orbital quantum number L are in effect due to this latter interaction.

In the actual problem, the calculation of the energy of an atom lacking an electron is analogous to that of an atom having an electron with quantum numbers n, l, and $j = 1 \pm \frac{1}{2}$. The fine structure consists of a doubling of the levels for which $l \neq 0$ (since for $l = 0$, $j = +\frac{1}{2}$, the total angular momentum which depends on j cannot be negative).

The diagram of the K, L, and M levels is presented in Fig. 75.1.

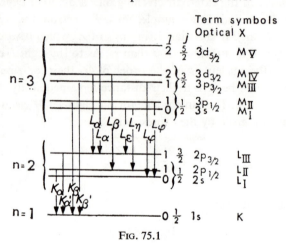

FIG. 75.1

The selection rules $\Delta l = \pm 1$ and $\Delta j = 0, \pm 1$ only permit the transitions indicated by the vertical arrows.

4. For the L level, of which there are three, $n = 2$. The differences between levels is accounted for by the corrective terms in equation (1). One finds for these terms

$$L_{\mathrm{I}} \left(l = 0, j = \frac{1}{2}, C'_x = 2 \right) : \frac{R\alpha^2 (Z-2)^4}{2^4} \left(\frac{2}{1} - \frac{3}{4} \right) = 116.5 \times 10^4 \text{ cm}^{-1},$$

$$L_{\mathrm{II}} \left(l = 1, j = \frac{1}{2}, C'_x = 3.5 \right) : \frac{R\alpha^2 (Z-3.5)^4}{2^4} \left(\frac{2}{1} - \frac{3}{4} \right) = 1.00 \times 10^4 \text{ cm}^{-1},$$

$$L_{\mathrm{III}} \left(l = 1, j = \frac{3}{2}, C'_x = 3.5 \right) : \frac{R\alpha^2 (Z-3.5)^4}{2^4} \left(\frac{2}{2} - \frac{3}{4} \right) = 0.20 \times 10^4 \text{ cm}^{-1}.$$

These values are added to the value of $\tilde{\nu}_L = 2156 \times 10^4$ cm^{-1} obtained in question II.1.

For given n and l, the differences in j are the origin of the spin doublet. For given n and j, the variation of the screening constant with l produces the screening doublet. For large Z, the spin doublet splitting is greater than the screening doublet splitting.

5. The K_{α_1} line is due to the $L_{III} \rightarrow K$ transition and the K_{α_2} line is due to the $L_{II} \rightarrow K$ transition. These lines form the spin doublet. One has $\left(n = 2, j = \frac{3}{2} \text{ or } \frac{1}{2}\right)$

$$\Delta\tilde{\nu} = \tilde{\nu}_{K_{\alpha_1}} - \tilde{\nu}_{K_{\alpha_2}} = \frac{R\alpha^2(Z - C_x')^4}{n^3}\left(\frac{1}{1} - \frac{1}{2}\right) = 0.80 \times 10^4 \text{ cm}^{-1}.$$

III

See § 11.3. The conservation of energy in the transformation of the kinetic energy, eV, of an electron into a single photon gives the frequency limit

$$h\nu_l = eV,$$

$$\tilde{\nu}_l = \frac{eV}{hc} = \frac{1.60 \times 10^{-19} \times 10^5}{3 \times 10^8 \times 6.5 \times 10^{-34}} = 82 \times 10^9 \text{ cm}^{-1},$$

$$\lambda = \frac{1}{\tilde{\nu}_l} = 1.22 \times 10^{-11} \text{ m} = 0.122 \text{ Å}.$$

PROBLEM 76

Mössbauer Effect

By radioactive transformation, the $^{57}_{27}$Co nucleus yields the $^{57}_{26}$Fe nucleus in an excited state with a mean lifetime $\tau = 1.45 \times 10^{-7}$ sec, 14.4 keV above the ground state. The decay of the excited state occurs with the emission of a γ-photon.

1. Find the width ΔW of the excited state as well as the relative natural width $\Delta\nu/\nu$ of the emitted γ-ray.

2. Assuming that the emitting nucleus is free and at rest, calculate the relative variation $(\Delta\nu/\nu)_r$ resulting from recoil during emission at the given frequency.

3. Natural iron contains 2% of $^{57}_{26}$Fe. This nucleus is capable, in principle, of absorbing the γ-radiation by a process analogous to optical resonance. If the absorbant like the emitter is free and at rest, is the resonant absorption observable?

4. To demonstrate this absorption, one places between the $^{57}_{27}$Co source and a γ-detector' an iron screen which can move either toward or away from the source. Calculate the relative velocity of the screen and the source which corresponds to the natural line width, the frequency variation $\Delta\nu$ which corresponds to a relative velocity of 1 mm/s, and the relative velocity which must be given to the screen in order to observe the resonance.

5. Calculate the relative spreading $(\Delta v/v)_T$ of the line due to thermal motions at $T = 300°C$ using the hypothesis of equal partition of thermal energy as well as the corresponding relative velocity.

6. In solid cobalt (mass density $\varrho = 7.8 \times 10^3$ kg/m³) the elastic waves transfer a mechanical motion at velocity $v = 3$ km/s and, in particular, that due to nuclear recoil. Estimate the number of atoms which participate in the recoil and show that the relative frequency variation $(\Delta v/v)_r$ is then negligible with respect to the natural line width of the v-line.

7. At the top of a 23-m tower one places a $^{57}_{27}$Co source which emits γ-photons toward the base where they are detected. Calculate the frequency variation $(\Delta v/v)_g$ as a function of H, c, and the gravitational acceleration $g = 9.81$ m/s². Is it detectable?

SOLUTION

1. Since the ground state is stable indefinitely, the energy uncertainty is due only to the limited lifetime of the excited state. The uncertainty relation relating time to energy gives (§ 12.9)

$$\Delta W \approx \frac{h}{\tau} = \frac{6.62 \times 10^{-34}}{1.45 \times 10^{-7}} = 4.56 \times 10^{-27} \text{ J} \tag{1}$$

or 2.85×10^{-8} eV. (If one takes $\Delta W \cdot \tau \approx \hbar$, $\Delta W = 4.5 \times 10^{-9}$ eV.)

The relative natural line width of the γ line is

$$\frac{\Delta v}{v} = \frac{h \Delta v}{hv} = \frac{\Delta W}{W} = \frac{2.85 \times 10^{-8}}{14.4 \times 10^3} = 1.98 \times 10^{-12} \tag{2}$$

(with $W = 4.5 \times 10^{-9}$ eV, $\Delta v/v = 3.14 \times 10^{-13}$).

2. The nucleus with mass m receives a recoil momentum p equal to the momentum hv/c carried by the emitted photon. The recoil kinetic energy $p^2/2m$ subtracts from the transition energy W, with the result that the energy balance is

$$W = hv + \frac{h^2 v^2}{2mc^2} = hv \left(1 + \frac{hv}{2mc^2}\right). \tag{3}$$

The relative frequency variation of the emitted photon is

$$\left(\frac{\Delta v}{v}\right)_r = \left(\frac{\Delta W}{W}\right)_r = -\frac{hv}{2mc^2}. \tag{4}$$

3. The preceding frequency with a change of sign is produced in the photon absorption which gives its momentum to the nucleus with the result that the energy available to modify the internal state in this transition from the ground state to an excited state is equal to the photon energy less the kinetic energy acquired by the nucleus. The relative separation between the absorbed and emitted frequency is then

$$\frac{hv}{mc^2} = \frac{14.4 \times 10^3 \times 1.6 \times 10^{-19}}{57 \times 1.67 \times 10^{-27} \times 9 \times 10^{16}} = 2.69 \times 10^{-7}. \tag{5}$$

This is more than 10^5 times larger than the natural relative width. As a result, the emission and absorption lines do not overlap and one cannot observe resonant absorption.

4. The motion of the iron screen leads to a variation in the frequency of absorption resulting from the Doppler effect. The frequency variation corresponding to a speed of 1 mm/s is

$$\Delta \nu = \frac{u}{c} \nu = u \frac{h\nu}{hc} = u \frac{\Delta W}{hc} = 11.60 \text{ MHz.}$$

To compensate for the frequency variation due to the recoil given by equation (5), the screen must approach the source with a velocity

$$u = 2c\left(\frac{\Delta \nu}{\nu}\right)_r = 3 \times 10^8 \times 2.69 \times 10^{-7} = 81 \text{ m/s.}$$

5. The spreading due to thermal motion is given by (§ 12.9)

$$\left(\frac{\Delta \nu}{\nu}\right)_T = 2 \sqrt{\frac{2kT \log 2}{mc^2}} = 2 \sqrt{\frac{2 \times 4.14 \times 10^{-21} \times 2.3 \times 0.301}{57 \times 1.67 \times 10^{-27} \times 9 \times 10^6}} = 16.36 \times 10^{-7}.$$

It is of the order of 10^6 times larger than the natural width. The relative velocity to which it corresponds is

$$u = c\left(\frac{\Delta \nu}{\nu}\right)_T = 3 \times 10^8 \times 16.36 \times 10^{-7} = 491 \text{ m/s.}$$

6. During the γ emission mean lifetime τ, the recoil motion takes place over a distance $v\tau$ in the metal and the momentum is transmitted to a volume

$$V = \tfrac{4}{3}\pi v^3 \tau^3.$$

The number of atoms per cubic metre is

$$\frac{\mathcal{N}\varrho}{A} = \frac{6 \times 10^{26} \times 7.8 \times 10^3}{57}.$$

The number which receive momentum is

$$N = \frac{4 \times 3.14 \times (3 \times 10^3 \times 1.45 \times 10^{-7})^3 \times 6 \times 7.8 \times 10^{29}}{3 \times 57} \approx 2.83 \times 10^{19}.$$

The momentum is then received by a mass equal to N times that of the emitter nucleus and the recoil kinetic energy, which arises in equation (3), is, as a result, negligible with respect to the energy $h\nu$. One can then observe resonance absorption, the target participating in the same way as the absorbing nucleus.

7. One of the fundamental principles of general relativity is the equivalence between a gravitational field and an inertial force field resulting from an accelerated motion. Now the earth's gravitational field produces, when it acts alone, a uniformly accelerated motion and the speed acquired by a mass m after falling through a vertical height H in this field is given by the kinetic energy theorem

$$\tfrac{1}{2}mv^2 = mgH,$$

hence

$$v^2 = 2gH.$$

For an observer situated on the earth, the time interval Δt becomes $\Delta t'$ in the system moving with velocity v and one finds (§ 9.9.2)

$$\Delta t = \Delta t' \sqrt{1 - \frac{v^2}{c^2}} = \Delta t' \sqrt{1 - \frac{2gH}{c^2}} \approx \Delta t'\left(1 - \frac{gH}{c^2}\right).$$

If the interval Δt represents the period of a clock (which can be an atomic nucleus), the relative variation due to changing the reference system has the value

$$\frac{T - T'}{T} = \frac{gH}{c^2} \, .$$

This is also the relative frequency variation $(\Delta v/v)_g$, which one calls the "gravitational Doppler effect".

With the given parameters

$$\frac{\Delta v}{v} = \frac{12 \times 9.81}{9 \times 10^{16}} \approx 1.3 \times 10^{-15}.$$

and this is difficult to observe.

PROBLEM 77

Vibrational and Rotational Spectrum of the Hydriodic Acid Molecule

Consider a diatomic hydriodic acid molecule HI at ordinary temperature. It has a rotational motion about an axis passing through the centre of gravity G.

I

1. Calculate the reduced mass μ of the molecule and its moment of inertia I about G. The interatomic distance is given as $r = 1.6 \times 10^{-10}$ m, the mass of the electron as $m = 9.11 \times 10^{-31}$ kg, and the respective mass of the two atoms as $m_H = 1836m$ and $M_I = 127m_H$.

2. Given the laplacian in spherical coordinates as

$$\Delta = \frac{1}{r^2} \frac{\partial}{\partial r}\left(r^2 \frac{\partial}{\partial r}\right) + \frac{1}{r^2 \sin\theta} \frac{\partial}{\partial\theta}\left(\sin\theta \frac{\partial}{\partial\theta}\right) + \frac{1}{r^2 \sin^2\theta} \frac{\partial^2}{\partial\phi^2} \tag{1}$$

write the time independent Schrödinger equation for the stationary states of rotation of the molecule. Calculate the values of its rotational energy W_J, given that the equation has a solution in the form $\psi = e^{jJ\phi}(\sin \theta)^J$.

II

1. What is the structure of the rotational spectrum of gaseous HI? What is the frequency of the radiation emitted or absorbed in a transition between state W_J and an adjacent state W_{J+1}? Find W_0, W_1, W_2, W_3, and the wavelengths emitted in the transitions $0 \leftrightarrows 1, 1 \rightleftarrows 2$, and $2 \rightleftarrows 3$. What frequency domain to they belong to? $h = 6.62 \times 10^{-34}$ joule-s.

2. Assume now that the molecule has a vibrational motion independent of the rotational motion. When the two atoms are separated by a distance r different from r_0 they are subject to a restoring force $F = -k_0 s$ where $s = |r-r_0|$. Making the change of variable $q = s\sqrt{b}$, Schrödinger's equation relative to this motion takes the form

$$\frac{d^2\psi}{dq^2} + \left(\frac{a}{b} - q^2\right)\psi = 0 \tag{2}$$

where

$$a = \frac{8\pi^2\mu}{h^2} W_v \quad \text{and} \quad b = \frac{4\pi^2\mu v_0}{h}.$$

In these expressions W_v represents the vibrational energy of the molecule and

$$v_0 = \frac{1}{2\pi}\sqrt{\frac{k_0}{\mu}} \tag{3}$$

its vibrational frequency.

(a) Given that (2) has acceptable solutions only if $a/b = 2v+1$ (where v is a positive integer), give the equation for the vibrational energy levels of the molecule. What is the asymptotic solution of (2)? To what quantum principle can one appeal to justify the fact that the minimal vibrational energy is not zero?

Recall that the dipole moment associated with all transitions $n \rightleftarrows m$ giving rise to radiation has the value

$$(D_q)_{nm} = e \int_{-\infty}^{+\infty} q\psi_n^*\psi_m \, dq.$$

Calculate $(D_q)_{01}$ and $(D_q)_{02}$ knowing that (to within a normalization constant)

$$\psi_0 = e^{-q^2/2}, \quad \psi_1 = qe^{-q^2/2}, \quad \psi_2 = (2q^2-1)\times e^{-q^2/2}.$$

Show that in this way one can verify the selection rule $\Delta v = \pm 1$ between stationary vibrational states.

3. Assume that the total energy of the molecule W_T is the sum of its rotational and vibrational energy. Give the value of W_T. What are the wave numbers of the radiation $\tilde{v}$ emitted and absorbed by the molecule assuming that only one vibrational level but a large

number of rotational levels are excited? This set of wave numbers can be put in the form $\tilde{\nu} = \tilde{\nu}_0 \pm m \, \Delta \tilde{\nu}$. Give the numerical value of $\Delta \tilde{\nu}$. Find $\tilde{\nu}_0$, ν_0, and k_0 [defined by (3)], given $\tilde{\nu} = 2332 \text{ cm}^{-1}$ for $m = 7$.

SOLUTION

I

1. Reduced mass:

$$\mu = \frac{m_H m_I}{m_H + m_I} = \frac{127}{128} m_H = \frac{127}{128} \times 1836 \times 9.11 \times 10^{-31} = 1.66 \times 10^{-27} \text{ kg.}$$

The moment of inertia (§ 14.2):

$$I = \mu r_0^2 = 1.66 \times 10^{-27} \times 2.56 \times 10^{-20} = 4.25 \times 10^{-47} \text{ kg/m}^2.$$

2. Schrödinger's equation relative to the stationary states of motion of the mass μ:

$$\Delta \psi + \frac{2\mu}{\hbar^2} (W - W_p) \psi = 0$$

here is, where $W_p = 0$, Δ is the given expression, and r_0 is constant during rotation,

$$\frac{1}{r_0^2 \sin \theta} \frac{\partial}{\partial \theta} \left(\sin \theta \frac{\partial \psi}{\partial \theta} \right) + \frac{1}{r_0^2 \sin^2 \theta} \frac{\partial^2 \psi}{\partial \phi^2} + \frac{2\mu}{\hbar^2} W \psi = 0$$

or

$$\frac{\partial^2 \psi}{\partial \theta^2} + \cot \theta \frac{\partial \psi}{\partial \theta} + \frac{1}{\sin^2 \theta} \frac{\partial^2 \psi}{\partial \phi^2} + \frac{2IW}{\hbar^2} \psi = 0. \tag{4}$$

Substituting the assumed solution in this latter equation, one has

$$J(J-1) \cos^2 \theta (\sin \theta)^{J-2} - J(\sin \theta)^J + J \cos^2 \theta (\sin \theta)^{J-2} - J^2 (\sin \theta)^{J-2} + \frac{2IW}{\hbar^2} (\sin \theta)^J = 0,$$

so that

$$W_J = J(J+1) \frac{\hbar^2}{2I}. \tag{5}$$

Note that the assumed solution

$$\psi = \exp (jJ\phi) \cdot (\sin \theta)^J$$

is a special case of the general solution of equation (1)

$$\psi = C \exp (jm\phi) P_J^m(\cos \theta)$$

when $J = m$. This can be verified using the definition of the associated Legendre polynomials P_J^m.

II

1. Hydriodic acid, as all polar diatomic molecules, has a dipole emission and absorption rotational spectrum. The selection rule for the permitted transitions is $\Delta J = \pm 1$. Using (5), the first four energy levels are given by

$$W_0 = 0, \quad W_1 = \frac{\hbar^2}{I}, \quad W_2 = \frac{3\hbar^2}{I}, \quad W_3 = \frac{6\hbar^2}{I}.$$

With

$$\frac{\hbar^2}{I} = \frac{(6.62)^2 \times 10^{-68}}{4 \times 9.87 \times 4.25 \times 10^{-47}} = 25.2 \times 10^{-22} \text{ J},$$

one has

$$W_0 = 0, \quad W_1 = 25 \times 10^{-22} \text{ J}, \quad W_2 = 75 \times 10^{-22} \text{ J}, \quad W_3 = 150 \times 10^{-22} \text{ J}.$$

The absorption frequency for a transition between an energy level and the next higher level is given by

$$\nu = \frac{W_{J+1} - W_J}{h} = \frac{h}{4\pi^2 I}(J+1).$$

hence

$$\lambda = \frac{c}{\nu} = \frac{4\pi^2 cI}{h} \cdot \frac{1}{J+1} = \frac{0.75 \times 10^{-3}}{J+1} \quad \text{m},$$

$$\lambda_{01} = 0.75 \text{ mm}, \quad \lambda_{12} = 0.37 \text{ mm}, \quad \lambda_{23} = 0.25 \text{ mm}.$$

These lines are in the microwave region. They have been widely studied since their discovery in 1945.

2. (a) Equation (2) is the harmonic oscillator expression (§ 14.4). One finds

$$2v + 1 = \frac{a}{b} = \frac{2W_v}{h\nu_0},$$

thus

$$W_v = (v + \tfrac{1}{2})h\nu_0.$$

The asymptotic solution of (2) for large values of r (and therefore of q) is found by neglecting a/b relative to q^2. The equation is written

$$\frac{d^2 f}{dq^2} - q^2 f = 0$$

and has the solution

$$f = A \exp(-q^2/2).$$

The existence of a residual energy $W_0 = \tfrac{1}{2}h\nu$ for $v = 0$ is a consequence of uncertainty (see Problem 57).

(b) Write the transition dipole expression between the first two energy levels and use the values for the wavefunctions given

$$(D_q)_{01} = e \int_{-\infty}^{+\infty} q\psi_0^*\psi_1 \, dq = e \int_{-\infty}^{+\infty} q^2 \exp(-q^2) \, dq.$$

The integration is done by parts:

$$\int_{-\infty}^{+\infty} q^2 \exp(-q^2) \, dq = \left[-q^2 \exp(-q^2) \right]_{-\infty}^{+\infty} + \tfrac{1}{2} \int_{-\infty}^{+\infty} \exp(-q^2) \, dq.$$

The quantity between the brackets is zero and the last integral is equal to $\sqrt{\pi}$. Thus

$$(D_q)_{01} = \frac{e\sqrt{\pi}}{2}.$$

The transition $0 \rightarrow 1$, with non-zero dipole moment is thereby allowed

$$(D_q)_{02} = e \int_{-\infty}^{+\infty} q(2q^2-1) \exp(-q^2) \, dq$$

$$= 2e \int_{-\infty}^{+\infty} q^3 \exp(-q^2) \, dq - e \int_{-\infty}^{+\infty} q \exp(-q^2) \, dq.$$

The last integral is zero and one finds

$$\int_{-\infty}^{+\infty} 2q^3 \exp(-q^2) \, dq = \left[-q^2 \exp(-q^2) \right]_{-\infty}^{+\infty} + 2 \int_{-\infty}^{+\infty} q \exp(-q^2) \, dq.$$

From what has been done above these integrals are zero. Thus $(D_q)_{02}$ is zero and the transition $0 \rightarrow 2$ is forbidden.

3. One has

$$W_T = W_J + W_v = \frac{\hbar^2}{2I} J(J+1) + \left(v + \frac{1}{2} \right) h\nu_0$$

hence, the wave numbers of the emitted and absorbed lines, using only the vibrational transition $v = 0 \rightarrow v = 1$

$$\tilde{\nu} = \frac{\Delta W_v + \Delta W_J}{ch} = \frac{\nu_0}{c} + [J'(J'+1) - J(J+1)] \frac{\hbar^2}{2Ic}.$$

If, in addition, one takes the selection rule $\Delta J = \pm 1$ into account, one finds
for $J' = J+1$:

$$\tilde{\nu} = \frac{\nu_0}{c} + (J+1) \frac{h}{4\pi^2 Ic},$$

for $J' = J-1$:

$$\tilde{\nu} = \frac{\nu_0}{c} - J \frac{h}{4\pi^2 Ic}.$$

All of the allowed values of $\tilde{\nu}$ can be represented by the expression

$$\tilde{\nu} = \tilde{\nu}_0 \pm m \frac{h}{4\pi^2 Ic} \qquad (m \text{ a positive integer})$$

Thus one has

$$\Delta\tilde{\nu} = \frac{h}{4\pi^2 Ic} = \frac{6.62 \times 10^{-34}}{4 \times 9.87 \times 4.25 \times 10^{-47} \times 3 \times 10^8} = 1.31 \times 10^3 \text{ m}^{-1}.$$

The problem gives $\tilde{\nu} = \tilde{\nu}_0 + 7 \, \Delta\tilde{\nu}$, thus

$$\tilde{\nu}_0 = 2332 - 7 \times 13.1 = 2230 \text{ cm}^{-1},$$
$$\nu_0 = c\tilde{\nu}_0 = 6.69 \times 10^{13} \text{ s}^{-1},$$
$$k_0 = 4\pi^2 \mu \nu_0^2 = 2.93 \times 10^2 \text{ N m}^{-1}.$$

PROBLEM 78

Calculation of the Velocity of Light

One precisely measures with a grating the wavelengths of the lines in the vibration–rotation band of carbon monoxide gas $^{12}C^{16}O$ found near 4.67 μ and gets for the six lines surrounding the centre of the band, the following wave numbers in cm^{-1} in vacuum: 2131.635, 2135.550, 2139.430, 2147.084, 2150.858, and 2154.599.

These values (and those of other lines in this band) are accurately described by the expression

$$\tilde{\nu} = \tilde{\nu}_0 \pm Bm - Cm^2, \tag{1}$$

where m is an integer ($m = 1, 2, 3, \ldots$), and B and C are constants.

In addition, electromagnetic millimeter waves produced by a klystron oscillator are absorbed when they are incident on the gas with the absorption frequency being $\nu = 114,737 \times 10^6$ Hz which corresponds to the first line in the pure rotational spectrum ($J = 0 \rightarrow 1$).

With this information find the velocity of light in vacuum.

SOLUTION

When one compares a diatomic molecule to an oscillator and assumes that the vibrational and rotational energies are independent, the wave numbers of a rotation–vibration band are given by the expression (§ 16.3)

$$\tilde{\nu} = \frac{\nu_v}{c} \pm m \frac{h}{4\pi^2 Ic} \tag{2}$$

where ν_v is the vibration frequency, m a positive integer, and I the moment of inertia about an axis passing through the center of mass of the molecule and normal to the internuclear line. Equation (1) in the problem statement is derived from (2) with $\tilde{\nu}_0 = \nu_v/c$ and $B = h/4\pi^2 Ic$. It also contains an additional term, Cm^2, due to the interaction of the rotations and vibrations.

Note that the first three given values differ in $\tilde{\nu}$ pairwise by 4 cm^{-1} as do the last three but that the difference between the third and the fourth is 8 cm^{-1}. The pure rotational line $\tilde{\nu}_0$ which does not appear in absorption (§ 16.35) is therefore situated between the third and fourth lines of the series. The values of m for the various lines can then be assigned. One finds

$$
\begin{aligned}
\tilde{\nu}_1 &= 2147.084 = \tilde{\nu}_0 + B - C, \\
\tilde{\nu}_2 &= 2150.858 = \tilde{\nu}_0 + 2B - 4C, \\
\tilde{\nu}_3 &= 2154.599 = \tilde{\nu}_0 + 3B - 9C, \\
\tilde{\nu}_{-1} &= 2139.430 = \tilde{\nu}_0 - B - C, \\
\tilde{\nu}_{-2} &= 2135.550 = \tilde{\nu}_0 - 2B - 4C, \\
\tilde{\nu}_{-3} &= 2131.635 = \tilde{\nu}_0 - 3B - 9C.
\end{aligned}
\tag{3}
$$

The first and the fourth equations give

$$
\tilde{\nu}_1 + \tilde{\nu}_{-1} = 2\tilde{\nu}_0 - 2C.
\tag{4}
$$

The second and fifth

$$
\tilde{\nu}_2 + \tilde{\nu}_{-2} = 2\tilde{\nu}_0 - 8C,
$$

hence

$$
C = \tfrac{1}{6}(\tilde{\nu}_1 + \tilde{\nu}_{-1}) - (\tilde{\nu}_2 + \tilde{\nu}_{-2}) = 0.0176.
\tag{5}
$$

One finds from (4)

$$
\tilde{\nu}_0 = \tfrac{1}{2}(\tilde{\nu}_1 + \tilde{\nu}_{-1}) + C = 2143.274,
\tag{6}
$$

also the first equation in (3) gives

$$
B = \tilde{\nu}_1 - \tilde{\nu}_0 + C = 3.8270.
\tag{7}
$$

Putting the values (5), (6), and (7) in (1) one gets the following values:

$$
\tilde{\nu}_3 = 2154.597 \qquad \tilde{\nu}_{-3} = 2131.635
$$

on good agreement with the given values.

Also, the frequency of the first rotation line of CO is given by (§ 16.2)

$$
\nu = \frac{W_1 - W_0}{h} = \frac{h}{4\pi^2 I} = cB.
$$

Hence

$$
c = \frac{\nu}{B} = \frac{11.4737 \times 10^{10}}{3.8270} = 2.99792 \times 10^{10} \text{ cm/s.}
$$

With more precise measurements using refinements not taken account of here, this method has been very effectively applied (Plyler *et al.*, *J. Opt. Soc. Amer.*, **45**, 1955).

PROBLEM 79

Spectroscopy and Specific Heat

The internal energy of a mole of a perfect gas can be thought of as being made up of four parts:

$$U = \mathcal{N}(\overline{W}_t + \overline{W}_r + \overline{W}_v + \overline{W}_e).$$

$\overline{W}_t$ is the mean value of the molecule's translational energy, $\overline{W}_r$ the mean rotational energy, $\overline{W}_v$ the vibrational energy, $\overline{W}_e$ the electronic energy and $\mathcal{N}$ is Avogadro's number. The molar heat capacity at constant volume is given by $C = (\partial U/\partial T)_v$ where T is the absolute temperature. One wants to find $U(T)$ and $C(T)$ in the region $0°K$ to $2500°K$ using spectroscopic data.

The mean value of an energy $\overline{W}_i$ can be found from the Boltzmann distribution law using the equation

$$\overline{W}_i = \frac{\sum\limits_0^\infty g_i W_i \exp\left(-\beta W_i\right)}{\sum\limits_0^\infty g_i \exp\left(-\beta W_i\right)}. \tag{1}$$

$\beta = 1/kT$ where k is Boltzmann's constant. The sum is taken over all the quantized energy states. g_i is the statistical weight, the number of distinguishable quantum states having the same energy. In the case where the number of energy levels in a given energy interval is large, the sums in (1) can be replaced by integrals.

I

Show that at a very low temperature, $1°K$ for example, the translational energy of a molecule W_t enclosed in a volume of the order of 1 cm³ has reached the value $\frac{3}{2}kT$, which is given it by the prequantum principle of equipartition of energy.

II

Consider a diatomic molecule AB made up of two different atoms A and B (complications due to symmetry arise in the case where A and B are the same). Derive the expression for the mean value $\overline{W}_r$ of the rotational energy. Take as a variable the dimensionless ratio $x = T_r/T$ where $T_r = \hbar^2/2kI$ ($I = $ moment of inertia of the molecule about an axis passing through its centre of mass normal to AB) is a characteristic rotational temperature and study the behaviour for the function $\overline{W}_r(T)$. Examine two limiting cases: first, $T \ll T_r$ where few rotational levels are excited and one can consider the first two alone, then secondly, where $T \gg T_r$ where a great many levels are excited. Calculate the relative number of molecules in the first eight levels when $T = 10T_r$. Give an expression for the rotational heat capacity and determine its behaviour with variation in temperature.

Numerical application. Calculate T_r for the molecules HD, $H^{35}Cl$, and $^{14}N^{16}O$ for which the values of $AB = r$ are respectively 0.75, 1.27, and 1.15 Å.

III

By comparing a diatomic molecule to a linear harmonic oscillator, derive the expression for the mean vibrational energy $\overline{W}_r$. Taking as a variable $y = T_v/T$ where $T_v = h\nu/k$ is the characteristic vibrational temperature, study the behaviour of the function $\overline{W}_v(T)$ and the corresponding molecular heat.

Numerical application. Calculate T_v for HD, HCl, and NO given that the wave numbers of their Raman vibrational lines are respectively 3630, 2886, and 1880 cm^{-1}.

IV

Given the first electronic excitations of HD, HCl, and NO as 90,400 cm^{-1}, 75,000 cm^{-1} and 45,000 cm^{-1}, respectively, show that it is not necessary to take $\overline{W}_e$ into account in the calculation of molecular heat up to 2500°K.

SOLUTION

I

The study of the translational motion of a particle confined in a given volume (§ 13.10) shows that the lowest energy level corresponds to a de Broglie wavelength of the order of the linear dimensions of the container. The quantum translational energy is:

$$W = \frac{h^2}{8mL} = \frac{44 \times 10^{-68}}{8 \times 1.6 \times 10^{-27} \times M \times 10^{-2}} \simeq \frac{4}{M} \times 10^{-39} \text{ J},$$

M being the mass number of the molecule. This energy is much less than $kT = 1.4 \times 10^{-23}$ J for $T = 1°K$. At this temperature a very large number of levels are excited and one can write for one of their three translational degrees of freedom, x for example, whose energy is $\frac{1}{2}mv^2$:

$$\overline{W} = \frac{\displaystyle\int_0^\infty \frac{1}{2} mv_x^2 \exp\left(-\beta \frac{mv_x^2}{2}\right) dv_x}{\displaystyle\int_0^\infty \exp\left(-\beta \frac{mv_x^2}{2}\right) dv_x}.$$

Taking $q^2 = \beta mv_x^2/2$, one gets:

$$\overline{W} = \frac{1}{\beta} \frac{\displaystyle\int_0^\infty q^2 \exp(-q^2) dq}{\displaystyle\int_0^\infty \exp(-q^2) dq} = \frac{1}{\beta} \frac{\frac{1}{2}\sqrt{\pi}}{\sqrt{\pi}} = \frac{1}{2\beta} = \frac{kT}{2}.$$

The total translational kinetic energy:

$$\frac{mv^2}{2} = \frac{m}{2}(v_x^2 + v_y^2 + v_z^2)$$

has the value $\overline{W}_t = \frac{3}{2}kT$ and the molar heat capacity corresponds to the constant value:

$$C_t = \mathcal{N} \cdot \frac{3}{2}k = \frac{3}{2}R.$$

II

The rotational energy of a free rotor which the diatomic molecule represents can take the values (§ 14.2):

$$W_J = J(J+1)\frac{\hbar^2}{2I} \qquad (J \text{ positive integer or zero})$$

The statistical weight of one of these states is $2J+1$, since the angular momentum $G = \sqrt{J(J+1)}\hbar$ can take $m = 2J+1$ different orientation with respect to a fixed axis. These values have the same energy in the absence of an external field acting on the molecule but can be separated in the presence of such a field. The selection rule for the rotational quantum number is $\Delta J = \pm 1$.

Thus, the expression for the mean rotational energy is:

$$\overline{W}_r = \frac{\sum\limits_0^\infty J(J+1)(2J+1)\dfrac{\hbar^2}{2I}\exp\left(-\dfrac{\beta J(J+1)\hbar^2}{2I}\right)}{\sum\limits_0^\infty (2J+1)\exp\left(-\dfrac{\beta J(J+1)\hbar^2}{2I}\right)}$$

or, taking $x = T_r/T$ and $T_t = \hbar^2/2Ik$,

$$W_r = kT_r \frac{\sum\limits_0^\infty J(J+1)(2J+1)\exp[-J(J+1)x]}{\sum\limits_0^\infty (2J+1)\exp[-J(J+1)x]}. \tag{2}$$

The rotational molar heat capacity is:

$$C_r = \mathcal{N}\frac{d\overline{W}_r}{dT} = R\frac{T_r^2}{T^2}\left\{\frac{\sum\limits_0^\infty J^2(J+1)^2(2J+1)\exp[-J(J+1)x]}{\sum\limits_0^\infty (2J+1)\exp[-J(J+1)x]}\right.$$

$$\left. -\left[\frac{\sum\limits_0^\infty J(J+1)(2J+1)\exp[-J(J+1)x]}{\sum\limits_0^\infty (2J+1)\exp[-J(J+1)x]}\right]^2\right\}. \tag{3}$$

For $T \ll T_r$, if one only considers the levels $J = 0$ and $J = 1$, one has:

$$\overline{W}_r = 6kT_r \exp\left(-2\frac{T_r}{T}\right),$$

$$C_r = 12R\left(\frac{T_r}{T}\right)\exp\left(-2\frac{T_r}{T}\right).$$

These expressions tend to zero with T.

For $T \gg T_r$, expression (2) becomes:

$$\overline{W}_r = kT_r\frac{\displaystyle\int_0^\infty 2J^3 \exp\left(-J^2 x\right)dx}{\displaystyle\int_0^\infty 2J \exp\left(-J^2 x\right)dx},$$

so that, taking $J^2 x = q$:

$$\overline{W}_r = kT\frac{\displaystyle\int_0^\infty q \exp\left(-q\right)dq}{\displaystyle\int_0^\infty \exp\left(-q\right)dq} = kT$$

and:

$$C_r = R.$$

One finds again the values corresponding to equipartition, because the linear rotor, which has only kinetic energy, has only two degrees of rotational freedom about axes normal to the line AB. In fact, the moment of inertia about the axis AB is not zero but only very small with the result that the corresponding quantum of rotation, inversely proportional to this moment of inertia, is so large that the molecule has no rotational energy about this axis except at very high temperature.

The variation of equation (3) is not simple. Starting at 0, it passes the value R for T near $0.6T_r$, approaches a maximum of the order of $1.1R$ for T near $0.8T_r$, then tends asymptotically toward the value R to which it is very close for $T = 2T_r$ (Fig. 79.1).

The relative number of molecules in a given state J is

$$\frac{(2J+1) \exp\left[-J(J+1)x\right]}{\displaystyle\sum_0^\infty (2J+1) \exp\left[-J(J+1)x\right]}. \tag{4}$$

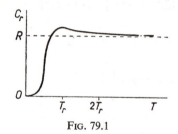

FIG. 79.1

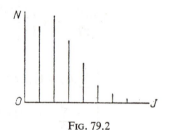

FIG. 79.2

For $x = 0.1$, the numerators N of (4) are as follows for J from 0 to 7:

$$J = 0 \qquad N = 1$$

$$1 \qquad 3 \exp(-0.2) = 2.457$$
$$2 \qquad 5 \exp(-0.6) = 2.75$$
$$3 \qquad 7 \exp(-1.2) = 2.10$$
$$4 \qquad 9 \exp(-2.0) = 1.22$$
$$5 \qquad 11 \exp(-3.0) = 0.55$$
$$6 \qquad 13 \exp(-4.2) = 0.19$$
$$7 \qquad 15 \exp(-5.6) = 0.05$$

Such a distribution explains the relative intensity of the pure rotational spectral lines or vibration-rotation spectra (Fig. 79.2).

Numerical application:

$$T_r = \frac{h^2}{8\pi^2 kI} = \frac{3.96 \times 10^{-46}}{I}, \quad I = \mu r^2, \quad \mu = \frac{m_1 m_2}{m_1 + m_2},$$

μ is the reduced mass.

For HD:

$$\mu = \tfrac{2}{3} \times 1.67 \times 10^{-27} \text{ kg}, \quad I = \tfrac{2}{3} \times 1.67 \times 10^{-27} \times (0.75)^2 \times 10^{-20} \text{ kg/m}^2,$$
$$T_r = 64°\text{K}.$$

For HCl:

$$\mu = \tfrac{35}{36} \times 1.67 \times 10^{-27} \text{ kg}, \quad I = \tfrac{35}{36} \times 1.67 \times 10^{-27} \times (1.27)^2 \times 10^{-20} \text{ kg/m}^2,$$
$$T_r = 15°\text{K}.$$

For NO:

$$\mu = \tfrac{224}{30} \times 1.6 \times 10^{-27} \text{ kg}, \quad I = \tfrac{224}{30} \times 1.6 \times 10^{-27} \times (1.15)^2 \times 10^{-20} \text{ kg/m}^2,$$
$$T_r = 2.4°\text{K}.$$

III

The quantized energy of the harmonic oscillator is (§ 14.4):

$$W_v = (v + \tfrac{1}{2})hv$$

v is the oscillation frequency and v a positive integer or zero. The mean value of the vibrational energy is

$$\overline{W}_v = \frac{\sum\limits_0^\infty (v + \tfrac{1}{2})hv \exp[-\beta(v+\tfrac{1}{2})hv]}{\sum\limits_0^\infty \exp[-\beta(v+\tfrac{1}{2})hv]} = \frac{\dfrac{hv}{2}\sum\limits_0^\infty \exp(-\beta vhv) + \sum\limits_0^\infty vhv \exp(-\beta vhv)}{\sum\limits_0^\infty \exp(-\beta vhv)}.$$

$$\overline{W}_v - \frac{hv}{2} = \frac{\sum\limits_0^\infty vhv \exp(-\beta vhv)}{\sum\limits_0^\infty \exp(-\beta vhv)}. \tag{5}$$

Note that the numerator of (5) is equal except for the sign, to the derivative of its denominator D with respect to β and thus one can write

$$\overline{W}_v - \frac{hv}{2} = -\frac{1}{D}\frac{\mathrm{d}D}{\mathrm{d}\beta} = -\frac{\mathrm{d}(\log D)}{\mathrm{d}\beta} = -\frac{\mathrm{d}}{\mathrm{d}\beta}\log\sum_0^\infty \exp(-\beta vhv).$$

D is the sum of a geometric series and has the value

$$D = \frac{1}{1-\exp(-\beta hv)},$$

hence

$$\overline{W}_v - \frac{hv}{2} = -\frac{\mathrm{d}}{\mathrm{d}\beta}\left(\log\frac{1}{1-\exp(-\beta hv)}\right) = \frac{hv}{\exp(\beta hv)-1} = \frac{kT_v}{\exp(T_v/T)-1}. \qquad (6)$$

For $T \ll T_v$, the one is negligible with respect to the exponential and one gets

$$\overline{W}_v - \frac{hv}{2} \simeq kT_v\exp\left(-\frac{T_v}{T}\right).$$

For $T \gg T_v$ one can expand the exponential as a series and one obtains

$$\overline{W}_v - \frac{hv}{2} = \frac{hv}{\beta hv\left(1+\dfrac{\beta}{2}hv+\ \dots\right)} \qquad (7)$$

which tends toward $1/\beta = kT$ for high temperatures. This latter value corresponds to the equipartition of energy because the linear oscillator has only one degree of freedom, but at the same time potential and kinetic energy as quadratic functions of the coordinates and velocity respectively.

One finds from (6) the equation for the molar vibrational heat capacity

$$C_v = \mathcal{N}\frac{\mathrm{d}\overline{W}_v}{\mathrm{d}T} = R\frac{T_v^2}{T^2}\times\frac{\exp(T_v/T)}{[\exp(T_v/T)-1]^2}. \qquad (8)$$

This expression tends to zero with T and to R at high temperatures. Figure 79.3 shows the behaviour of the function.

Numerical application:

$$T_v = \frac{hv}{k} = \frac{hc}{k}\tilde{v} = 1.44\times10^{-2}\tilde{v}.$$

For HD: $T_v = 1.44\times10^{-2}\times3630\times10^2 = 5227°K.$

For HCl: $T_v = 1.44\times10^{-2}\times2886\times10^2 = 4266°K.$

For NO: $T_v = 1.44\times10^{-2}\times1880\times10^2 = 2607°K.$

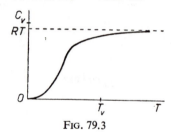

FIG. 79.3

IV

Look at the ratio N_2/N_1 of the number of molecules excited into the first electronic level to the number in the ground state by applying the Boltzmann distribution and assuming the statistical weights of the levels are unity (since this can only introduce a change of the order of unity):

$$\frac{N_2}{N_1} = \exp\left(-\frac{W_2-W_1}{kT}\right).$$

One has $kT = 1.38\times10^{-23}\times25\times10^2 = 3.45\times10^{-20}$ J,

$$W_2-W_1 = hc\tilde{\nu} = 19.96\times10^{-26}\tilde{\nu}.$$

For HD: $W_2-W_1 = 182\times10^{-20}$, $\dfrac{N_2}{N_1} = \exp(-52) = 0.26\times10^{-22}$.

For HCl: $W_2-W_1 = 150\times10^{-20}$, $\dfrac{N_2}{N_1} = \exp(-43) = 0.21\times10^{-18}$.

For NO: $W_2-W_1 = 90\times10^{-20}$, $\dfrac{N_2}{N_1} = \exp(-26) = 0.51\times10^{-11}$.

The number of excited molecules is thus very small and the electronic energy does not contribute to the specific heat of the molecules under consideration.

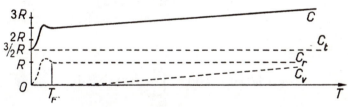

FIG. 79.4

The general behavior of $C(T)$ is given by the solid line in Fig. 79.4, and the translational, rotational, and vibrational contributions by the dotted curves. At ordinary temperatures the limiting value of the vibrational heat is not reached except for heavier molecules than those considered here.

APPENDIX A

THE FOURIER TRANSFORMATION

THE Fourier transformation is a mathematical operation which is frequently applied in optics. This integral arises in many different problems (spatial coherence, temporal coherence, diffraction, structure factors for X-rays, uncertainty relations, ...).

This treatment will not be rigorous. It is presented simply as a tool for the opticist in order to help simplify his calculation.

A. Definitions and General Properties of the Fourier Transformation

I. *Notation and definitions*

Let x be a real variable lying between $-\infty$ and $+\infty$, and $f(x)$ a function of x having real or complex values. $f(x)$ must be a summable function, that is, it must never go to ∞ for $x \to \infty$. *This is always the case in optics.*

By definition:

$$\text{F.T.}[f(x)] = F(u) = \int_{-\infty}^{+\infty} f(x)\, e^{2\pi j u x}\, dx. \tag{1}$$

One writes

$$f(x) \xrightarrow{\text{F.T.}} F(u). \tag{2}$$

One says that $F(u)$ is the Fourier transform (F.T.) of $f(x)$ or the spectrum of $f(x)$. u and x are called conjugate variables.

Consider, for example, the propagation equation of:

electromagnetic waves: $E = E_m \exp\left[2\pi j(vt - \sigma x)\right]$,

$$\left.\begin{array}{l} v \text{ and } t \\ \sigma \text{ and } x \end{array}\right\} \text{ are conjugate variables;} \tag{3}$$

the wave associated with a particle,

$$\psi = \psi_m \exp\left[\frac{2\pi j}{h}(Wt - px)\right],$$

$$\left.\begin{array}{l} W \text{ and } t \\ p \text{ and } x \end{array}\right\} \text{ are conjugate variables.} \tag{4}$$

340

II. *Reciprocal property of the F.T.*

If $F(u)$ is a known function, one can obtain $f(x)$ by the following operation:

$$f(x) = \int_{-\infty}^{+\infty} F(u)\, e^{-2\pi jux}\, du. \tag{5}$$

(Note the change in sign of the exponential in equation (5) compared to that in equation (1).)
The equation with dimensions of (5) is

$$[f] = [F]\,[u];$$

while for equation (1) one has

$$[F] = [f]\,[x];$$

hence

$$[u]\,[x] = 1. \tag{6}$$

This latter relation leads to an easy introduction of the Heisenberg uncertainty relations.
Note. Some other authors write these equations in the form

$$F(u) = \frac{1}{\sqrt{2\pi}} \int_{-\infty}^{+\infty} f(x)\, e^{jux}\, dx$$

and

$$f(x) = \frac{1}{\sqrt{2\pi}} \int_{-\infty}^{+\infty} F(u)\, e^{-jux}\, du.$$

III. *Properties*

1. *Linearity*

If one lets the two functions $f_1(x)$ and $f_2(x)$ have for their F.T., $F_1(u)$ and $F_2(u)$, respectively, and if a_1 and a_2 are constants, one finds:

$$\int_{-\infty}^{+\infty} [a_1 f_1(x) + a_2 f_2(x)]\, e^{2\pi jux}\, dx = a_1 \int_{-\infty}^{+\infty} f_1(x)\, e^{2\pi jux}\, dx + a_2 \int_{-\infty}^{+\infty} f_2(x)\, e^{2\pi jux}\, dx \tag{7}$$

or

$$[a_1 f_1(x) + a_2 f_2(x)] \xrightarrow{\text{F.T.}} [a_1 F_1(u) + a_2 F_2(u)]. \tag{8}$$

The F.T. of a linear combination of functions is the linear combination of the F.T. of these functions.

2. *Translation*

Translate the function $f(x)$ by the constant x':

$$\int_{-\infty}^{+\infty} f(x-x')\, e^{2\pi jux}\, dx = \int_{-\infty}^{+\infty} f(X)\, e^{2\pi ju(X+x')}\, dX, \tag{9}$$

by taking $X = x - x'$, (9) can be written

$$e^{2\pi j u x'} \int_{-\infty}^{+\infty} f(X) e^{2\pi j u X} dX = e^{2\pi j u x'} F(u), \tag{10}$$

$$f(x - x') \xrightarrow{\text{F.T.}} F(u) e^{2\pi j u x'}. \tag{11}$$

If one translates $f(x)$ by the constant amount x', its F.T. is multiplied by $e^{2\pi j u x'}$.

3. *Symmetry property*

Taking the F.T. of $f(x) e^{-2\pi j u' x}$ with u' constant:

$$\int_{-\infty}^{+\infty} f(x) e^{-2\pi j u' x} e^{2\pi j u x} dx = \int_{-\infty}^{+\infty} f(x) e^{2\pi j (u - u') x} dx = F(u - u'), \tag{12}$$

$$e^{-2\pi j u' x} f(x) \xrightarrow{\text{F.T.}} F(u - u'). \tag{13}$$

Note the analogy between equations (11) and (13).
These results can be applied to various examples.

$$f(x + x') + f(x - x') \qquad\qquad F(u) e^{-2\pi j u x'} + F(u) e^{+2\pi j u x'} = 2F(u) \cos 2\pi u x', \tag{14}$$

$$f(x - x') - f(x + x') \qquad\qquad F(u) e^{+2\pi j u x'} - F(u) e^{-2\pi j u x'} = 2jF(u) \sin 2\pi u x', \tag{15}$$

$$2f(x) - f(x - x') - f(x + x') \qquad 2F(u)[1 - \cos 2\pi u x'] = 4F(u) \sin^2 \pi u x', \tag{16}$$

$$f(x) \cos 2\pi u' x = \tfrac{1}{2} f(x) [e^{2\pi j u' x} + e^{-2\pi j u' x}] \quad \tfrac{1}{2}[F(u + u') + F(u - u')], \tag{17}$$

$$f(x) \sin 2\pi u' x = \frac{1}{2j} f(x) [e^{2\pi j u' x} - e^{-2\pi j u' x}] \quad \frac{1}{2j} [F(u + u') - F(u - u')], \tag{18}$$

$$f(x) \sin^2 \pi u' x = \tfrac{1}{2} f(x) [1 - \cos 2\pi u' x] \quad \tfrac{1}{4}[2F(u) - F(u + u') - F(u - u')]. \tag{19}$$

4. *Expansion*

Let a be a real constant. The F.T. of $f(ax)$ is desired. Make the change of variable $y = ax$.
if $a > 0$:

$$\int_{-\infty}^{+\infty} f(ax) e^{2\pi j u x} dx = \frac{1}{a} \int_{-\infty}^{+\infty} f(y) e^{2\pi j u y / a} dy = \frac{1}{a} F\left(\frac{u}{a}\right) \tag{20}$$

if $a < 0$:

$$\frac{1}{a} \int_{+\infty}^{-\infty} f(y) e^{2\pi j u y / a} dy = -\frac{1}{a} F\left(\frac{u}{a}\right). \tag{21}$$

In general one can write

$$f(ax) \xrightarrow{\text{F.T.}} \frac{1}{|a|} F\left(\frac{u}{a}\right). \tag{22}$$

In the special case where $a = -1$, equation (22) is written:

$$f(-x) \xrightarrow{\text{F.T.}} F(-u).$$

5. *Symmetries*

Taking the F.T. of $f^*(x)$, one finds

$$\int_{-\infty}^{+\infty} f^*(x)\, e^{2\pi jux}\, dx = \left[\int_{-\infty}^{+\infty} f(x)\, e^{-2\pi jux}\, dx\right]^* = F^*(-u). \tag{23}$$

$$f^*(x) \xrightarrow{\text{F.T.}} F^*(-u). \tag{24}$$

One often has occasion to examine functions with a special kind of symmetry. Assume that $f(x)$ is made up of an even function $p(x)$ and an odd function $i(x)$.

One can write

$$f(x) = p(x) + i(x) \tag{25}$$

where $p(x)$ and $i(x)$ may be complex.

The F.T. of $f(x)$ reduces to

$$F(u) = 2 \int_0^{+\infty} p(x) \cos 2\pi ux\, dx + 2j \int_0^{+\infty} i(x) \sin 2\pi ux\, dx. \tag{26}$$

The following general results are found:

$$f(x) \text{ real, even} \qquad \xrightarrow{\text{F.T.}} F(u) \text{ real, even}; \tag{27}$$

$$f(x) \text{ real, odd} \qquad \xrightarrow{\text{F.T.}} F(u) \text{ imaginary, odd}; \tag{28}$$

$$f(x) \text{ imaginary, even} \xrightarrow{\text{F.T.}} F(u) \text{ imaginary, even}. \tag{29}$$

The following table summarizes these results. (Re designates the real part of and Im the imaginary part of.)

$$f(x) = p(x) + i(x) = \text{Re}\,[p(x)] + j\,\text{Im}\,[p(x)] + \text{Re}\,[i(x)] + j\,\text{Im}\,[i(x)] \tag{30}$$

$$f(u) = P(u) + I(u) = \text{Re}\,[P(u)] + j\,\text{Im}\,[P(u)] + \text{Re}\,[I(u)] + j\,\text{Im}\,[I(u)]. \tag{31}$$

The arrows indicate the correspondences between the F.T.

IV. *Extension to two variables*

Taking $F(u, v)$ as the F.T. of $f(x, y)$, one states

$$F(u, v) = \int_{-\infty}^{+\infty} \int_{-\infty}^{+\infty} f(x, y)\, e^{2\pi j(ux+vy)}\, dx\, dy \tag{32}$$

reciprocally,

$$f(x, y) = \int_{-\infty}^{+\infty} \int_{-\infty}^{+\infty} F(u, v)\, e^{-2\pi j(ux+vy)}\, du\, dv. \tag{33}$$

The functions f and F play symmetric roles, one being the spectrum of the other.

These relationships arise, for example, in Huygen's principle where, if one has an amplitude distribution on a wave surface, the F.T. allows one to calculate the spectrum $F(u, v)$ of $f(x, y)$ and thereby to get the diffraction pattern. Inversely, if one knows a diffracted amplitude $F(u, v)$, one can calculate the structure of the wave surface which gave rise to it.

V. *Various useful F.T.*

In the table at the end of this appendix is given a limited number of F.T. which the reader will encounter (see examples 1 to 9).

In graphical representation the functions are normalized.

B. Convolution

I. *Definition*

Let the two functions $f(x)$ and $g(x)$ be limited and summable (Figs. A.10a and A.10b). The convolution of these two functions is $h(x)$:

$$h(x) = \int_{-\infty}^{+\infty} f(y)\, g(x-y)\, dy. \tag{34}$$

This is often written using the notation

$$h(x) = f(x) \otimes g(x). \tag{35}$$

Figures A.10c and A.10d illustrate the operations which give the convolution: the function $g(-y)$ is translated by an amount x. The product $f(y)\, g(x-y)$ is then formed. The ordinate $h(x)$ in Fig. A.10d is then equal to the shaded area in Fig. A.10c.

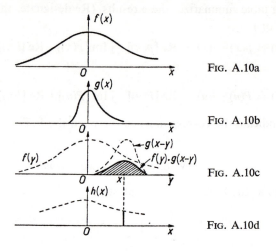

FIG. A.10a

FIG. A.10b

FIG. A.10c

FIG. A.10d

II. *Properties*

A. *The convolution product is commutative*

In equation (34) let $x-y = Y$

$$h(x) = \int_{-\infty}^{+\infty} f(x-Y)\, g(Y)\, d(-Y) = \int_{-\infty}^{+\infty} f(x-Y)\, g(Y)\, dY$$

$$h(x) = g \otimes f. \tag{36}$$

B. *Fourier transform of the convolution product*

$$f(x) \xrightarrow{\text{F.T.}} F(u)$$
$$g(x) \xrightarrow{\text{F.T.}} G(u).$$

(37)

Equation (34) can be written

$$h(x) = \int_{-\infty}^{+\infty} f(y) \int_{-\infty}^{+\infty} G(u) e^{-2\pi j u(x-y)} \, du \, dy$$

(38)

thus, by reversing the order of integration,

$$h(x) = \int_{-\infty}^{+\infty} G(u) e^{-2\pi j u x} \int_{-\infty}^{+\infty} f(y) e^{2\pi j u y} \, dy \, du$$

$$h(x) = \int_{-\infty}^{+\infty} F(u) \cdot G(u) e^{-2\pi j u x} \, du.$$

(39)

In summary one has the following reciprocal theorem:

$$\boxed{\begin{array}{l} f \otimes g \xrightarrow{\text{F.T.}} F \cdot G \\ f \cdot g \xrightarrow{\text{F.T.}} F \otimes H \end{array}}$$

(40)

This theorem is known as Parseval's theorem.

III. *Special cases*

(a) If $x = 0$ in equation (34), one gets

$$h(0) = \int_{-\infty}^{+\infty} f(y) g(-y) \, dy = \int_{-\infty}^{+\infty} F(u) \cdot G(u) \, du.$$

(41)

For $f = g$, one finds

$$\int_{-\infty}^{+\infty} f(x) f(-x) \, dx = \int_{-\infty}^{+\infty} F^2(u) \, du.$$

(42)

(b) *Correlation*
Taking

$$h'(x) = f(x) \otimes g^*(-x) = \int_{-\infty}^{+\infty} f(y) g^*(y-x) \, dy.$$

(43)

Now

$$f(x) \xrightarrow{\text{F.T.}} F(u),$$
$$g^*(-x) \xrightarrow{\text{F.T.}} G^*(u).$$

(44)

Equation (43) becomes

$$h'(x) = \int_{-\infty}^{+\infty} F(u) \cdot G^*(u) e^{-2\pi j u x} \, du.$$

(45)

In the special case where $x = 0$

$$h'(0) = \int_{-\infty}^{+\infty} F(u)\, G^*(u)\, du. \tag{46}$$

(c) *Autocorrelation*

$$f(x) = g(x) \tag{47}$$

$$h'(x) = f(x) \otimes f^*(-x). \tag{48}$$

Equation (48) becomes

$$h'(x) = \int_{-\infty}^{+\infty} |F(u)|^2\, e^{-2\pi jux}\, du. \tag{49}$$

The convolution $f(x) \otimes f^*(-x)$, called the autocorrelation function of $f(x)$ its F.T. is $|F(u)|^2$.

For $x = 0$:

$$\int_{-\infty}^{+\infty} |f(x)|^2\, dx = \int_{-\infty}^{+\infty} |F(u)|^2\, du. \tag{50}$$

This theorem expresses the conservation of energy, independent of the plane where this is applied. (Rayleigh's theorem.)

Applications of the autocorrelation

As a standard example, find the autocorrelation of the slit function and then the F.T. of this autocorrelation function. $f(x)$ is a real, even function. One has

$$f \otimes f \xrightarrow{\text{F.T.}} F^2(u). \tag{51}$$

The various functions are shown in Fig. A.11.

The autocorrelation function of the pupil, known as the *transfer function*, plays a very important role in optical instruments illuminated with incoherent light (see Problems 35 and 37).

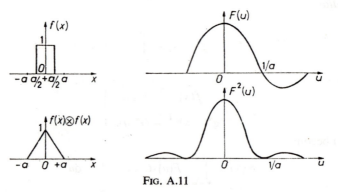

Fig. A.11

One can represent the general distribution of luminance of an object by a superposition of an infinity of sinusoidal variations, each of which is characterized by

a characteristic direction, that of the wave vector;

a spatial frequency proportional to the inverse of the wavelength (this is the frequency of the sinusoidal component under consideration);

an amplitude and a phase.

These various sinusoidal components are transmitted through the optical instrument. They are then acted upon by a filtering law given by the transfer function.

The transfer function specifies the quality of the instrument. This provides information on all spatial frequencies. For this reason it is preferable to characterize an instrument by its transfer function rather than by its limit of resolution which gives the limiting frequency transmitted by the instrument, but no information regarding intermediate frequencies (see Problems 35 and 37).

C. Dirac Distribution. Poisson Distribution

Certain functions such as $f(x) = 1$, $f(x) = \cos x$, $\ldots$ do not satisfy the conditions for application of the F.T. In these cases it is only possible to define a F.T. by a limiting process.

The Fourier series can only be put within the framework of the F.T. through use of distributions. We are not going to deal with the theory of distributions here, but only give some useful definitions and properties.

I. Dirac distribution

1. Definition

Consider an impulse $\delta(x)$ with very narrow width and very large height such that its area is unity:

$$f(x) = 0 \quad \text{for} \quad x = 0 \tag{52}$$

$$\int_{+\infty}^{-\infty} \delta(x)\,dx = 1. \tag{53}$$

2. Representation

The impulse $\delta(x)$ is represented by a spike with its height normalized to one (Fig. A.12a).

3. Convolution

One can write

$$\int_{-\infty}^{+\infty} \delta(x) f(x)\,dx = f(0) \tag{54}$$

or

$$\int_{-\infty}^{+\infty} \delta(x-a) f(x)\,dx = f(a) \tag{55}$$

$$\int_{-\infty}^{+\infty} \delta(x) f(x-a)\,dx = f(-a). \tag{56}$$

Extending this to convolution one gets

$$\int_{-\infty}^{+\infty} \delta(y) f(x-y)\, dy = \int_{-\infty}^{+\infty} \delta(x-y) f(y)\, dy = f(x) \tag{57}$$

$$\delta(x) \otimes f(x) = f(x) \otimes \delta(x) = f(x). \tag{58}$$

The Dirac function is the unit element for convolution (just as zero is the unit element for addition and one is the unit element for multiplication).

4. Translation

Starting with the preceding equation, one can write

$$f(x-a) = f(x) \otimes \delta(x-a). \tag{59}$$

Translation can be thought of as a convolution operation.

5. Fourier transformation

Call $\Delta(u)$ the F.T. of $\delta(x)$:

$$\delta(x) \xrightarrow{\text{F.T.}} \Delta(u). \tag{60}$$

Applying the convolution theorem to (58) gives

$$\Delta(u) \times F(u) = F(u). \tag{61}$$

hence

$$\Delta(u) = 1. \tag{62}$$

In summary:

$$\delta(x) \xrightarrow{\text{F.T.}} \Delta(u) = 1 \qquad \text{(Fig. A.12.b).} \tag{63}$$

with

$$\delta(x) = \int_{-\infty}^{+\infty} e^{-2\pi j u x}\, du. \tag{64}$$

If one translates the Dirac function by an amount a, one finds

$$\delta(x-a) \xrightarrow{\text{F.T.}} e^{2\pi j u a}. \tag{65}$$

6. Properties

$$\delta(ax) = \frac{1}{|a|}\, \delta(x), \tag{66}$$

$$\delta(-x) = \delta(x), \tag{67}$$

$$f(x) \delta(x) = f(0)\, \delta(x). \tag{68}$$

Or

$$f(x)\, \delta(x-a) = f(a)\, \delta(x-a). \tag{69}$$

Taking $f(x) = x$, one finds

$$\int_{-\infty}^{+\infty} x\delta(x) = 0 \quad \text{or} \quad x\delta(x) = 0. \tag{70}$$

II. *Fourier series*

1. *Fourier transform of a Poisson distribution (or a Dirac "comb" series)*

We state, without proof, that the F.T. of a Poisson distribution of period p is a Poisson distribution of period $1/p$ (Fig. A.13), in other words that

$$\sum_{k=-\infty}^{+\infty} \delta(x-kp) \xrightarrow{\text{F.T.}} \sum_{-\infty}^{+\infty} \delta\left(u-\frac{k}{p}\right). \tag{71}$$

2. *F.T. of an unbounded periodic function*

Let $h(x)$ be an unbounded function of period p.

One can assume that $h(x)$ is obtained by translating by integral multiples of p the simple convergent function $f(x)$ (Figs. A.14a and A.14a′). Since translation is a convolution process, one can take

$$h(x) = f(x) \otimes \sum_{k=-\infty}^{+\infty} \delta(x-kp). \tag{72}$$

Taking the F.T. of each side, one gets

$$H(u) = F(u) \sum_{k=-\infty}^{+\infty} \delta\left(u-\frac{k}{p}\right) \tag{73}$$

u can only take on discrete values of k/p so that

$$H(u) = \sum_{-\infty}^{+\infty} F\left(\frac{k}{p}\right) \times \delta\left(u-\frac{k}{p}\right) \tag{74}$$

(Figs. A.14b and A.14b′).

In summary:

the F.T. of an unbounded periodic function is a distribution;

if the period of $h(x)$ is p, the period of $H(u)$ is $1/p$;

the uniformly spaced Dirac weightings are equal to $F(k/p)$, where $F(u)$ is the F.T. of $f(x)$ and $F(k/p)$ is the value of $F(u)$ at point $u = k/p$.

Reversibility. By substituting equation (73) in equation (72), one can see that the F.T. of and unbounded periodic distribution is an unbounded periodic function. This is only the case where the Dirac weightings have equal weight so that the F.T. of a distribution is a distribution.

Note. The special case of a function (or distribution) which is periodic and bounded can be easily treated by using previously obtained results. One can always assume that the bounded function (or distribution) is the product of a function $g(x)$ with a periodic unbounded function (or distribution) (Fig. A.15). Most commonly $g(x)$ is a slit function.

If

$$h'(x) = h(x) \times g(x) \qquad \text{(the order of the factors must be retained)} \tag{75}$$

$$H'(u) = H(u) \otimes G(u), \tag{76}$$

$$H'(u) = F(u) \cdot \sum_{-\infty}^{+\infty} \delta\left(u - \frac{k}{p}\right) \cdot G(u). \tag{77}$$

TABLES

1. SLIT FUNCTION

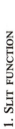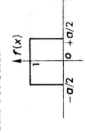

Fig. A.1a

$f(x) = 1$
for $-a/2 < x < +a/2$
$f(x) = 0$
for $x < -a/2$ and $x > a/2$

$$\Delta x = a$$

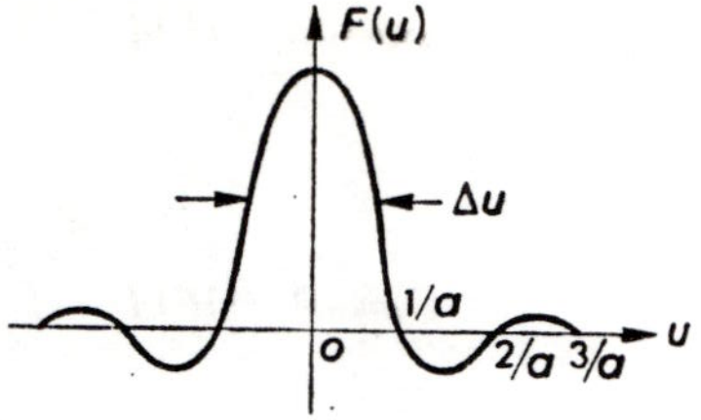

Fig. A.1b

$$F(u) = \frac{\sin \pi u a}{\pi u a}$$

$$\Delta u = \frac{1}{a}$$

Influence of the spatial magnitude of the source (Problem 3)

$$\Delta\sigma \times \Delta x = 1$$

Influence of the magnitude of the slit on the diffraction image (Problem 35)

$$\Delta x \times \Delta u = 1$$

Influence of the width of the energy level on the lifetime of a wave packet (Problem 71)

$$\Delta W \times \Delta t = 1$$

Influence of the uncertainty in momentum on the position of the particle (Problem 57)

$$\Delta p \times \Delta x = h$$

2. GAUSSIAN FUNCTION

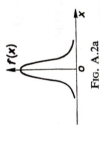

Fig. A.2a

$$f(x) = e^{-\pi x^2}$$

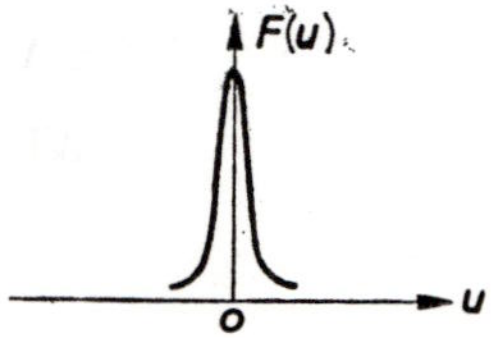

Fig. A.2b

$$F(u) = \int_{-\infty}^{+\infty} e^{-\pi x^2} e^{+j2\pi u x} \, dx$$

The preceding applications can also be treated with a gaussian profile

For a rigorous proof see the mathematics references. One takes $Z = x - ju$ and assumes j to be a constant

$$F(u) = e^{-\pi u^2} \int_{-\infty}^{+\infty} e^{-\pi Z^2} \, dZ$$

by making the change of variables $\pi Z^2 = X^2$

$$F(u) = e^{-\pi u^2} \int_{-\infty}^{+\infty} e^{-X^2} \frac{dX}{\sqrt{\pi}}.$$

Since X and Y are independent variables, one can state:

$$[F(u)]^2 = \frac{e^{-2\pi u^2}}{\pi} \int_{-\infty}^{+\infty} \int_{-\infty}^{+\infty} e^{-(X^2 + Y^2)} \, dX \, dY$$

and, treating the surface integral in polar coordinates

$$[F(u)]^2 = \frac{e^{-2\pi u^2}}{\pi} \int_0^\infty \int_0^{2\pi} e^{-\varrho^2} \varrho \, d\varrho \, d\theta = e^{-2\pi u^2} \int_0^\infty e^{-\varrho^2} \, d(\varrho)^2$$

$$= -e^{-2\pi u^2} \left[e^{-\varrho^2} \right]_0^\infty = e^{-2\pi u^2}$$

$$f(x) = e^{-\pi x^2} \qquad\qquad F(u) = e^{-\pi u^2}.$$

In the most general sense:

$$f(x) = e^{-\pi x^2/a^2} \qquad\qquad F(u) = e^{-\pi a^2 u^2}.$$

3. BOUNDED COSINUSOID

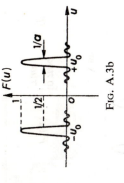

FIG. A.3a

$$f(x) = \cos 2\pi u_0 x$$

for $-a/2 < x < +a/2$

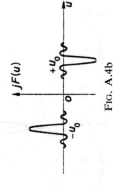

FIG. A.3b

$$F(u) = \frac{\sin \pi (u+u_0)a}{\pi(u+u_0)a} + \frac{\sin \pi (u-u_0)a}{\pi(u-u_0)a}$$

A perfectly monochromatic wavetrain having finite duration gives a spectral line of finite width

The line is centred on frequency v_0 and has width $1/\tau$ where τ is the lifetime of the wave train (Problem 3). Likewise, if one considers the width Δx of the wave train, one finds $\Delta \sigma \times \Delta x = 1$

4. BOUNDED SINUSOID

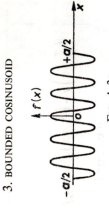

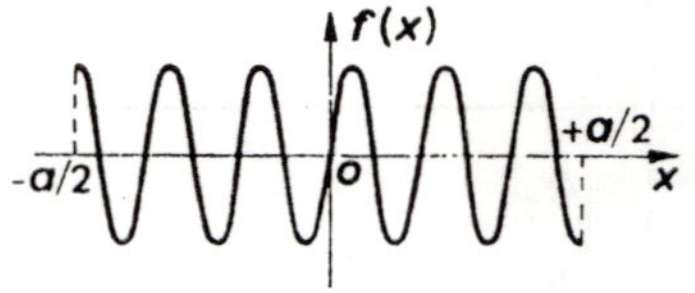

FIG. A.4a

$$f(x) = \sin 2\pi u_0 x$$

for $-a/2 < x < +a/2$

FIG. A.4b

$$F(u) = \frac{1}{j} \left[\frac{\sin \pi(u+u_0)a}{\pi(u+u_0)a} - \frac{\sin \pi(u-u_0)a}{\pi(u-u_0)a} \right]$$

The sinusoidal or cosinusoidal systems only give two spectra (Problem 35)

5. $f(x) = 1 + \cos 2\pi u_0 x$
for $-a/2 < x < +a/2$.

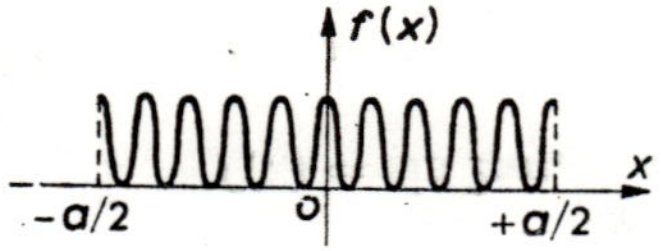

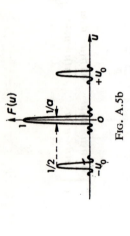

Fig. A.5a

$$F(u) = \left[\frac{\sin \pi u a}{\pi u a} + \frac{1}{2}\frac{\sin \pi(u+u_0)a}{\pi(u+u_0)a} + \frac{1}{2}\frac{\sin \pi(u-u_0)a}{\pi(u-u_0)a}\right]$$

Fig. A.5b

Transmission grating system $f(x)$
gives three unequal spectra

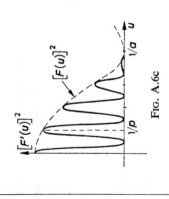

Fig. A.6c

$$[F'(u)]^2 = \frac{1}{2}[F(u)]^2(1+\cos 2\pi p u)$$

6. YOUNG'S SLITS

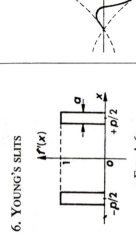

Fig. A.6a

2 identical slits of width a and
separation p

$$f'(x) = f\left(x+\frac{p}{2}\right) + f\left(x-\frac{p}{2}\right)$$

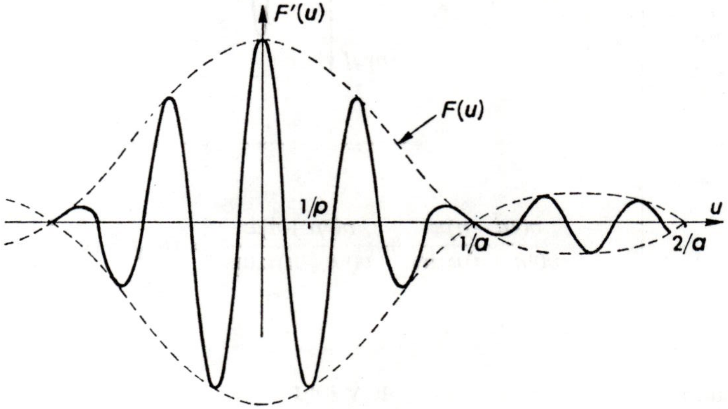

Fig. A.6b

$$F'(u) = \frac{\sin \pi u a}{\pi u a}\cos \pi p u$$

7. THREE IDENTICAL SLITS

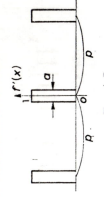

FIG. A.7a

$$f'(x) = f(x) + f(x+p) + f(x-p)$$

$$[F'(u)]^2$$

FIG. A.7b

$$F'(u) \simeq F(u)\,[1 + 2\cos 2\pi p u]$$

8. N IDENTICAL SLITS (GRATING)

$$f'(x) = \sum_{n=0}^{N-1} f(x+np)$$

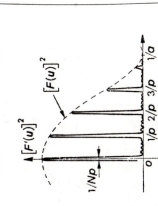
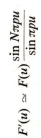

FIG. A.8

$$F'(u) \simeq F(u)\,\frac{\sin N\pi p u}{\sin \pi p u}$$

One can compare the three graphs representing $[F'(u)]^2$ from 6, 7, and 8

the modulation by $F(u)$ is the same,
the position of the spectra is the same,
the width of the spectra depends on the number of slits

9. CIRCULAR FUNCTION

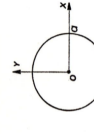

Fig. A.9a

$$f(r) = \frac{1}{\pi a^2}$$

for $\quad x^2 + y^2 = r^2 < a^2$

AIRY FUNCTION

If $\quad u^2 + v^2 = \varrho^2$

$$F(\varrho) = 2\,\frac{\mathcal{J}_1(2\pi\varrho a)}{2\pi\varrho a}$$

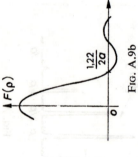

$\frac{1.22}{2a}$

Fig. A.9b

See Problem 36

10. DIRAC IMPULSE

$\delta(x)$

Fig. A.12a

$\Delta(u)$

1

Fig. A.12b

Uncertainty relationships:

If one variable is determined absolutely, the conjugate variable is completely undeterminable. One can re-examine the examples given with figure 1

11. DIRAC SERIES

$\Sigma\,\delta(x-kp)$

$o\ p\ 2p\ 3p$

Fig. A.13a

$\Sigma\,\delta\left(u-\frac{k}{p}\right)$

$o\ 1/p\ 2/p\ 3/p\ \ u$

Fig. A.13b

Grating $\left\{\begin{array}{l}\text{infinitely}\\\text{narrow slits}\\\text{spacing } p\\\text{width } \infty\end{array}\right.$ → Spectrum $\left\{\begin{array}{l}\text{infinite}\\\text{number}\\\text{spacing } 1/p\\\text{infinitely}\\\text{sharp}\end{array}\right.$

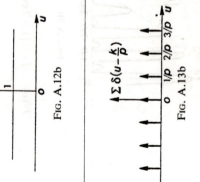

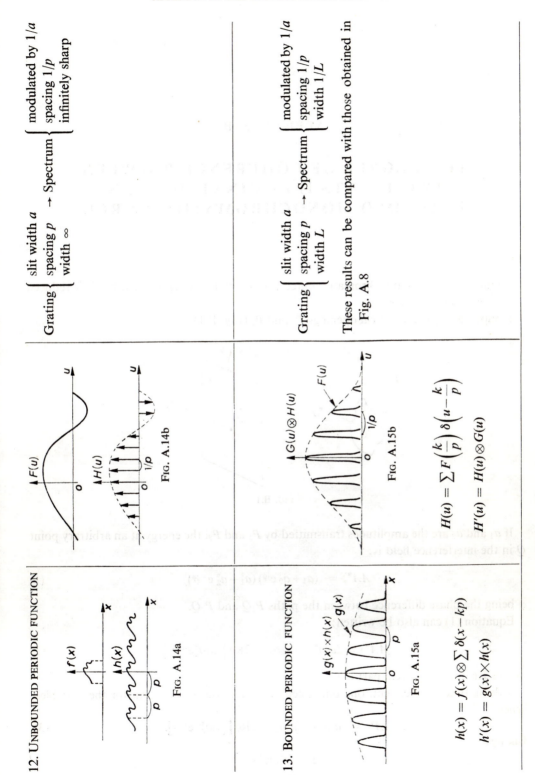

12. UNBOUNDED PERIODIC FUNCTION

$f(x)$

$h(x)$

Fig. A.14a

$F(u)$

$H(u)$ $1/p$

Fig. A.14b

Grating $\begin{cases} \text{slit width } a \\ \text{spacing } p \\ \text{width } \infty \end{cases}$ → Spectrum $\begin{cases} \text{modulated by } 1/a \\ \text{spacing } 1/p \\ \text{infinitely sharp} \end{cases}$

13. BOUNDED PERIODIC FUNCTION

$g(x) \times h(x)$

$g(x)$

Fig. A.15a

$G(u) \otimes H(u)$

$F(u)$

$1/p$

Fig. A.15b

$$H(u) = \sum F\left(\frac{k}{p}\right) \delta\left(u - \frac{k}{p}\right)$$

$$H'(u) = H(u) \otimes G(u)$$

$$h(x) = f(x) \otimes \sum \delta(x - kp)$$

$$h'(x) = g(x) \times h(x)$$

Grating $\begin{cases} \text{slit width } a \\ \text{spacing } p \\ \text{width } L \end{cases}$ → Spectrum $\begin{cases} \text{modulated by } 1/a \\ \text{spacing } 1/p \\ \text{width } 1/L \end{cases}$

These results can be compared with those obtained in Fig. A.8

THE DEGREE OF COHERENCE BETWEEN TWO POINTS ILLUMINATED BY AN EXTENDED MONOCHROMATIC SOURCE

TAKE:

a finite, extended, monochromatic source whose points are represented by their reduced coordinates u and v;

a pupil having two identical openings P_1 and P_2 (Fig. B.1).

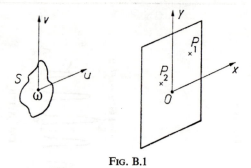

FIG. B.1

If a_1 and a_2 are the amplitudes transmitted by P_1 and P_2, the energy at an arbitrary point Q in the interference field is:

$$\langle AA^* \rangle = \langle (a_1 + a_2\, e^{j\phi})\,(a_1^* + a_2^*\, e^{-j\phi}) \rangle \tag{1}$$

ϕ being the phase difference between the paths P_1Q and P_2Q.

Equation (1) can also be written:

$$\langle AA^* \rangle = \langle a_1 a_1^* \rangle + \langle a_2 a_2^* \rangle + 2\mathrm{Re}\,[\langle a_1 a_2^* e^{-j\phi} \rangle] \tag{2}$$

(where Re implies the real part of).

ϕ being time independent, one only needs to make use of the mean for the variables, hence:

$$\langle AA^* \rangle = \langle a_1 a_1^* \rangle + \langle a_2 a_2^* \rangle + 2\mathrm{Re}\,[\langle a_1 a_2^* \rangle\, e^{-j\phi}]. \tag{3}$$

Taking:

$$a_1 a_2^* = |a_1 a_2^*|\, e^{-j\theta};$$

equation (3) becomes:

$$\langle AA^* \rangle = \langle a_1 a_1^* \rangle + \langle a_2 a_2^* \rangle + 2\langle |a_1 a_2^*| \rangle \cos(\phi + \theta). \tag{4}$$

One finds:

maximal illumination (for $\phi + \theta = 2k\pi$),

$$I_{max} = \langle a_1 a_1^* \rangle + \langle a_2 a_2^* \rangle + 2\langle |a_1 a_2^*| \rangle \tag{5}$$

minimal illumination (for $\phi + \theta = (2K+1)\pi$),

$$I_{min} = \langle a_1 a_1^* \rangle + \langle a_2 a_2^* \rangle - 2\langle |a_1 a_2^*| \rangle. \tag{6}$$

If one takes as the definition of the contrast

$$\Gamma = \frac{I_{max} - I_{min}}{I_{max} + I_{min}}, \tag{7}$$

Γ can be written as:

$$\Gamma = \frac{2\langle |a_1 a_2^*| \rangle}{\langle a_1 a_1^* \rangle + \langle a_2 a_2^* \rangle}. \tag{8}$$

Consider an atom having coordinates u and v which emits a vibration $a(t)$. These vibrations falling on P_1 and P_2 can be written respectively as:

$$a(t)\,e^{-j2\pi(ux_1 + vy_1)} \quad \text{and} \quad a(t)\,e^{-j2\pi(ux_2 + vy_2)}. \tag{9}$$

Characterize by the index i the various atoms in the source. Their contribution to the fields at P_1 and P_2 is:

$$\begin{aligned} a_1 &= \Sigma a_i(t)\,e^{-j2\phi(u_i x_1 + v_i y_1)} \\ a_2 &= \Sigma a_i(t)\,e^{-j2\pi(u_i x_2 + v_i y_2)}. \end{aligned} \tag{10}$$

One can take:

$$\langle a_1 a_2^* \rangle = \Sigma a_i(t)\,e^{-j2\pi(u_i x_1 + v_i y_1)}\,\Sigma a(t)\,e^{+j2\pi(u_j x_2 + v_j y_2)}. \tag{11}$$

One distinguishes between the products relative to a single atom and those relative to two different atoms (these latter are zero since the atoms involved will radiate incoherently). One has:

$$\langle a_1 a_2^* \rangle = \Sigma\langle a_i a_i^* \rangle\,e^{-j2\pi[u_i(x_1-x_2) + v_i(y_1-y_2)]}. \tag{12}$$

Since the density of atoms is large, the sum can be replaced by an integral and:

$$\langle a_1 a_2^* \rangle = \iint_S I(u,\,v)\,e^{-j2\pi[u(x_1-x_2) + v(y_1-y_2)]}\,du\,dv, \tag{13}$$

where $I(u,\,v)$ is the energy contributed by the element of the source characterized by $u,\,v$. Additionally,

$$\langle a_1 a_1^* \rangle = \langle a_2 a_2^* \rangle = \iint_S I(u,\,v)\,du\,dv. \tag{14}$$

The degree of partial coherence between P_1 and P_2 can thus be written:

$$\Gamma(x_1-x_2, y_1-y_2) = \frac{\iint\limits_S I(u, v)\,e^{-j2\pi[u(x_1-x_2)+v(y_1-y_2)]}\,du\,dv}{\iint I(u, v)\,du\,dv}.$$ (15)

Van Cittert–Zernike theorem

The degree of coherence between a fixed point P_1 and a variable point P_2 illuminated by an extended monochromatic source is equal to the complex amplitude, normalized at P_1, of a diffraction pattern centred on P_2. This artificial diffraction pattern is obtained by replacing the source by a pupil having the same dimensions and form as the source and with an amplitude distribution in the pupil equal to the intensity distribution in the source.

INDEX

OTHER TITLES IN THE SERIES IN NATURAL PHILOSOPHY